Sources and Studies in the History of Mathematics and Physical Sciences

Sources and Studies in the
History of Mathematics and Physical Sciences

Continued after Subject Index

Giovanni Ferraro

The Rise and Development
of the Theory of Series up to the
Early 1820s

 Springer

Giovanni Ferraro
Università del Molise
Dipartimento STAT
c. da Fonte Lappone
86090 Pesche
Isernia, Italy
giovanni.ferraro@unimol.it

Sources and Series Editor:
Jesper Lützen
Institute for Mathematical Sciences
University of Copenhagen
DK-2100 Copenhagen
Denmark

ISBN: 978-0-387-73467-5   e-ISBN: 978-0-387-73468-2

Library of Congress Control Number: 2007939827

Mathematics Subject Classification (2000): 01Axx 01A50 26-03 40xx

Printed on acid-free paper

9 8 7 6 5 4 3 2 1

springer.com

A Pina, Maria Grazia, Serena, Giuseppe

Preface

The theory of series in the 17th and 18th centuries poses several interesting problems to historians. Indeed, mathematicians of the time derived numerous results that range from the binomial theorem to the Taylor formula, from the power series expansions of elementary functions to trigonometric series, from Stirling's series to series solution of differential equations, from the Euler–Maclaurin summation formula to the Lagrange inversion theorem, from Laplace's theory of generating functions to the calculus of operations, etc. Most of these results were, however, derived using methods that would be found unacceptable today, thus, if we look back to the theory of series prior to Cauchy without reconstructing internal motivations and the conceptual background, it appears as a corpus of manipulative techniques lacking in rigor whose results seem to be the puzzling fruit of the mind of a magician or diviner rather than the penetrating and complex work of great mathematicians.

For this reason, in this monograph, not only do I describe the entire complex of 17th- and 18th-century procedures and results concerning series, but also I reconstruct the implicit and explicit principles upon which they are based, draw attention to the underlying philosophy, highlight competing approaches, and investigate the mathematical context where the series theory originated. My aim is to improve the understanding of the framework of 17th- and 18th-century mathematics and avoid trivializing the complexity of historical development by bringing it into line with modern concepts and views and by tacitly assuming that certain results belong, in some unproblematic sense, to a unified theory that has come down to us today.

The initial and final points of my monograph require some clarification. The point of departure is the publication of a paper by Viète, *Variorum de rebus mathematicis responsorum. Liber VIII* (1593), where geometrical series are discussed and π is expressed in the form of an infinite product. Even though previous tracks of infinite series can be found, Viète's paper, when considered in the context of the new rising symbolic algebra, appears to be a step forward in a path –very slow to begin with, but that developed much more rapidly after 1650– that has made series an essential instrument in mathematics. The point of arrival is the early 1820s when Cauchy published *Cours d'analyse* and *Résumé des leçons données à l'École Royale Polytechnique sur le calcul infinitésimal*, which can be considered to mark the definitive abandonment of the 18th-century formal approach to the series theory.

My main arguments can be summarised as follows. The mathematicians who first used series were interested in their capacity to represent geometrical quantities and had an intuitive idea of convergence. They thought that a series represented a quantity and had a quantitative meaning if, and only if, it was convergent to this quantity. However, a distinction between finite and infinite sums was lacking, and this gave rise to formal procedures consisting of the infinite extension of finite procedures. In the works of mathematicians such as Newton and Leibniz, the quantitative and the formal aspect co-existed and formal manipulations were a tool for deriving convergent series.

As from the 1720s, several results began to upset the previously established balance between the quantitative and the formal. Mathematicians introduced recurrent series, which stressed the law of formation of coefficients, independently of the convergence of series. The attempt to improve the acceleration of series subsequently led to the emergence of asymptotic series, which showed the possibility of using divergent series to obtain appropriate approximations. Furthermore, the investigation of continued fractions and infinite products and certain applications of series (for instance, in numerical analysis and in number theory) increasingly stressed the formal aspects.

In this context, Euler offered a unitary interpretation of the complex of results concerning series, which even allowed the acceptance of those findings that did not form part of the early theory. A series was thought to be the result of a formal transformation of an analytical quantity expressed in a closed form. This transformation was considered sufficient to give a meaning to the series, even when the latter was not convergent. However, mathematicians were not free to invent transformations by a free creative act. They limited themselves to using the same transformations that were used in the original theory or at least were compatible with it. This seemed to guarantee that the new more formal conception was a generalization of earlier conception, which remained the essential basis from which all the parts of the series theory were subsequently generated.

The more formal Eulerian approach was widely predominant during the second part of the 18th-century for two main reasons. First, mathematicians who were critical of it were not able to eliminate the formal aspects of the early concept and found a really new theory: They always used the formal methodology that had led to asymptotic series and to the combinatorial use of series. Second, the formal concept of series contributed to the growth of mathematics. It led to many new discoveries and even to a new branch of analysis: the calculus of operations.

The formal approach became unsuited to most advanced mathematical research toward the end of the 18th century and the beginning of the 19th century. Applied mathematics encouraged investigations and introduction of new functions in analysis, but formal methodology was unable to treat quantities that were not elementary quantities and series that were not power series. The need to use trigonometric series to enable the analytical investigation of heat led Fourier to reject the formal concept of series and assert an entirely quantitative notion of series. Similarly, the need to introduce hypergeometric and gamma functions into analysis and to have an adequate analytical theory of them forced Gauss to highlight the quantitative meaning of the sum of series and to reject formal manipulations. The new approach based only upon convergence was the basis of Cauchy's treatises.

Given the purposes of this book, I cannot avoid dealing with some topics that are closely connected to series theory and are crucial to an understanding of its historical evolution: Not only do these include other infinite processes (continued fractions and infinite products) but also certain basic mathematical notions (quantity, numbers, functions) and the 18th-century concept of analysis.

This book is divided into four parts. The first part starts with a chapter devoted to the use of series prior to the rise of the calculus (Chapter 1), where I deal principally with Viète, Grégoire de Saint-Vincent, Mengoli, Wallis, and Gregory. I then move on to investigate the conception of the founders of the calculus (Leibniz in Chapter 2; Newton in Chapter 4). On the basis of this examination, and after discussing the contributions of Johann and Jacob Bernoulli (Chapters 3 and 5) and the notion of a quantity and of a number (Chapter 7), I offer an interpretative scheme of the early theory of series in Chapter 8. The first part also includes the appearance of Taylor series in Newton and Taylor (Chapter 6) and the rise of the problem of the sum of a divergent series in one of Grandi's writings and the ensuing debate in Leibniz, Varignon, Daniel Bernoulli, and Goldbach (Chapter 9).

In the second part, I illustrate the development of series theory from the 1720s to the 1750s. De Moivre's recurrent series and Bernoulli's method for solving equations are the subject of Chapter 10. Chapter 11 deals with the attempt to improve the acceleration of series and Stirling's series, the first example of asymptotic series. Chapter 12 examines the geometric conception of Colin Maclaurin. Most of the second part is devoted to Euler, "the master of all us," to use an expression that Libri [1846, 51] ascribes to Laplace. From 1730 to 1750, Euler obtained many important results, which I examine in Chapters 13 to 17. In particular, I shall concentrate on the problem of interpolation and some of Euler's first findings (Chapter 13), on Euler's derivation of the Euler–Maclaurin summation formula (Chapter 14), on issues connected to the interpretation of asymptotic series (Chapter 15), on the theory of infinite products and continued fractions (Chapter 16), and on the application of series to number theory (Chapter 17). Chapter 18 is a digression on some basic principles of analysis during the period from the 1740s to the 1810s, which is essential for understanding series theory in the second half of the 18th century. In particular, the relationship between analysis and geometry, the notion of a function, and the principle of generality of algebra are examined. In Chapter 19, I discuss some criticisms of certain procedures and how Euler rejected them by giving a merely formal interpretation of the notion of the sum.

The third part is devoted to the period when formal conception held undisputed sway. I begin by illustrating some of the greatest successes of the formal approach during the second part of the 18th century: the La-

grange inversion theorem, which is discussed in Chapter 20, the calculus
of operations, examined in Chapter 21, and Laplace's theory of generating
functions (the subject of Chapter 22). The problem of the representation
of transcendental quantities and their analytical investigation is treated in
Chapters 23, 24, and 25.

Integration by series and series solutions to differential equations were
already known by the beginning of the calculus, but they underwent a remarkable development after Euler: Some examples from Euler, Laplace and
Legendre are given in Chapter 26. I then deal with trigonometric series for
which mathematicians applied the same procedure as that used for power
series. This prevented them from being fully understood (Chapter 27).

The attempts to prove the binomial theorem, Lagrange's view of the
Taylor theorem, and other significant developments that took place between
the end of the 18th century and the beginning of 19th century are the subject
matter of Chapter 28.

Chapters 29 and 30 focus on the problematic attempt of Legendre to
enlarge the realm of accepted functions and to the emergence of techniques
of inequalities in d'Alembert's and Lagrange's work.

The fourth and final part is devoted to the crisis in formal methods. It
deals with Fourier's investigations of Fourier series (Chapter 31), Gauss's
work on hypergeometric and gamma functions (Chapter 32), and Cauchy's
contributions on series during the early 1820s (Chapter 33). The conceptions of these mathematicians differ from all other mathematicians discussed
in this book since they belong to a new historical phase. However, the discussion of their approach allows me to illustrate some hypotheses about the
abandonment of 18th-century series theory.

In order to write this monograph I have drawn on various papers of mine,
in particular:

Some parts of "True and fictitious quantities in Leibniz's theory of series",
published in *Studia Leibnitiana*, 32 (2000), pp. 43–67 (copyright Franz
Steiner Verlag GmbH, Stuttgart) are reproduced in Chapters 2, 3, and
9.

Some parts of "Functions, functional relations and the laws of continuity
in Euler", published in *Historia Mathematica*, 27 (2000), pp. 107–
132 (copyright Elsevier), and "Analytical symbols and geometrical figures. Eighteenth century analysis as nonfigural geometry", published
in *Studies in History and Philosophy of Science Part A*, 32 (2001), pp.
535–555 (copyright Elsevier), are reproduced in Chapter 18.

Some parts of "Some aspects of Euler's theory of series. Inexplicable functions and the Euler–Maclaurin summation formula", published in *Historia Mathematica*, 25 (1998), pp. 290–317 (copyright Elsevier), are
reproduced in Chapters 13, 14, and 24.

Some parts of "The value of an infinite sum. Some observations on the Eulerian theory of series", published in *Sciences et Techniques en Perspective*, 4 (2000), pp. 73–113, are reproduced in Chapters 15 and 19.

Some parts of "Convergence and formal manipulation in the theory of series from 1730 to 1815", published in *Historia Mathematica*, 34 (2007) pp. 62–88 (copyright Elsevier), are reproduced in Chapters 26, 27 and 31.

Some parts of "The foundational aspects of Gauss's work on the hypergeometric, factorial and digamma functions", published in *Archive for History of Exact Sciences*, 61 (2007), 457-518 (copyright Springer-Verlag) are reproduced in Chapters 29, 30, and 32.

I would like to thank *Studia Leibnitiana, Sciences et Techniques en Perspective, Studies in History and Philosophy of Science, Historia Mathematica*, and *Archive for History of Exact Sciences* for their permission to include material from the above-mentioned articles.

Finally, I would like to thank Craig Fraser and Jesper Lützen for their suggestions that have been helpful in the preparation of this volume.

Afragola, Italy
May 2007

Contents

Part I

From the beginnings of the 17th century to about 1720: Convergence and formal manipulation

In the first part of the present book, I examine the emergence of series theory and its development up to around 1720.

Series were introduced in mathematics mainly to solve geometric problems. Their use, which was initially rather sporadic, began to take on importance around 1650 and was crucial to the rise of the calculus. In Newton and Leibniz's times many results were obtained: They came to form an organic corpus of knowledge that constituted the early theory of series.[1]

From the very beginning mathematicians had an intuitive idea of convergence and they thought that convergent series[2] could represent geometric quantities.[3] However, convergence was not considered as a preliminary condition for handling series. Mathematicians did not distinguish between operations on infinite series and operations on finite series, and they formally manipulated infinite series by applying the same rules that were employed for finite sums. Therefore, series theory had a twofold aspect, the first based upon convergence, the second upon formal manipulation.

The formal aspect was not caused by the imprecision or vagueness of certain formulations and certain concepts; rather it was rooted in the basic notion of 17th- and 18th-century mathematics, the notion of quantity. Sometimes, especially in the first years, the formal aspect was almost hidden by the simplicity of the employed series and from the immediacy of geometric reference. At other times, the formal aspect appeared. This occurred mainly when the attention was focused on the problem of finding the development of given quantities. In early series theory, mathematicians thought

---

[1] The term "theory" is, of course, not intended in the sense of a formal theory but instead as an organic set of principles, rules, methods, and logical deductions concerning a specific subject.

[2] Two terminological specifications are necessary. First, during the 17th and 18th centuries, mathematicians used the term "series" to denote both series and sequences. This could give rise to confusion. I shall therefore distinguish a series $\sum_{n=1}^{\infty} a_n$ from a sequence $\{a_n\}_{n=1}^{\infty}$. Second, prior to Cauchy, the term "convergence" usually denoted that the sequence a_n was decreasing and tended toward 0. I prefer to employ the term "convergence" in Cauchy's sense, namely a series is convergent if it has a finite sum, except for quotation and some particular cases that are explicitly indicated.

[3] I shall discuss the notion of quantity later on. At this moment I use the term "geometric quantity" to refer to lines or other geometrical objects connected to a curve, such as ordinate, abscissa, arc length, subtangent, normal, area between curves and axes, etc.

1

that the coexistence of the two aspects of series theory was guaranteed

(a) by the assumption that the expansion into series of a given quantity was convergent at least for an interval of values of the variable,

(b) by the possibility of postponing the investigation of convergence to the phase of application of a certain series to specific geometrical, mechanical or numerical problems.

Only in a few isolated cases did mathematicians recognize tensions or difficulties between convergence and formal manipulation.

Part I is divided into nine chapters. In the first chapter, I examine the earliest researches on series, infinite products, and continued fractions mainly by examining the works of Viète, Grégoire de Saint-Vincent, Mengoli, Wallis, Mercator, and James Gregory. In the following five chapters, I explore the early theory of series with particular attention to the relationship between convergence and formal manipulation and to the geometrical context in which the theory was originated. I concentrate upon the writings of Leibniz (Chapters 2 and 3), Johann Bernoulli (Chapter 3), Newton (Chapter 4), Jacob Bernoulli (Chapter 5), and Taylor (Chapter 6). This investigation provides the basis for an analysis of the notion of quantity and a comprehensive interpretation of early theory series (Chapters 7 and 8). Finally, Chapter 9 is devoted to the question of Grandi's series and the early debate on divergent series.

1 Series before the rise of the calculus

Even though series were occasionally found earlier, it is only from the 17th century that they began to be a topic of importance in mathematics. Their use mainly arose in the context of the problem of quadratures and rectifications of curves. During the 17th century, mathematicians attempted to find new methods for squaring curved lines, which avoided the difficulty of the so-called method of exhaustion.[4] This method, which had been one of the greatest successes of Greek geometry, made it possible to determine the area A of a given figure by means of a complex procedure divisible in two phases.

1. One or two sequences of polygons were constructed so that the areas of these polygons approximated to the given figure and suggested that the sought area A was equal to a certain area P.

2. One proved $A = P$ by means of a double *reductio ad absurdum* (namely, one showed that neither $A > P$ nor $A < P$ was true).

A classic example is the quadrature of the parabolic segment obtained by Archimedes[5]. As the first step in the proof, one considers the triangle ABC with area F, which is greater than one-half of the parabolic segment ACB with area P (see Fig. 1). Then, one considers the diameters $B_1 V_1$ and $B_1 V_2$ such that $AV_1 = AV_2 = \frac{AH}{2}$ and constructs the triangles $AB_1 C$ and $BB_2 C$. These triangles are greater than one-half of the corresponding parabolic segments $AB_1 C$ and $BB_2 C$. Moreover, both the triangles $AB_1 C$ and $BB_2 C$ are equal to $\frac{1}{8}F$ and, consequently, their sum is $\frac{1}{4}F$. The process can be continued so as to construct a sequence H_n such that

- H_n is a polygon formed by the sum of the triangles,

- at the nth step, the area of H_n is $\frac{1}{4^{n-1}}F$,

- the polygons H_n exhaust (namely, fill up entirely) the segment,

- the sum S_n of the areas of all the triangles up to the nth step is given by the finite geometric progression

$$S_n = F + \frac{F}{4} + \frac{F}{16} + \frac{F}{64} + \ldots + \frac{F}{4^{n-1}}. \tag{1}$$

After having shown that

$$S_n + \frac{1}{3}\frac{F}{4^{n-1}} = \frac{4}{3}F, \tag{2}$$

Archimedes proved that the area of the parabolic segment is $\frac{4}{3}F$ by reasoning as follows.

---

[4] The name is due to Grégoire de Saint-Vincent [1647, 740].

[5] See Archimedes [QA, 233–252].

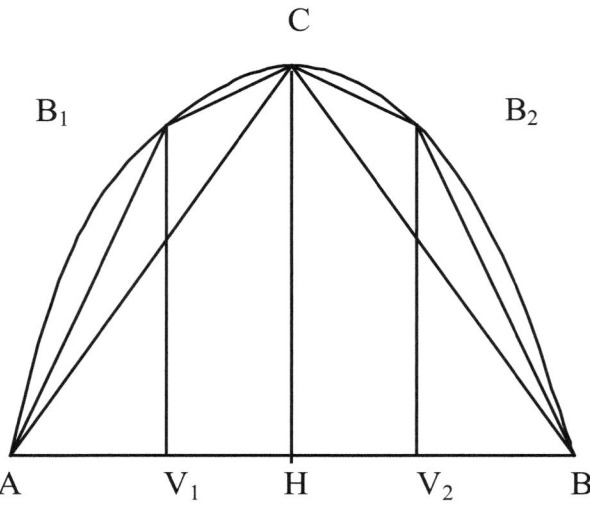

Fig. 1

- If $P > \frac{4}{3}F$, then $P - \frac{4}{3}F > 0$ and one can continue the exhaustion process until one obtains a sum S_n such that $P - S_n < P - \frac{3}{4}F$. Hence, $S_n > \frac{3}{4}F$, which contradicts formula (2).

- If $P < \frac{4}{3}F$, then $\frac{4}{3}F - P > 0$. Since the triangles constructed become increasingly smaller, at a certain step n, the area $\frac{F}{4^{n-1}}$ of the polygon H_n becomes less than $\frac{4}{3}F - P$. From (2), one obtains

$$\frac{4}{3}F - S_n = \frac{1}{3}\frac{F}{4^{n-1}} < \frac{F}{4^{n-1}} < \frac{4}{3}F - P.$$

Hence, $S_n > P$, which is impossible.

During the 17th century, the method of exhaustion was always considered as a model of a rigorous mathematical reasoning, although it was thought to be too difficult, especially because of the double *reductio ad absurdum*. It was also thought to be too particular, since it was connected to specific properties of certain geometrical figures and the reasoning used in a specific case could not be used in others. In effect, the method of exhaustion was not a method of finding or discovery, but rather it was a method of justification of known results. Consequently, mathematicians searched for new methods that were easier and had a more general application. This led in a very natural way to the consideration of series and even infinite products and continued fractions. For instance, in the above-mentioned quadrature of the parabola, it is possible to avoid the double *reductio ad absurdum* by using the series $\sum_{n=0}^{\infty} \frac{1}{4^n}$. It is no wonder that series are found in many 17th-century works concerning the quadratures of curves and almost all the precursors of the calculus run up against series. In particular, the attempt to merge Cavalieri's geometrical method of indivisibles with the emerging use of algebra led to the investigation of several series.

*   *   *

Geometric series played a crucial role in earlier research on series. In the 1590s, geometric series[6] were mentioned in a work by Francois Viète, *Variorum de rebus mathematicis responsorum*, in which he tackled the problem of the quadrature of circle. In this paper Viète determined the sum of a geometric series $\sum_{i=1}^{\infty} a_i$. His starting point was Proposition 12 in Book 5 of Euclid's *Elements*: If any number of magnitudes are proportional, then one of the antecedents is to one of the consequents as the sum of the antecedents is to the sum of the consequents (see Euclid [E]). In modern symbols, if $s_n = \sum_{i=1}^{n} a_i$, then

$$a_1 : a_2 = (s_n - a_n) : (s_n - a_1).$$

Hence,

$$\frac{a_1 - a_2}{a_1} = \frac{a_1 - a_n}{s_n - a_n}.$$

By assuming that the terms of the geometric series were decreasing, Viète obtained

$$\frac{a_1 - a_2}{a_1} = \frac{a_1}{s}, \qquad (3)$$

where $s = \sum_{i=1}^{\infty} a_i$. He justified (3) by stating that the magnitudes a_n were changed into nothing (*in nihil*) when the series was continued *ad infinitum*.[7] As an example, Viète considered the series

$$\sum_{n=0}^{\infty} \frac{1}{4^n} = \frac{4}{3}$$

and explicitly noted that it fitted the Archimedean quadrature of the parabola.

---

[6]It worthwhile pointing out that geometric series had already appeared earlier. N. Oresme dealt with the nature and summation of geometric series in a manuscript, the *Quaestiones super geometriam Euclidis,* which was only published in 1961 (On Oresme's treatment of series, see Mazet [2003]). Oresme's results seem to have had little influence on the rise of series theory in 17th century.

[7]See Viète [1593, 397–398].

A few decades later, Grégoire de Saint-Vincent made geometric series a crucial instrument of his method of quadratures.[8] He wrote a remarkable treatise, the *Opus geometricum,* devoted to the quadrature of conics, which was published in 1647 though its essential aspects dated back to before 1625. Grégoire observed that the classic problems inherited from the Ancients had not been solved after many centuries; he therefore thought that new techniques and new methods needed to be discovered (*unde novas artes et methodo novas iudicam excogitandas*) to fill the lacunae of ancient geometry [1647, 51–52]. Such new methods were grounded precisely on infinite geometric series, which he discussed at length in the second book of the *Opus geometricum.*

Saint-Vincent defined a geometric series to be "a finite quantity divided by an uninterrupted sequence according to a given ratio" and distinguished series from progressions [1647, 54]. He used the term "progression" to mean both a finite sequence of the terms of a geometric series (which he understood as infinite) and the sum of this finite sequence. Saint-Vincent used the term "limit" to denote the sum of a geometric series and stated that the "limit" of a progression was the end of the series that the progression did not reach –even if it continued indefinitely; however, the progression could approach this limit more than any given quantity [1647, 54].

Saint-Vincent, as well Viète, had an intuitive but clear idea of what the sum of series was (whatever words they used to denote the sum). By using more recent terminology, we could state that, in their opinion, a series $\sum_{k=0}^{\infty} a_k$ had a sum S if the sequence of nth sums $S_n = \sum_{k=0}^{n} a_k$ was convergent to S; namely, if it approached S indefinitely when n increased so that the difference between S_n and S (in absolute value) became less than any given quantity. As we shall see below, this idea of the sum lay at the heart of the series theory during both the 17th century and, in a more complicated form, the 18th century.

Basing his argument on the concept of the sum, Saint-Vincent examined the famous paradox of Achilles and the turtle. He showed that Achilles gains on the turtle according to a decreasing geometric series, which has a finite sum. Therefore, Achilles does reach the turtle and one can also determine the point where the turtle is reached by summing the series [1647, 97–98].

Saint-Vincent obtained several results by applying geometric series.[9] One of the most interesting is the following proposition concerning the quadrature of the hyperbola:

> *Let AY and AX be the asymptote of the hyperbola HKM* [see Fig. 2]. *If the segment AX is divided into segments AB, AC,*

---

[8]On Grégroire de Saint-Vincent, see Dhombres [1995].

[9]Saint-Vincent, in particular, determined the sums of $\sum_{n=0}^{\infty} q^n$ and $\sum_{n=0}^{\infty} q^{kn}$, for an integer k [1647, 115–149] and constructed two geometric series with different n-th terms but with the same sum [1647, 97–98].

CD, DE that are in continued proportion, then the areas BCKH,
CDLK,DEMN are equal[10] (Saint Vincent [1647, 586]).

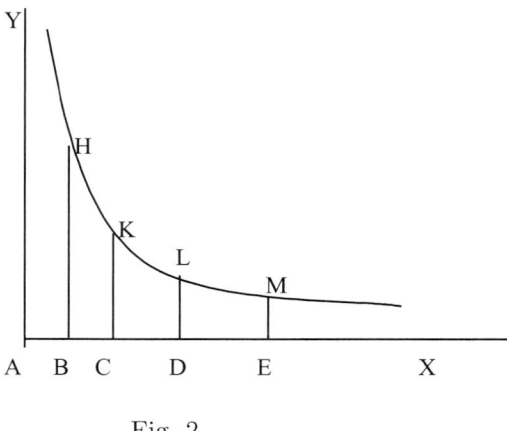

Fig. 2

Saint-Vincent did not employ the term "logarithm". This term had already been introduced at that time although it was used with a meaning that
differed considerably from the modern one. Indeed, the word "logarithms"
denoted the terms of an arithmetical progression that were matched with
the terms of a geometric progression in sequence[11]. By using the term logarithm in this sense, Saint-Vincent's theorem can be formulated by stating
that the areas $BCKH, CDLK, DEMN$ are the logarithms of the abscissas
of the hyperbola HKM. This formulation was made explicit by de Sarasa
in 1649.[12]

*   *   *

Pietro Mengoli was another mathematician who made a remarkable contribution to the rising theory of series. He was taught mathematics by
Cavalieri and was influenced by Saint-Vincent and Torricelli[13]. In 1650,

---

[10]In other words, if the abscissa are in a geometric progression, then the areas are in an
arithmetic progression.

[11]See Burn [2001, 4].

[12]As regards different historical interpretations of the actual contributions of Saint-
Vincent and de Serasa to the study of natural logarithms, see Burn [2001].

[13]I point out that Torricelli gave a geometric proof of the sum of a geometric series in his
De dimensione Parabolae [1644]. For Torricelli's proof, I refer to Panza [1992, 307–308].
A similar geometrical proof, given by Leibniz, is discussed in Chapter 2.

Mengoli published several results concerning series in *Novae quadraturae arithmeticae, seu de additione fractionum,* a treatise that stemmed from the examination of the Archimedean quadrature of parabola, as he stated in the introduction.[14] Mengoli based his argument upon two axioms:

1. If infinite magnitudes have an infinite extension, then one can take a certain number of these magnitudes such that they exceed any finite extension (In modern terms, if the sum of a series is infinite, then the partial sums become greater than any positive number) (Mengoli [1650, 18]).

2. If infinite magnitudes have a finite extension and if they are thought of as being arranged and gathered together to form another extension, then these two extensions are equal (that is to say, if a series with positive terms[15] converges to a finite number, then any rearrangement of the series converges to the same number) (Mengoli [1650, 19]).

From these axioms Mengoli derived various properties of the series of magnitudes. In particular,

a. if the sum of any number of a sequence of infinite quantities is bounded, then the series has a finite extension (in modern words, if the partial sums of a series are bounded, the series is convergent) (Mengoli [1650, 18]);

b. if a series has the finite extension S and A is a quantity less than S, then there is a finite number of the given magnitudes such that their sum exceeds A [namely, there exists a partial sum S_n of the series such that $S_n < A\ (< S_{n+1})$] (Mengoli [1650, 19]).

Mengoli applied these axioms and properties to the determination of the sum of various numerical series by conceiving the numbers present in such series as specific values of geometric quantities. He represented the terms, partial sums and remainder of series by means of segments. In order to sum the series

$$\sum_{n=1}^{\infty} \frac{1}{n(n+1)},$$

Mengoli employed a relation, which he had proved in his *Novae quadraturae arithmeticae* [1650, 9] and which, using modern symbols, can be written as

$$\frac{a_2 - a_1}{a_1 a_2} + \frac{a_3 - a_2}{a_2 a_3} + \frac{a_4 - a_3}{a_3 a_4} + \ldots + \frac{a_n - a_{n-1}}{a_{n-1} a_n} = \frac{a_n - a_1}{a_1 a_n},$$

---

[14]On Mengoli's contribution to series theory, see Agostini [1941].

[15]Since Mengoli referred to geometrical quantities, he tacitly assumed that the terms of series were positive.

This formula makes it possible to establish that the partial sums of $\sum_{n=1}^{\infty} \frac{1}{n(n+1)}$ are

$$S_n = \frac{1}{1 \cdot 2} + \frac{1}{2 \cdot 3} + \ldots + \frac{1}{n(n+1)} = \frac{n}{n+1}, \tag{4}$$

Since $\frac{n}{n+1} < 1$, the series has a finite extension S. This extension is precisely equal to 1. Indeed, if $S > 1$, then there should exist a partial sum S_n such that $S_n > 1$, which is impossible. Now, let $S < 1$ be. Since the numbers $\frac{n}{n+1}$ approach 1 indefinitely when n increases, the partial sums

$$S_n = \frac{n}{n+1}$$

would become greater than S when n is large enough. This is also impossible. Consequently, $S = 1$.

Similarly, Mengoli obtained the sum of many other series, such as

$$\sum_{n=1}^{\infty} \frac{1}{n(n+2)} = \frac{3}{4},$$

$$\sum_{n=1}^{\infty} \frac{1}{n(n+3)} = \frac{11}{18},$$

$$\sum_{n=1}^{\infty} \frac{1}{n(n+1)(n+2)} = \frac{1}{4},$$

$$\sum_{n=1}^{\infty} \frac{1}{(2n+1)(2n+3)(2n+5)} = \frac{1}{12}.$$

Moreover, in the introduction to *Novae quadraturae arithmeticae*, Mengoli showed that the harmonic series did not converge.[16] In modern terms, his proof can be formulated as follows. Since

$$\frac{1}{n-1} + \frac{1}{n} + \frac{1}{n+1} > \frac{3}{n},$$

one has

$$\begin{aligned}
S &= 1 + \frac{1}{2} + \frac{1}{3} + \frac{1}{4} + \frac{1}{5} + \frac{1}{6} + \frac{1}{7} + \frac{1}{8} + \frac{1}{9} + \frac{1}{10} + \ldots \\
&= 1 + \left(\frac{1}{2} + \frac{1}{3} + \frac{1}{4} \right) + \left(\frac{1}{5} + \frac{1}{6} + \frac{1}{7} \right) + \left(\frac{1}{8} + \frac{1}{9} + \frac{1}{10} + \right) \ldots \\
&> 1 + \frac{3}{3} + \frac{3}{6} + \frac{3}{9} + \frac{3}{12} + \ldots \\
&= 1 + 1 + \frac{1}{2} + \frac{1}{3} + \frac{1}{4} + \ldots \\
&= 1 + S.
\end{aligned}$$

---

[16]This result was not new (see Oresme [A]).

Consequently, S cannot be a finite quantity.

In the introduction to *Novae quadraturae arithmeticae,* Mengoli also took the series

$$\sum_{n=0}^{\infty} \frac{1}{n^2}$$

into consideration. He failed to calculate the sum of such a series, but this problem was subsequently tackled by Jakob Bernoulli, and became known as the Basel problem. It was considered a very interesting problem of pure mathematics and its solution was one of Euler's most important successes.

Mengoli also wrote *Geometriae speciosae elementa*[17] [1659] and *Circolo* [1672], where he rediscovered an infinite product expansion for $\pi/2$, which had already been found by Wallis in a way that I shall now go on to examine.

*   *   *

The use of series was worked on extensively by John Wallis. In *Arithmetica infinitorum* [1656] he tried to provide an arithmetical version of the method of indivisibles; this led him to deal with a large number of series by means of a peculiar methodology that had an enormous influence on later mathematicians.[18] As Maierù [1994, 118–119] noted, Wallis's treatment of series developed in a number of particular cases and makes use of the specific geometric properties of particular figures. To illustrate Wallis's method,[19] consider the problem of finding the area under the curves $y = x^k$ $(k = 1, 2, \dots)$ and over the segment $[0, a]$ (see Fig. 3, where the curve $y = x^k$ is represented by means of PSR, $PQ = AB = a$, and $RQ = BC = a^k$). Following Cavalieri, Wallis regarded the figure PQR as consisting of an infinite number of parallel lines, every one of them having length equal to x^k. Therefore, if one divides the segment $PQ = AB = a$ into n pieces of length $h = \frac{a}{n}$, where n is infinite, the sum of these infinite lines is of the type

$$0^k + h^k + (2h)^k + (3h)^k + \ldots + (nh)^k, \ k = 1, 2, \ldots. \tag{5}$$

Similarly, the area of the rectangle is

$$a^k + a^k + a^k + \ldots + a^k = (nh)^k + (nh)^k + (nh)^k + \ldots + (nh)^k, \ k = 1, 2, \ldots.$$

The ratio between the parabola PQR and the rectangle $ABCD$ is

$$\frac{\textit{Area} \text{ parabola } PSR}{\textit{Area} \text{ rectangle } ABCD} = \frac{0^k + 1^k + 2^k + 3^k + \ldots + n^k}{n^k + n^k + n^k + n^k + \ldots + n^k}, \ k = 1, 2, \ldots. \tag{6}$$

This procedure led Wallis to consider the problem of determining the values

---

[17]For this work, I refer to Massa [1997].

[18]On Wallis's method of quadrature, see Scott [1938], Panza [1995, 135–176], and Maierù [1994], [1995], and [2000].

[19]See Wallis [1656, 1–52].

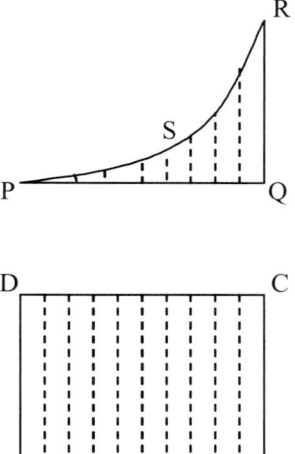

Fig. 3

of

$$\frac{0^k + 1^k + 2^k + 3^k + \ldots + n^k}{n^k + n^k + n^k + n^k + \ldots + n^k},\tag{7}$$

for $n = \infty$ and $k = 1, 2, 3, \ldots$[20] He stated

> The simplest method of investigation ... is to consider a certain number of individual cases, and to observe the emergent ratios, and to compare these with one another, so that a universal proposition may be established by induction. (Wallis [1656, 1])

He first considered the case $k = 1$ and observed that

$$
\begin{array}{ll}
\frac{0+1}{1+1} & = \frac{1}{2} \\
\frac{0+1+2}{2+2+2} & = \frac{1}{2} \\
\frac{0+1+2+3}{3+3+3+3} & = \frac{1}{2} \\
\frac{0+1+2+3+4}{4+4+4+4+4} & = \frac{1}{2}.
\end{array}
$$

---

[20]In modern terms, he sought

$$\lim_{n \to \infty} \frac{1^k + 2^k + 3^k + \ldots + n^k}{(n+1)n^k}.$$

The divergent series $\sum_j j^k$ appears in Wallis's work as intermediate steps during the analytical manipulation of geometrical entities.

By induction, Wallis asserted that

$$\frac{0+1+2+3+\ldots+n}{n+n+n+n+\ldots+n} = \frac{1}{2}.$$

Wallis then considered the case $k=2$. Since

$$
\begin{array}{rcl}
\frac{0+1}{1+1} & = \frac{1}{2} & = \frac{1}{3} + \frac{1}{6} \\
\frac{0+1+4}{4+4+4} & = \frac{5}{12} & = \frac{1}{3} + \frac{1}{6\cdot2} \\
\frac{0+1+4+9}{9+9+9+9} & = \frac{14}{36} & = \frac{1}{3} + \frac{1}{6\cdot3} \\
\frac{0+1+4+9+16}{16+16+16+16+16} & = \frac{30}{80} & = \frac{1}{3} + \frac{1}{6\cdot4},
\end{array}
$$

he stated

$$\frac{0^2 + 1^2 + 2^2 + 3^2 + \ldots + n^2}{n^2 + n^2 + n^2 + n^2 + \ldots + n^2} = \frac{1}{3} + \frac{1}{6n}.$$

The ratio approached $\frac{1}{3}$ as the number of terms increased, and

$$\frac{0^2 + 1^2 + 2^2 + 3^2 + \ldots + n^2}{n^2 + n^2 + n^2 + n^2 + \ldots + n^2} = \frac{1}{3}$$

for $n=\infty$. In case $k=3$, Wallis proceeded in a similar way and found

$$\frac{0^3 + 1^3 + 2^3 + 3^3 + \ldots + n^3}{n^3 + n^3 + n^3 + n^3 + \ldots + n^3} = \frac{1}{4} + \frac{1}{4n}$$

and

$$\frac{0^3 + 1^3 + 2^3 + 3^3 + \ldots + n^3}{n^3 + n^3 + n^3 + n^3 + \ldots + n^3} = \frac{1}{4}$$

for $n=\infty$. By generalizing these results, Wallis asserted[21]

$$\frac{0^k + 1^k + 2^k + 3^k + \ldots + n^k}{n^k + n^k + n^k + n^k + \ldots + n^k} = \frac{1}{k+1}. \tag{8}$$

Wallis did not stop here. He continued to generalize in order to give a meaning to (8) even when $k \neq 1, 2, 3, \ldots$. He first stated that if the value 0 was assigned to k, then one obtained

$$\frac{0^0 + 1^0 + 2^0 + 3^0 + \ldots + n^0}{n^0 + n^0 + n^0 + n^0 + \ldots + n^0} = \frac{1}{1}.$$

He then sought to justify the assignment of fractional values to k in the following way. If we denote[22] the series $0^k + 1^k + 2^k + 3^k + \ldots + n^k$ by A_k,

---

[21] Of course, from this formula and (6), one can deduce that the area under the parabola $y = x^k$ from 0 to a is $\frac{a^{k+1}}{k+1}$ $\left(= \sum_{j=0}^{n} j^k \right)$.

[22] The symbolism is mine.

and the reciprocal of their sums (the corresponding ratio, in Wallis's terms) by $b_k (= k + 1)$, then formula (8) can be written in the form

$$\frac{A_k}{n^k + n^k + \ldots + n^k} = \frac{1}{b_k}.$$

Wallis observed that

$$\sqrt{0^4} + \sqrt{1^4} + \sqrt{2^4} + \sqrt{3^4} + \ldots = 0^2 + 1^2 + 2^2 + 3^2 + \ldots .$$

The terms $n^2 = \sqrt{n^4}$ of A_2 are the square roots of the terms of A_4 and, therefore, A_2 can be viewed as the series "interpolating" A_0 and A_4. The corresponding ratios of A_0, A_2, and A_4 are the numbers $b_0 = 1$, $b_2 = 3$, and $b_4 = 5$, which are in arithmetic progression (see table below).

$A_0 = 0^0 + 1^0 + 2^0 + 3^0 + \ldots + n^0$	$->$	$b_0 = 1$
$A_2 = 0^2 + 1^2 + 2^2 + 3^2 + \ldots + n^2$	$->$	$b_2 = 3$
$A_4 = 0^4 + 1^4 + 2^4 + 3^4 + \ldots + n^4$	$->$	$b_4 = 5$

At this point Wallis considered $A_{\sqrt{}} = \sqrt{0} + \sqrt{1} + \sqrt{2} + \sqrt{3} + \ldots$ and stated that it was the series interpolating A_0 and A_1 since it behaved with respect to A_0 and A_1 as A_2 behaved with respect to A_0 and A_4. By analogy, the value of

$$\frac{\sqrt{0} + \sqrt{1} + \sqrt{2} + \sqrt{3} + \ldots + \sqrt{n}}{\sqrt{n} + \sqrt{n} + \sqrt{n} + \sqrt{n} + \ldots + \sqrt{n}}$$

ought to be a number $\frac{1}{b}$ such that $b_0 = 1$, b and $b_1 = 2$ (namely, the corresponding ratios of $A_0 = 1$, $A_{\sqrt{}}$, and $A_1 = 2$) were in arithmetic progression. Hence, $b = \frac{1}{2} + 1$. By observing that for $k = \frac{1}{2}$, formula (8) becomes

$$\frac{0^{1/2} + 1^{1/2} + 2^{1/2} + 3^{1/2} + \ldots + n^{1/2}}{n^{1/2} + n^{1/2} + n^{1/2} + n^{1/2} + \ldots + n^{1/2}} = \frac{1}{\frac{1}{2} + 1},$$

Wallis concluded that $n^{\frac{1}{2}} = \sqrt{n}$.

Similarly, Wallis observed that

$$A_1 = \sqrt[3]{0^3} + \sqrt[3]{1^3} + \sqrt[3]{2^3} + \sqrt[3]{3^3} + \ldots = 0 + 1 + 2 + 3 + \ldots,$$

$$A_2 = \left(\sqrt[3]{0^3}\right)^2 + \left(\sqrt[3]{1^3}\right)^2 + \left(\sqrt[3]{2^3}\right)^2 + \left(\sqrt[3]{3^3}\right)^2 + \ldots = 0^2 + 1^2 + 2^2 + 3^2 + \ldots,$$

and that the terms $n = \sqrt[3]{n^3}$ of A_1 were the cube roots of the terms of A_3 and the terms $n^2 = \left(\sqrt[3]{n^3}\right)^2$ of A_2 were the squares of cube roots. For this reason A_1 and A_2 could be viewed as the series interpolating A_0 and A_3. The corresponding ratios of A_0, A_1, A_2, A_3 ($b_0 = 1$, $b_1 = 2$, $b_2 = 3$, $b_3 = 4$) were in arithmetical progression. Wallis then considered

$$\frac{\sqrt[3]{0} + \sqrt[3]{1} + \sqrt[3]{2} + \sqrt[3]{3} + \ldots + \sqrt[3]{n}}{\sqrt[3]{n} + \sqrt[3]{n} + \sqrt[3]{n} + \sqrt[3]{n} + \ldots \sqrt[3]{n}} = \frac{1}{r} \qquad (9)$$

and

$$\frac{\left(\sqrt[3]{0}\right)^2 + \left(\sqrt[3]{1}\right)^2 + \left(\sqrt[3]{2}\right)^2 + \left(\sqrt[3]{3}\right)^2 + \ldots + \left(\sqrt[3]{n}\right)^2}{\left(\sqrt[3]{n}\right)^2 + \left(\sqrt[3]{n}\right)^2 + \left(\sqrt[3]{n}\right)^2 + \left(\sqrt[3]{n}\right)^2 + \ldots \left(\sqrt[3]{n}\right)^2} = \frac{1}{q} \tag{10}$$

and assumed that

$$\sqrt[3]{0} + \sqrt[3]{1} + \sqrt[3]{2} + \sqrt[3]{3} \quad \text{and} \quad \left(\sqrt[3]{0}\right)^2 + \left(\sqrt[3]{1}\right)^2 + \left(\sqrt[3]{2}\right)^2 + \left(\sqrt[3]{3}\right)^2$$

behaved with respect to

$$A_0 \quad \text{and} \quad A_1$$

in the same way as A_1 and A_2 behaved with respect to A_0 and A_3. By analogy he concluded that b_0, r, q, b_1 had to be in arithmetical progression as b_0, b_1, b_2, b_3. Therefore,

$$r = \frac{4}{3} \quad \text{and} \quad q = \frac{5}{3}.$$

This made it possible to write (9) and (10) as

$$\frac{0^{1/3} + 1^{1/3} + 2^{1/3} + 3^{1/3} + \ldots + n^{1/3}}{n^{1/3} + n^{1/3} + n^{1/3} + n^{1/3} + \ldots n^{1/3}} = \frac{1}{1 + \frac{1}{3}}$$

and

$$\frac{0^{2/3} + 1^{2/3} + 2^{2/3} + 3^{2/3} + \ldots + n^{2/3}}{n^{2/3} + n^{2/3} + n^{2/3} + n^{2/3} + \ldots n^{2/3}} = \frac{1}{1 + \frac{2}{3}}.$$

Consequently, $\sqrt[3]{n}$ was equal to $n^{1/3}$ and $\left(\sqrt[3]{n}\right)^2$ was equal to $n^{2/3}$. In this way Wallis was able to find the meaning of the power x^α, where α was a rational number ($n^{\frac{l}{k}} = \sqrt[k]{n^l}$). He even considered the case in which α was an irrational and a negative number.

    Wallis's analogical procedure (later known as Wallis's interpolation) was of great importance in the 18th century. It can be considered as an answer to the following problem:

> *Given a sequence P_k, defined for integral values of k, find the meaning of P_α where α is a nonintegral number.*

In the case Wallis considered, P_k were the sequences x^k and

$$\frac{0^k + 1^k + 2^k + 3^k + \ldots + n^k}{n^k + n^k + n^k + n^k + \ldots + n^k},$$

and the problem was reduced to the interpolation of the number sequence

$$\frac{1}{k+1}, \quad k = 0, 1, 2, \ldots.$$

From a modern point of view, this problem is meaningless. A modern mathematician attributes meaning to new operations, formulas, or symbols using appropriate definitions. Operations, formulas, and symbols do not have a "natural" meaning. Thus, if x^n is defined only for an integer value of n, then any meaning can be assigned to a new symbol such as $x^{1/2}$.

Wallis viewed the matter differently. New combinations of symbols, such as $x^{1/2}$ and x^0, were not introduced arbitrarily. Mathematical objects were not given by definition, but they existed in nature (or were an idealization of natural objects). It seemed obvious to them that $x^{1/2}$ and x^0 had a "natural" meaning and that mathematicians had to discover it. When a new symbol or a new object had to be introduced, mathematicians asked "What is the value (or the meaning) of the symbol?" and not "How shall we define it?"

For Wallis, interpolating x^n required investigating the objects x, x^2, ... and reconstructing the "nature" of these objects just as one reconstructed the nature of a physical phenomenon by interpolating physical data. When he met with the undefined symbolic notation $x^{1/2}$, he did not take $x^{1/2} = \sqrt{x}$ by a useful but arbitrary definition; rather he "discovered" that the true meaning[23] of $x^{1/2}$ was $\sqrt{x}$.

In *Arithmetica infinitorum*,[24] Wallis reduced the problem of the quadrature of the circle to determining the corresponding ratio of the series whose general term is $\zeta_p = \sqrt{R^2 - p^2 a^2}$. To do this, he considered the series whose general terms are

$$(R^2 - p^2 a^2)^0, \ (R^2 - p^2 a^2)^1, \ (R^2 - p^2 a^2)^2, \ (R^2 - p^2 a^2)^3, \ \ldots \qquad (11)$$

which have for their corresponding ratios

$$1, \ \frac{2}{3}, \ \frac{8}{15}, \ \frac{48}{105}, \ \ldots.$$

If the series $\zeta_p = \sqrt{R^2 - p^2 a^2}$ is interpolated between the first and second terms of (11), the corresponding ratio of

$$\zeta_p = \sqrt{R^2 - p^2 a^2}$$

---

[23]See also Ferraro [1998, 291–293].
[24]See Wallis [1656, 89–182].

is given by the interpolated value of 1, $\frac{2}{3}$, $\frac{8}{15}$, $\frac{48}{105}$, ... between 1 and $\frac{2}{3}$. Wallis introduced the symbol $\square$ to denote the sought-after number and constructed several numerical tables such as

1		1		1		1	
	$\square$						
1		2		3		4	
1		3		6		10	
1		4		10		20	

where

- the numbers in the first row and column are 1,

- those in the second row and column are the natural numbers $\{n\}_{n=1,2,...,\,\infty}$,

- those in the third row and column are $\left\{\frac{n(n+1)}{1\cdot2}\right\}_{n=1,2,...,\infty}$ (triangular numbers),

- those in the fourth row and column are $\left\{\frac{n(n+1)(n+2)}{1\cdot2\cdot3}\right\}_{n=1,2,...,\infty}$ (triangular pyramidal number),

- ...

After a long sequence of calculations, he succeeded in expressing $\square$ as the infinite product

$$\square = \left(\frac{4}{\pi}\right) = \frac{3\cdot3\cdot5\cdot5\cdot7\cdot7\ldots}{2\cdot4\cdot4\cdot6\cdot6\cdot8\ldots}. \tag{12}$$

* * *

Formula (12) was not the first infinite product to be found in the history of mathematics. In his *Variorum de rebus mathematicis responsorum* [1593], Viète had already squared the circle by means of an infinite product. He assumed the circle to be a polygon with infinite sides and considered regular inscribed polygons of 4, 8, 16, ... sides. By using geometric properties of these polygons he represented π in the form[25]

$$\frac{2}{\pi} = \sqrt{\frac{1}{2}}\sqrt{\frac{1}{2}+\frac{1}{2}\sqrt{\frac{1}{2}}}\sqrt{\frac{1}{2}+\frac{1}{2}\sqrt{\frac{1}{2}+\frac{1}{2}\sqrt{\frac{1}{2}}}}\ldots. \tag{13}$$

---

[25]See Viète [1593, 400].

In the *Arithmetica infinitorum*,[26] Wallis also published an expansion of $\frac{4}{\pi}$ into continued fractions. He had submitted (12) to Lord Brouncker, who expressed $\frac{4}{\pi}$ in the form

$$\frac{4}{\pi} = 1 + \frac{1}{2+} \frac{9}{2+} \frac{25}{2+} \frac{49}{2+}. \tag{14}$$

This formula was published by Wallis in the *Arithmetica Infinitorum*, Proposition 191. On this occasion Wallis introduced the term "continued fraction". However, he did not expound the procedure used by Lord Brouncker to derive (14).

It should be emphasised that when Brouncker obtained (14), continued fractions were already known, at least since 1613 when Cataldi had shown how a root $\sqrt{p}$ could be expanded into a continued fraction. Earlier, in his *Algebra* [1572, 37–38], Bombelli had published a procedure for calculating the approximate value of a root which can be interpreted *a posteriori* as a procedure for developing numbers into continued fractions.

To compute the value of $\sqrt{13}$, Bombelli first observed that 3 is the greatest integer less than $\sqrt{13}$. Then he considered the difference $\sqrt{13} - 3 = x$ (for the sake of simplicity, I use the letter x to denote this difference, though Bombelli did not use symbols of this kind). The first approximation of x (say x_1) is given by $\frac{2}{3}$ because $13 - 3^2 = 4$ and

$$x_1 = \frac{4}{2 \cdot 3} = \frac{4}{6} = \frac{2}{3}.$$

To find a second approximation x_2, he set

$$x_2 = \frac{4}{6 + \frac{2}{3}} = \frac{3}{5}.$$

The third approximation is

$$x_3 = \frac{4}{6 + \frac{3}{5}} = \frac{20}{33}.$$

Similarly, he found

$$x_4 = \frac{66}{109}, \quad x_5 = \frac{109}{180}, \quad x_6 = \frac{720}{1189}.$$

The approximation can be improved as desired.

In modern terms, Bombelli's procedure can be described as follows. If one sets

$$\sqrt{p} = n + x,$$

---

[26]See Wallis [1665, 181–193].

where n is the maximum integer such that $n^2 \leq p$, then one has

$$p - n^2 = 2nx + x^2.$$

By neglecting x^2, one obtains the first approximation

$$x_1 = \frac{r}{2n},$$

where $r = p - n^2$. By approximating $2n + x$ to $2n + x_1$, the equation $r = (2n + x)x$ can be written as $r = (2n + x_1)x_2$. Hence,

$$x_2 = \frac{r}{2n + x_1},$$

and so on. In this way, one derives a repeating continued fraction

$$\sqrt{p} = n + \cfrac{r}{2n + \cfrac{r}{2n + \cfrac{r}{2n + \cdots}}}$$

The absence of appropriate symbolism makes it hard to state whether Bombelli had grasped the idea of continued fractions. This idea is clear in Cataldi, who expounded the same procedure in his *Trattato* [1613]. He considered "the manner of finding roots, by adding step by step to the denominator of the fraction which is the last in the preceding root, a fraction equal the first one" [1613, 70]. As an example, he expressed the square root of 18 as a continued fraction and used the following symbols:

$$4 \& \frac{2}{8} \, \& \frac{2}{8} \, \& \frac{2}{8}$$

and

$$4 \& \frac{2}{8} \, \& \frac{2}{8} \, \& \frac{2}{8}$$

During the 17th century, other mathematicians used continued fractions. For instance, Christiaan Huygens used the convergents of continued fractions to approximate the correct design for the toothed wheels of a planetarium.[27] Wallis himself returned to continued fractions later in his *Algebra* [1685], where he expounded some of the now-familiar properties of convergents. However, it was only during the 18th century that new important results concerning continued fractions were obtained.

$$* \quad * \quad *$$

----

[27]See Huygens [P, 627–631].

In 1668, Nicholas Mercator published a book, entitled *Logarithmotech-nia*, in which he illustrated a technique for facilitating the calculation of logarithms. This technique was based on the series expansion of the natural logarithm[28]

$$\log(1 + x) = x - \frac{x^2}{2} + \frac{x^3}{3} - \dots \tag{15}$$

which he derived by using the results of Saint-Vincent and de Sarasa (namely, the fact that the logarithm is the area under the hyperbola $y = \frac{1}{1+x}$) and by applying Wallis's method of quadrature (see Mercator [1668, 28–33]). Mercator's procedure can be summarized as follows.[29] To compute the area under the hyperbola $\frac{1}{1+x}$ and above the segment $[1, A]$, one divides this segment into an equal number of parts (*in aequales partes innumeras*). Denoting each of the parts by h, the area of the hyperbola can be thought of as the sum of the ordinates

$$\frac{1}{1 + h}, \quad \frac{1}{1 + 2h}, \quad \frac{1}{1 + 3h}, \quad \dots, \quad \frac{1}{1 + nh},$$

where $nh = A$. By expanding these fractions by long division (I shall clarify the meaning of this term a few lines below), one obtains

$$\frac{1}{1 + h} = 1 - h + h^2 - h^3 + \dots,$$

$$\frac{1}{1 + 2h} = 1 - 2h + 4h^2 - 8h^3 + \dots,$$

$$\frac{1}{1 + 3h} = 1 - 3h + 9h^2 - 27h^3 + \dots,$$

$$\dots.$$

By summing column by column, the sought-after area comes to be equal to

$$(1 + 1 + 1 + \dots) \tag{16}$$
$$- (h + 2h + 3h + \dots)$$
$$+ (h^2 + 4h^2 + 9h^2 + \dots)$$
$$- (h^3 + 8h^3 + 27h^3 + \dots)$$
$$+ \dots.$$

The series $\sum_n (na)^k$, $k = 0, 1, 2, \dots$, in formula (16) are the same series used by Wallis and can be thought of as representing the areas of the parabolas

---

[28] The expansion of logarithm had already been found by Newton (see p. 56). However, Mercator was the first to publish it in 1668.

[29] Here I follow the simplified version of Mercator's proof that was expounded by Wallis in the same year (1668) in a letter to Brouncker published in *Philosophical Transactions* (see Wallis [1668, 753–754]).

$y = x^k$ between 0 to A. Therefore,

$$\sum_n (na)^k = \frac{A^{k+1}}{k+1}, \qquad k = 0, 1, 2, \ldots$$

and

$$\log(1 + A) = A - \frac{A^2}{2} + \frac{A^3}{3} - \ldots.$$

Mercator's result was improved by Gregory [1668, 9–13] and Halley [1695, 58–67], who provided the expansion

$$\log \frac{1+x}{1-x} = 2x + \frac{2x^3}{3} + \frac{2x^5}{5} + \ldots,$$

which is faster and more useful in calculations.

In his *Logarithmotechnia*, Proposition 7, Mercator used the method of long division, today know as one of Mercator's rules.[30] This method was based on the idea that the usual algebraic operations could be used to generate series and that *one could operate on a series in the same manner as one operated on a closed analytical expression*. In the specific case of *long division*, the usual algorithm to find the quotient between two polynomials was continued endlessly to obtain an infinite series. For instance, consider the fraction

$$\frac{1}{1+x}.$$

By dividing 1 by $1 + x$, one obtains the quotient 1 and the remainder $-x$. By dividing such a remainder by $1 + x$, one obtains the quotient x and the remainder $-x^2$. By continuing *in infinitum*, one obtains the series $1 - x + x^2 - \ldots$.

$$* \quad * \quad *$$

The last mathematician I shall discuss in this chapter is James Gregory. He made a number of remarkable contributions to series theory and some of his results overlapped with the findings of Newton. In his *Vera Circuli et Hyperbolae Quadratura* [1667], while investigating the areas of conic sections, he introduced the expression "convergent series" [1667, 10]. However, his definition is rather different from the modern one, even if it contains the basic idea of quantities approaching a limit. Gregory did not give the name of 'convergent series' to a series or sequence, but to a pair of sequences a_n and b_n so that

---

[30]The other is the extraction of roots (see Chapter 4). On Leibniz's treatment of long division, see Chapter 2.

a. they were defined by recurrence,[31] namely

$$a_n = f(a_{n-1}, b_{n-1}) \text{ and } b_n = g(a_{n-1}, b_{n-1}), \ n > 1$$

b. $|a_n - b_n| < |a_{n-1} - b_{n-1}|$.

He also gave the name "convergent terms" to the pair of terms (a_n, b_n) of the convergent sequences.

An example of convergent sequences in Gregory's sense is the following. Given the positive number a, b, c, d, e, with $b > a$, $c > d$, $c > e$, consider the sequences a_n and b_n thus defined:

$$a_1 \ = \ a, \qquad b_1 = b,$$
$$a_n \ = \ a_{n-1} + \frac{d}{c}(b_{n-1} - a_{n-1}), \qquad b_n = b_{n-1} - \frac{e}{c}(b_{n-1} - a_{n-1}).$$

It is not difficult to prove that

$$a_{n-1} < a_n < b_{n-1} \quad \text{and} \quad a_{n-1} < b_n < b_{n-1}.$$

Hence,

$$|b_n - a_n| < |b_{n-1} - a_{n-1}|.$$

The two sequences therefore approach each other more and more. Since

$$\frac{b_{n-1} - a_{n-1}}{b_n - a_n} = \frac{b_1 - a_1}{b_2 - a_2},$$

the ratio $\frac{b_{n-1} - a_{n-1}}{b_n - a_n}$ is a constant value (< 1) and the differences $|b_n - a_n|$ becomes less than each given quantity. Gregory remarked

> Then, if we imagine that the series is continued in infinitum, we can imagine the *last* convergent terms to be equal. We call the equal terms the termination of the series. (Gregory [1667, 18–19, my emphasis])

In *Vera Circuli et Hyperbolae Quadratura*, Gregory applied his notion of convergence to the investigation of the problem of the quadrature of the circle by means of appropriate sequences of regular n-polygons, inscribed and circumscribed to a sector of a conic section.

---

[31] Condition a mainly depended on the special context of Gregory's research (the quadrature of conic sections).

In a letter written on November 23, 1670 to Collins,[32] Gregory provided the formula of interpolation, which today is named after him and Newton:[33]

$$f(x_0 + ct) = f(x_0) + t\Delta f(x_0) + \frac{t(t-1)}{2!}\Delta^2 f(x_0)$$
$$+ \frac{t(t-1)(t-2)}{3!}\Delta^3 f(x_0) + \dots, \tag{17}$$

This formula gives the interpolated value of a function $f(x)$ at the point $x_0 + ct$ by using the forward differences[34] $\Delta^h f(x_0)$, $h = 1, 2, 3, \dots$ at the point x_0. Gregory expressed (17) in the form

$$\alpha\gamma = \frac{ad}{c} + \frac{bf}{c} + \frac{kh}{c} + \frac{li}{c} + \text{etc.},$$

where $\alpha\gamma$ is the interpolated value of the ordinate of a given curve ABH at the point $\frac{a}{c}$, the letters d, f, k, $\dots$ are the first, second, third, etc. differences taken at the point $x_0 = 0$, and $\frac{b}{c}$, $\frac{k}{c}$, $\dots$ denote

$$\frac{a(a-c)}{c \cdot 2c}, \quad \frac{a(a-c)(a-2c)}{c \cdot 2c \cdot 3c}, \quad \dots,$$

respectively. He assumed that the ordinate of the curve at the point 0 is 0.

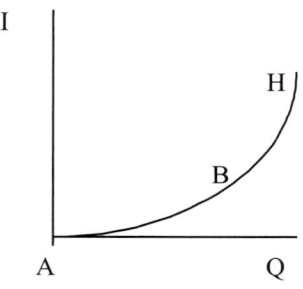

Fig. 4

Gregory gave no proof of his formula and observed

> This method, as I apprehend, is both more easie and universal than either Briggs or Mercator's,[35] and also performed without table. (Turnbull [GT, 131])

---

[32] See Turnbull [GT, 131].

[33] As regards the use of this formula before Gregory, see the first chapter of Goldstine [1977].

[34] Forward differences are defined as follows:
$$\Delta f(x_0) = f(x_0 + h) - f(x_0) \quad (h \text{ is a constant}),$$
$$\Delta^n f(x_0) = \Delta^{n-1} f(x_0 + h) - \Delta^{n-1} f(x_0), n = 2, 3, 4, \dots.$$

[35] Gregory referred to Briggs's *Arithmetica logarithmica* and Mercator's *Logarithmotech-nia*.

Gregory applied the interpolation formula to logarithms. Indeed, he considered the problem:

> *Given b, d, $e = \log b$ and $e + c = \log(b + d)$, find the number whose logarithm is $e + a$*

and stated the the sought-after number is[36]

$$b + \frac{a}{c}d + \frac{a(a-c)}{c \cdot 2c}\frac{d^2}{b} + \frac{a(a-c)(a-2c)}{c \cdot 2c \cdot 3c}\frac{d^3}{b^2} + \ldots \qquad (18)$$

To obtain (18), one can observe that

$$
\begin{aligned}
e + a &= \log b + \frac{a}{c}(e + c - e) = \log b + \frac{a}{c}(\log(b + d) - \log e) \\
&= \log\left(b\left(1 + \frac{b}{d}\right)^{\frac{a}{c}}\right)
\end{aligned}
$$

and then the sought-after number is $b\left(1 + \frac{b}{d}\right)^{\frac{a}{c}}$. If one considers the differences

$$
\begin{aligned}
\Delta(1+p)^z &= p(1+p)^z, \\
\Delta^2(1+p)^z &= p^2(1+p)^z, \\
\Delta^3(1+p)^z &= p^3(1+p)^z, \\
&\cdots
\end{aligned}
$$

of the polynomial $(1+p)^z$, one obtains

$$[\Delta^n(1+p)^z]_{z=0} = p^n.$$

By applying (17) to $(1+p)^z$ for $p = \frac{d}{b}$ and $z = \frac{a}{c}$, one finds[37] (18).

In another letter to Collins written on February 15, 1671,[38] Gregory referred to seven power series expansions for the quantities which, using modern symbols terminology, would be written as

$$
\arctan x, \qquad \tan x, \qquad \sec x, \qquad \log \sec x,
$$
$$
\log \tan\left(\frac{x}{2} + \frac{\pi}{4}\right), \qquad \sec^{-1}\left(\sqrt{2}e^x\right), \qquad 2\arctan\left(\tanh x\right).
$$

---

[36] One could note that expansion (18) includes the binomial theorem. Newton became aware of this from the winter of 1664–65. On the priority of the discovery, see Pensivy [1987–88, 45–48].

[37] Cf. Pensivy [1987–88, 43] and Goldstine [1977, 76].

[38] See Turnbull [GT, 170–171].

The first three expansions were written by Gregory in the form

$$a = t - \frac{t^3}{3r^2} + \frac{t^5}{5r^4} - \frac{t^7}{7r^6} + \frac{t^9}{9r^9},$$

$$t = a + \frac{a^3}{3r^2} + \frac{2a^5}{15r^4} + \frac{17a^8}{315r^6} + \frac{3233a^9}{181440r^8},$$

$$s = r + \frac{a^2}{2r} + \frac{5a^4}{24r^3} + \frac{61a^6}{720r^5} + \frac{277a^8}{8064r^7},$$

where r is the radius of a circle, a the arc of this circle cut off by the angle θ, $\frac{t}{r}$ the tangent of θ, and $\frac{s}{r}$ the secant of θ.

In his *James Gregory Tercentenary Memorial Volume* [GT, 26], Turnbull hypothesized that Gregory might have used a procedure corresponding to the use of the Taylor formula to obtain some of his expansions. He based his argument upon an error of calculation of the series for arctangent (in effect the fifth term is $\frac{3968a^9}{181440r^8}$ and not $\frac{3233a^9}{181440r^8}$). In his *James Gregorie on Tangents and the "Taylor" Rule of Series Expansions* [1993], Malet states convincingly that Gregory calculated the coefficient of any of the expansions sent to Collins by means of the following recursive procedure:

> Let
>
> $$y = y_0 + A_1^* x + A_2^* \frac{x^2}{r} + A_3^* \frac{x^3}{r^2} + \text{etc.}$$
>
> be any of Gregorie's [expansions]. [...] The constant r and the term r^i are introduced to preserve the homogeneity of variable y [...] the y always were the ordinates of some curves [...] y_0 is the value taken of y at a given point, say x_0, and the coefficients A_i^* are the numerical values taken by the variables A_i at the point x_0. In order to find the variable A_1, Gregory applied his new method of tangent [...] to the curve the ordinate of which are y, thereby determining its subtangent, say T [...] A_1 was then determined as $\frac{y}{T}r$ [...] Next Gregorie would assume A_1 to represent the ordinates of a curve having the same axis as the y-curve and calculate its subtangent [...] and so on. (Malet [1993, 100–101])

2 Geometrical quantities and series in Leibniz

Leibniz began to study higher mathematics in 1672 following Huygens' prompting. In a short time he became interested in infinite series. It is well-known that the investigation of number sequences and their difference and sum sequences was of great importance in his discovery of the calculus.[39] He himself asserted

> I arrived at the method of inassignables through the method of infinite increments in the series of numbers, as the nature of the things requires. (Leibniz [GMS, 4:413])

Between 1675 and 1676, he wrote a treatise, entitled *De quadratura arithmetica circuli ellipseos et hyperbolae cujus corollarium est trigonometria sine tabulis* [KQA], which aimed to provide the quadrature of certain curves by means of series. Leibniz wrote at least six versions of this treatise, which nevertheless remained unpublished through his life. Only recently has Eberhard Knobloch published its last and most extensive version, which consists of 51 propositions and many scholia.[40] In the years that followed, Leibniz published many of the results of the *De quadratura arithmetica* (but not the proofs and the solution methods) in *De vera proportione circuli ad quadratum circumscriptum in numeris rationalibus expressa* [1682] and in *Quadratura arithmetica communis sectionum conicarum* [1691].

Then Leibniz wrote other important papers concerning series, in particular *Supplementum geometriae practicae sese ad problemata transcendentia extendens, ope novae methodi generalissimae per series infinitas* [1693] and *Epistola ad V. Cl. Christianum Wolfium Grandi* [1710].

This chapter is divided into two sections. The first is devoted to the investigation of Leibniz's notion of convergence and the way in which he manipulated series. In the second section, I shall examine how power series were employed in the geometrical context of the early calculus. Other aspects of Leibniz's conception (Leibniz's derivation of the Bernoulli series, Leibniz's analogy, the rise of the question of divergent series) are discussed in Chapters 3 and 9.

2.1 The capacity of series to express quantities and their manipulation

In *De quadratura arithmetica*, Leibniz formulated the widelyknown theorem on the sum of a geometric series as follows:

> *The greatest term of an infinite geometric series is the mean proportional between the greatest sum and the greatest difference.*
> (see Leibniz [KQA, 71])

---

[39]See Bos [1974, 13] and Guicciardini [1999, 137–138].
[40]For an analysis of the manuscript, see Knobloch [1989] and [1991].

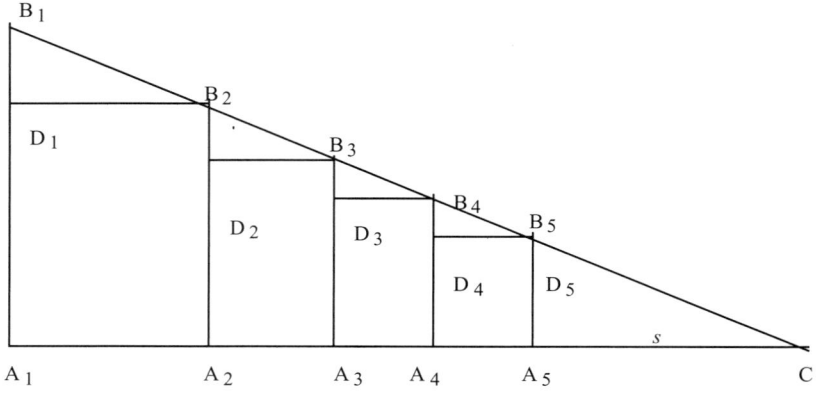

Fig. 5

Leibniz's proof is geometric and runs as follows. Given a decreasing geometric series a_n and a straight line s, one draws a segment $A_1 B_1$ so that it equals the first term a_1 of the series and is perpendicular s (see Fig. 5). One takes A_2 on the line s such that $A_1 A_2 = A_1 B_1$ and draws the segment $A_2 B_2$ so that it equals the second term a_2 of the series and is perpendicular to s. Then one takes A_3 on s such that $A_2 A_3 = A_2 B_2$ and so on. The point B_n fall on the line $B_1 B_2$, which intersects s in C. Leibniz states that the segment $A_1 C$ is the sum of the series. Indeed the triangles $D_n B_n B_{n+1}$ are similar to $A_n B_n C$, and therefore the segments $A_n C$ are proportional to

$$D_n B_{n+1} = A_n A_{n+1} = A_n B_n.$$

It is true that no A_n coincides with C; however, the segments $A_n B_n$ become smaller than any quantity and the sequence A_n approaches C closer and closer, with an error less than any assignable quantity. Since

$$A_1 C : A_1 B_1 = A_1 B_1 : D_1 B_1,$$

one has

$$S : a_1 = a_1 : (a_1 - a_2),$$

where S is the sum of the series (see Leibniz [KQA, 71–73]).

Leibniz, like the other mathematicians considered in chapter 1, thought that the sum of a series was a determinate quantity (the segment $A_1 C$) to which the partial sums of the series approached increasingly. Leibniz's notion of the sum resembles the modern one; however, it presents some aspect that makes these two notions different.[41] These aspects concern:

1. the relationship between a series and the quantity expressing this series,

2. the way Leibniz handled series.

As regards the first point, I would emphasise that, according to Leibniz, a series could express exact relationships and not merely approximations; however, *a series expressed an exact relationship insofar as it is considered a whole and not a limit process*. Indeed, in *De vera proportione*,[42] Leibniz justified that $\frac{\pi}{4}$ is equal to

$$1 - \frac{1}{3} + \frac{1}{5} - \frac{1}{7} + \dots$$

by observing that

- if we take the first term of this series, then $\frac{\pi}{4}$ is approximated with an error less than $\frac{1}{3}$,

- if we take the first two terms of this series $1 - \frac{1}{3}$, the error is less than $\frac{1}{5}$,

- if we take the first three terms of this series $1 - \frac{1}{3} + \frac{1}{5}$, the error is less than $\frac{1}{7}$,

- etc.

If the series is continued, the error becomes less than any given quantity. However, only the whole series (*tota series*), i.e., the actual infinity of the terms of series, contains all (*omnes*) approximations and expresses the exact value (*Quare tota series exactum exprimit valorem*).[43]

Leibniz went on to state that although one could express the sum of the series by means of no (rational) number and this series was produced *in infinitum*, one, however, conceived the whole series in his mind, since the law of progression was known (Leibniz [GMS, 5:120]). In other terms, Leibniz thought that *the sum of a series was achieved when the aggregate of the series* (namely, the infinite terms of the series) *was all together taken into account*. He illustrated this concept of the sum by Fig.6.

---

[41] Such aspects can also be noted in the works of previous mathematicians. (See, e.g., Wallis's interpolation, p. 10, Mercator's rule, p. 20, and Gregory's reference to the last term, p. 21.) In Leibniz they appear in a clearer way.

[42] See Leibniz [1682, 44].

[43] See Leibniz [GMS, 5:120].

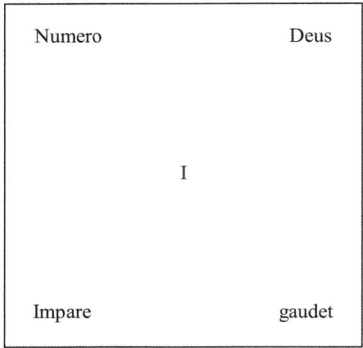
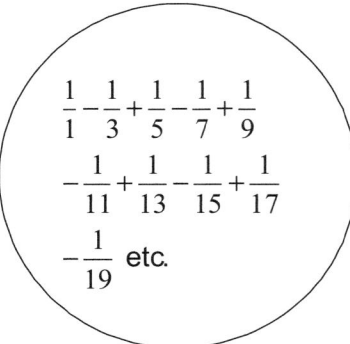

Fig. 6

Leibniz explained that a circle is equal to an infinite series in the same way as the segment of length 1 is equal to the sum of the segment

$$\frac{1}{2} + \frac{1}{4} + \frac{1}{8} + \dots$$

since (see Fig.7) the segment $AB = 1$ could be obtained by the juxtaposition of the segments

$$AC = \frac{1}{2}, \quad CD = \frac{1}{4}, \quad DE = \frac{1}{8}, \quad \dots$$

(see Leibniz [GMS, 5:120–121]).

A   C   D   E   B

Fig. 7

While an infinite series was equal to a quantity exactly only if it was conceived globally, the limit process of the partial sums provided approximations of the quantity represented by the series. Leibniz stressed the importance of series as the instrument that made it possible for numerical results to be computed with as small a margin of error as desired (Leibniz [KQA, 79]). According to Leibniz, if one moved from theoretical observations to practice, the results of *De quadratura arithmetica circuli ellipseos et hyperbolae cujus corollarium est trigonometria sine tabulis* (the title[44] gives

---

[44]It means: "On the arithmetical quadrature of the circle, the ellipse, and the hyperbola. A corollary is a trigonometry without tables."

a significant indication of the main scope of the treatise) enabled trigonometric operations to be performed without tables with as small a margin
of error as desired; he argued that this was an incredible sign of the power
of geometry[45] (Leibniz [KQA, 79]). Although the partial sums were fundamental for practical applications, they did not define the sum in the same
manner as it was thought that

$$1.4, \quad 1.41, \quad 1.414, \quad 1.4142, \quad 1.41421, \quad \ldots$$

approximated $\sqrt{2}$ but did not define the square root of 2, which was the
ratio of two incommensurable quantities.[46] The partial sums could even be
used to find the sum of a series or to prove a result; however, in order to sum
a series, one had to reach *the ultimate value* of series. It was not necessary
that the ultimate value was conceived of as a real object, a really existing
entity: It was sufficient that one conceived the possibility of it. According to
Leibniz, the ultimate term was a fictitious quantity. For instance, he stated:

> [Infinite numbers, infinitely small numbers, and ultimate terms
> of series] are nothing but fictions. Every number is finite and
> assignable . . . and infinite and infinitely small [magnitudes] mean
> nothing but magnitudes that can be assumed as large or as small
> as desired so that it is demonstrated without doubt that error is
> less that any given number. (Leibniz [D, 1:107])

I shall discuss Leibniz's notion of fictitious quantities in the following
section.

∗   ∗   ∗

As concerns the second above-mentioned aspect –exactly how Leibniz
handled series–, I first of all observe that, in modern mathematics, given the
series

$$\sum_{n=1}^{\infty} \frac{1}{2n+1},$$

---

[45] In a letter to Gallois written in December 1678, Leibniz wrote "I left my manuscript
on the arithmetical quadrature in Paris. The theorems which are contained in it are
considerable in theory and very useful in practice. Since if one memorises just two very
simple progressions which I give and which are almost impossible to forget, once one has
learnt them, all problems of trigonometry could easily be solved without tables, without
instruments, and without books, with the exactness one wishes [. . .] Having some tables is
a convenience, but not being able to solve problems without tables is a defect of science,
for which I claim to have found a remedy" (Leibniz [GMS 1:186]).

[46] On the notion of number, see Section 7.2.

one can write

$$\frac{1}{2n+1} = \frac{1}{2n-1} - 2\frac{1}{(2n)^2 - 1}$$

or

$$\sum_{n=1}^{\infty} \frac{1}{2n+1} = \sum_{n=1}^{\infty} \left(\frac{1}{2n-1} - 2\frac{1}{(2n)^2 - 1} \right).$$

The last equality merely means that the nth term of the series is a linear combination of the nth terms of the series $\sum_{n=1}^{\infty} \frac{1}{2n-1}$ and $\sum_{n=1}^{\infty} \frac{1}{(2n)^2-1}$. Instead Leibniz considered the relation between the terms of the series as a relation between the sums of series and assumed

$$\sum_{n=1}^{\infty} \frac{1}{2n+1} = \sum_{n=1}^{\infty} \frac{1}{2n-1} - 2\sum_{n=1}^{\infty} \frac{1}{(2n)^2 - 1}.$$

Indeed, in *De quadratura arithmetica* [KQA, 82], he set

$$\frac{1}{1} + \frac{1}{3} + \frac{1}{5} + \frac{1}{7} + \ldots = A,$$

$$\frac{1}{3} + \frac{1}{15} + \frac{1}{35} + \frac{1}{63} + \ldots = B,$$

$$\frac{2}{3} + \frac{2}{15} + \frac{2}{35} + \frac{2}{63} + \ldots = 2B.$$

Subtracting term by term, he derived

$$C = \frac{1}{3} + \frac{1}{5} + \frac{1}{7} + \frac{1}{9} + \ldots = A - 2B = A - 1.$$

At this juncture Leibniz did not merely consider $A - 2B$ and $A - 1$ as two symbols to denote two ways for deriving the terms of series C by operating upon the terms of A and B. He handled $A - 2B$ and $A - 1$ as if A and B were numbers and $A - 2B = A - 1$ an ordinary equation in the unknown B; thus, he derived $B = \frac{1}{2}$ and

$$\sum_{n=1}^{\infty} \frac{1}{(2n)^2 - 1} = \frac{1}{2}. \tag{19}$$

Leibniz often used this method.[47] For instance, he set[48]

$$\sum_{n=1}^{\infty} \frac{1}{n} = A$$

and

$$\sum_{n=1}^{\infty} \frac{2}{n(n+1)} = 2B$$

and derived

$$\sum_{n=2}^{\infty} \frac{1}{n} = A - B = A - 1.$$

Hence, $B = 1$ and[49]

$$\sum_{n=1}^{\infty} \frac{2}{n(n+1)} = 2$$

(see Leibniz [KQA, 83]).

Nowadays we realize that this method worked well because Leibniz employed series $\sum_{n=1}^{\infty} a_n$ such that $\lim_{n \to \infty} a_n = 0$ and transformed them into a series with the nth term of the type

$$c_n = a_n - a_{n+s},$$

where $s \geq 1$ is a fixed integer. Under this constraint, one has

$$\sum_{n=1}^{\infty} c_n = \sum_{n=1}^{\infty} (a_n - a_{n+s}) = \sum_{n=1}^{s} a_n - \lim_{k \to \infty} \sum_{n=1}^{s} a_{k+n} = \sum_{n=1}^{s} a_n.$$

Leibniz was able to use this procedure correctly and, at least intuitively, understood that the numerical result depended on $a_n = 0$.[50] This could give

---

[47]In *Rencensio libri*, which was written at the height of the controversy with Leibniz, Newton [OO, 4:459] dealt with the presumed difficulty of Leibniz's results rather ironically and showed how to obtain $\sum_{n=1}^{\infty} \frac{2}{n(n+1)} = 2$ by a procedure conceptually identical to Leibniz's one. Indeed, he wrote

$$1 = 1 - \frac{1}{2} + \frac{1}{2} - \frac{1}{3} + \frac{1}{3} - \frac{1}{4} + \frac{1}{4} - \frac{1}{5} + \frac{1}{5} + \cdots$$
$$= \frac{1}{1 \times 2} + \frac{1}{2 \times 3} + \frac{1}{3 \times 4} + \frac{1}{4 \times 6} + \cdots$$

[48]Leibniz knew that the harmonic series was divergent (see Leibniz [KQA, 103–104]).
[49]Note the difference with Mengoli's derivation of $\sum_{n=1}^{\infty} \frac{1}{n(n+1)} = 1$ (see p. 9).
[50]This condition was explicitly formulated by Jacob Bernoulli (see Chapter 5) and by de Moivre [1730, 129].

the idea that the use of divergent series was due to difficulties in symbolism or an inadequate formalization of the notion of convergence. In reality, Leibniz thought that series could be handled independently of any preliminary meaning that might be attributed to them. For instance, in *De vera proportione* [1682, 121], Leibniz disregarded the fact that

$$\sum_{k=0}^{\infty} \frac{1}{1+4k}$$

and

$$\sum_{k=0}^{\infty} \frac{1}{3+4k}$$

were divergent and stated

$$\frac{\pi}{4} = \frac{1}{1} + \frac{1}{5} + \frac{1}{9} + \frac{1}{13} + \cdots - \left(\frac{1}{3} + \frac{1}{7} + \frac{1}{11} + \frac{1}{15} + \cdots\right). \quad (20)$$

A proof of this relation is found in *De quadratura arithmetica* [KQA, 81] and merely consists in rearranging the series $1 - \frac{1}{3} + \frac{1}{5} - \frac{1}{7} + \frac{1}{9} - \ldots$ in order to obtain (20). Equation (20) can be interpreted in the sense that the difference between

$$Q_1(i) = \sum_{k=0}^{i} \frac{1}{1+4k} \quad \text{and} \quad Q_2(i) = \sum_{k=0}^{i} \frac{1}{3+4k}$$

approaches $\pi/4$ closer and closer and it is ultimately

$$Q_1(i) - Q_2(i) = \frac{\pi}{4}.$$

Today we should say that the sums $Q_1(i)$ and $Q_2(i)$ have the same asymptotic behavior (with respect to $\log i$). This idea was widely used by Euler in the 1730s.[51]

Another interesting example is the theorem known today as Leibniz's convergence criterion for alternating series:

> *Given a decreasing sequence $a_n > 0$, if a_n goes to 0 when $n \to \infty$, then the series*
>
> $$\sum_{n=1}^{\infty} (-1)^{n+1} a_n$$
>
> *is convergent.*

---

[51] See Section 13.3.

In a letter to Johann Bernoulli of October 25, 1713, Leibniz formulated it thus: If the terms of a series are continuously decreasing and, alternatively, positive and negative, then the series is "advergent" (namely, convergent) (see Leibniz [GMS, 3: 926])[52].

To prove this criterion, Leibniz set

$$S = \sum_{n=1}^{\infty} (-1)^{n+1} a_n$$

and observed that

$$
\begin{aligned}
s_{2n-1} &= a_1 - a_2 + a_3 - a_4 + \ldots + a_{2n-1} \\
&> (a_1 - a_2 + a_3 - a_4 + \ldots + a_{2n-1}) \\
&\quad + (-a_{2n} + a_{2n+1}) + (-a_{2n+2} + a_{2n+3}) + \ldots \\
&= S
\end{aligned}
$$

(s_n is the nth sum) since $a_n \geq a_{n+1}$.

Similarly

$$
\begin{aligned}
s_{2n} &= a_1 - a_2 + a_3 - a_4 + \ldots + a_{2n} \\
&< (a_1 - a_2 + a_3 - a_4 + \ldots + a_{2n}) \\
&\quad + (a_{2n-1} - a_{2n+2}) + (a_{2n+3} - a_{2n+4}) + \ldots \\
&= S.
\end{aligned}
$$

Since

$$s_{2n} < S < s_{2n+1},$$

he derived that S was finite. Furthermore, since

$$S - s_{2n} < a_{2n+1} \quad \text{and} \quad s_{2n-1} - S < a_{2n},$$

the difference between the finite series s_n and the infinite series S could be made smaller than a_n, namely it became as small as desired, provided one considered a sufficient number of terms.

---

[52]The proof I give here was sent to Bernoulli on January 10, 1714. An initial version of Leibniz's criterion is found in *De quadratura arithmetica* [KQA, 115], where it is formulated as follows (apart from the use of modern symbols): If the quantity S is equal to the series $\sum_{n=1}^{\infty} (-1)^{n+1} a_n$, then

$$\left| S - \sum_{n=1}^{m-1} (-1)^{n+1} a_n \right| < a_m$$

(see also Leibniz [GM 4:273]). The formulation of 1713–1714 shows traces of the problem of divergent series, which had surfaced at the beginnings of the 18th century (see Chapter 9). On Leibniz's criterion, see Knobloch [1993].

From a modern point of view, the first part of Leibniz's demonstration consists of the statement: If the sum exists, it is finite; the second part is a vicious circle. Today, by setting $S = \sum_{n=1}^{\infty}(-1)^{n+1}a_n$, we mean precisely that the difference between s_n and S is less than any quantity when n goes to infinity. To understand this proof, we must admit that, according to Leibniz,

$$S = \sum_{n=1}^{\infty}(-1)^{n+1}a_n$$

is not defined by

$$S = \lim_{m \to \infty}\sum_{n=1}^{m}(-1)^{n+1}a_n,$$

namely S does not denote the value to which the series $a_1 - a_2 + a_3 - a_4 + \ldots$ approaches closer and closer.

The previous examples show that, according to Leibniz, the symbol $a_1 - a_2 + a_3 - a_4 + \ldots$ could be used independently of the possibility of associating it with a finite number or a finite quantity or before proving that it was possible to associate a number with certain series. He operated upon these series by assuming that if an algebraic operation (and also differentials or integrals, if one considers power series, as we shall see later) could be performed upon finite series then it could be performed upon infinite series.

As concerns the nature of the above series, Knobloch [by referring in particular to (19)] stated:

> [A] is a symbol, sufficiently known by the known law of formation of the relative series. *A* does not depend on summation step by step: Leibniz takes the whole expression as a fictive quantity like signs for imaginary numbers. (Knobloch [1991, 276])

I substantially agree. In the quotation on p. 29 we already saw that Leibniz justified infinitesimals, infinitely large numbers, and ultimate terms of series in this way. Leibniz's use of fictitious entities is largely known; for instance, Bos stated: "[Leibniz] could not invoke the existence of infinitesimals in answer to objections to the validity of the calculus. Instead, he had to treat the infinitesimals as 'fictions' which need not correspond to actually existing quantities, but which nevertheless can be used in the analysis of problems" (Bos [1974, 54–55]).[53]

An observation about this point is appropriate. In a letter written on February 2, 1702, to Varignon, Leibniz refused to reduce the science of the

---

[53]On Leibniz's conception, see Horvàth [1982 and 1986] and, above all, a very stimulating paper by Knobloch published in 1999 (Knobloch [1999]).

infinite to a "fiction". In this letter, however, the word "fiction" is used to refer solely to invention, in other words, creation at our own discretion, such as in the rules of a game. According to Leibniz, imaginary numbers, infinite numbers, infinitesimals, the powers whose exponents were not "ordinary" numbers and other mathematical notions are not mere inventions; they are auxiliary and ideal quantities that have their foundation in nature and serve to shorten the path of thought:

> Without worry one can use infinitely small and large lines as ideal concepts — even though they do not exist as real objects in the metaphysically rigorous sense — as a means to shorten calculation, just as the imaginary roots in ordinary analysis, such as for example $\sqrt{-2}$. Regardless of whether one calls these "imaginary", they are nonetheless useful and sometimes even indispensable, in order to express real magnitudes analytically; so, for example, it is impossible, without using them, to give an analytical expression for a line segment, which divides a given angle into three equal parts. Just so, one could not elaborate our calculus of transcendental curves, without talking about differences, which are in the act of vanishing, and introducing once and for all the concept of incomparably small magnitudes ... It is always in the same way that one knows the dimensions above three, and the same is true for the powers whose exponents were not ordinary numbers; all this establishes ideas which are suitable for shortening reasoning and are based on reality. Nevertheless one must not imagine that the science of the infinite is degraded for this reason and reduced to fiction; because it always remains a syncateromatic infinite, to use the term used in the School, and it remains true, for example, that 2 is as much as $\frac{1}{1} + \frac{1}{2} + \frac{1}{4} + \frac{1}{8} + \frac{1}{16} + \frac{1}{32}$ etc., which is an infinite series, where all the fractions are included at the same time, though one employs nothing but ordinary numbers and though one considers no infinitely small fraction, or fractions whose denominator is an infinite number. Also the imaginary numbers have their foundation in reality (*fundamentum in re*). When I pointed out to the late Mr. Huygens that $\sqrt[2]{1 + \sqrt{-3}} + \sqrt[2]{1 - \sqrt{-3}}$ is equal to $\sqrt[2]{6}$, he was so amazed and he answered that, for him, there is something incomprehensible in this. But just so, one can say, that the infinite and infinitely small have such a solid basis, that all results of geometry, and even the processes of Nature, behave as if both were complete realities ... because everything obeys the Rule of Reason.[54] (Leibniz [GMS 4:92–93])

---

[54]See also Leibniz's *Quid sit idea* [GP, 7:263–264], where he stated that certain expressions have their foundation in nature, others in our discretion.

The use of fictitious quantities could lead to the erroneous idea of objects whose existence is assured by their very definition and therefore to ascribing a modern conception to Leibniz. In reality, what finds its foundation in Nature cannot be created by the human mind by means of a definition[55]. Modern mathematical theories do not make direct comparisons with nature but sound out our thought. They describe the logical consequences of certain axioms by means of clearly defined rules of inference: A new object or symbol is always introduced by an (explicit or implicit) definition and its meaning derives entirely from that definition and the subject matter of the search is circumscribed *a priori* by the given axioms. Thus, one can manipulate an object O only after having defined O and the operations that can be performed upon O. However, Leibniz believed that objects that have their foundation in nature have a meaning in their own right. This meaning could be not entirely clear so that these objects could be only partially known; however, one could manipulate the symbols $\sqrt{-1}$ or $1 + 1/2 + 1/3 + \ldots$ This was possible because Leibnizian fictitious quantities were analogical extensions of ordinary quantities and were manipulated by analogical extensions of the rules of ordinary quantities. Thus, one could operate upon $\sqrt{-1}$ in the same manner as one operated upon $\sqrt{2}$, and one could handle $a_1 - a_2 + a_3 - a_4 + \ldots$ in the same manner as one handled the finite sum $a_1 - a_2 + a_3 - a_4 + \ldots + a_n$.

I shall return to the notion of fictitious quantities in Chapter 7, where I shall show that it can be used to improve our understanding of 18th-century mathematics.

2.2   Power series

Although Leibniz considered certain numerical series interesting in their own right, he concentrated above all on what is referred to today as power series. However, his investigations originated in the field of geometry and this gave a peculiar aspect to series theory. Leibniz's analysis was not based upon the notion of function (either in the modern or in the 18th-century sense) but upon curves. He did not consider a function $y = f(x)$ and represented it diagrammatically; instead, he considered a certain figure F and adapted an appropriate system of coordinates (origins, unit, axes) to the figure F. The difference is very important. Today, given a function $y = f(x)$, we know from the start the constants, the dependent variable and the independent variable. By contrast, given a figure F, Leibniz had some geometric quantities connected to the figure F and, according to the specific circumstances, chose what were to be considered as variables and what were to be considered as constants. To compute an area or length connected

---

[55] "Definitions, according to Leibniz, are no mere conventions but have to show the possibility of the idea defined" (Otte [1989, 15]). For instance, see Leibniz [1684, 537–542].

to the figure F, the series that expressed the area or length had to be convergent, but convergence could be obtained simply by changing constants into variables, modifying the value of the constant in an appropriate way, or choosing between different expansions regarding the given quantities. For instance, let us consider the method of the long division.[56] In *De quadratura arithmetica*, Leibniz expounded it as follows. He observed that

$$\frac{a}{b+c} = \frac{a}{b} - \frac{ac}{b^2 + bc}.$$

If one replaces

$$ac, \quad b^2, \quad bc$$

by

$$a_1, \quad b_1, \quad c_1,$$

then $\frac{a_1}{b_1+c_1} = \frac{a_1}{b_1} - \frac{a_1 c_1}{(b_1)^2 + b_1 c_1}$ and

$$\frac{a}{b+c} = \frac{a}{b} - \frac{ac}{b^2} + \frac{a_1 c_1}{(b_1)^2 + b_1 c_1}.$$

In a similar way,

$$\frac{a}{b+c} = \frac{a}{b} - \frac{ac}{b^2} + \frac{ac^2}{b^3} - \frac{a_2 c_2}{(b_2)^2 + b_2 c_2}.$$

The remainder of the series decreases as desired (*quantumlibet decrescentes*); if one continues the series to infinity, then one finds the expansion

$$\frac{a}{b} - \frac{ac}{b^2} + \frac{ac^2}{b^3} - \dots \tag{21}$$

of $\frac{a}{b+c}$. In reality, this is true only if $c < b$; however, if one reverses b by c, then one obtains a different expansion of $\frac{a}{b+c}$:

$$\frac{a}{c} - \frac{ab}{c^2} + \frac{ab^2}{c^3} - \dots$$

(see Leibniz [KQA, 76–77]).[57] Therefore, if the ratio of certain geometrical quantities is $a : (b+c)$, then it can always be expanded, except the isolated case $b = c$. There is no reason to refuse one of two different expansions preliminarily. Only, *a posteriori* can one choose between two different developments according to the geometric situation. This approach is entirely different from the modern expansion of the function $f(c) = \frac{a}{b+c}$, where the

---

[56]See Chapter 1, p. 20.
[57]Of course, Leibniz assumed that the quantities a, b, and c are positive.

fact the c is a variable and a and b are constants is established *a priori*. In this case, it is not possible to change c into b without changing the function $f(c)$ (which can be expanded only if $|c| < b$) into the function $g(b) = \frac{a}{b+c}$ (which can be expanded only if $|b| < c$).

To make this point clearer, now let us investigate how Leibniz squared the hyperbola in *De quadratura arithmetica* [KQA, 88–90]. He observed that, given the hyperbola GCH, the area of the square $ABCD$ is equal to

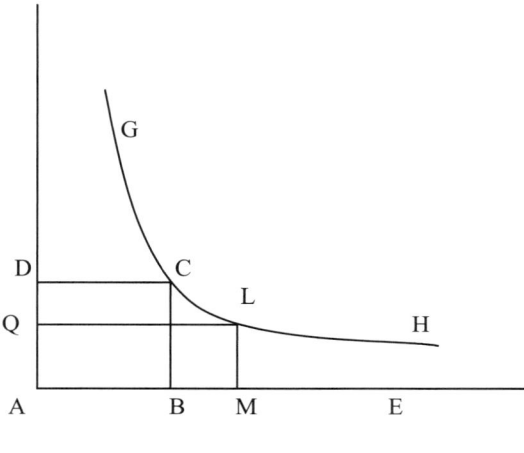

Fig. 8

the rectangle $AMLQ$ by the nature of the hyperbola (see Fig. 8). Therefore,

$$ML = \frac{AB^2}{AM} = \frac{AB^2}{AB + BM}.$$

By expanding, he had

$$ML = AB - BM + \frac{BM^2}{AB} - \frac{BM^3}{AB^2} + \frac{BM^4}{AB^3} - \frac{BM^5}{AB^4} + \cdots$$

under the condition $BM < AB$. Leibniz applied some theorems that he had proved earlier in *De quadratura arithmetica* and showed that the area of the figure $BEHC$ ($=$ to the sum all ML from BC to EH) was given from the sum of the areas of the curves having ordinate equal to $\frac{BM^n}{AB^m}$ (with M varying from B to E). Since AB is constant, the area AB is $AB \cdot AB$; the area of BM is a triangle of sides BE and BE, namely $\frac{BE^2}{2}$, and so on. He, *de facto*, integrated term by term[58] and obtained[59]

$$\text{Area } BEHC = AB \cdot BE - \frac{BE^2}{2} + \frac{BE^3}{3AB} - \frac{BE^4}{4AB^2} + \frac{BM^5}{5AB^3} - \frac{BE^6}{6AB^4} + \dots . \quad (22)$$

If one chooses BE appropriately so that $BE < AB$, then (22) makes it possible for any finite part of the given hyperbola to be squared: This means that, although the interval of convergence[60] is finite, one can compute any finite part of the area by means of a change in the system of coordinates. The condition of convergence is not an *a priori* condition in the sense that it does not restrict the possibility of formal procedures. Indeed, even if Leibniz knew that (22) converged only if $BE < AB$, he posed $AB = BE$ and obtained

$$\text{Area } BEHC = AB^2 - \frac{AB^2}{2} + \frac{AB^2}{3} - \frac{AB^2}{4} + \frac{AB^2}{5} - \frac{AB^2}{6} + \dots$$

and

$$\text{Area } BEHC = \left(AB^2 - \frac{AB^2}{2} \right)$$
$$+ \left(\frac{AB^2}{3} - \frac{AB^2}{4} \right) + \left(\frac{AB^2}{5} - \frac{AB^2}{6} \right) + \dots$$
$$= \frac{AB^2}{2} + \frac{AB^2}{12} + \frac{AB^2}{30} + \dots$$

Hence, he found that the area of $BEHC$ for $BE = \frac{1}{2}$ is equal to

$$\frac{1}{4} \log 2 = \frac{1}{8} + \frac{1}{48} + \frac{1}{120} + \dots .$$

---

[58]In a modern form, B being the origin of the axes, if $AB = b$, $BE = c$, and $BM = x$, Leibniz's reasoning is equivalent to considering the hyperbola $y = \frac{b^2}{b+x}$, expanding it into the power series $b - x + \frac{x^2}{b} - \frac{x^3}{b^2} + \frac{x^4}{b^3} - \frac{x^5}{b^4} + \dots$ and integrating it term by term in order to obtain

$$\text{Area } BEHC = \int_0^c \frac{b^2}{b+x} dx = bc - \frac{c^2}{2b} + \frac{c^3}{3b^2} - \frac{c^4}{4b^3} + \frac{c^5}{5b^4} - \frac{c^6}{6b^4} + \dots$$

[59]Leibniz derived several developments from expansion (22). For instance, for $BE < AB$ and $AB = 1$, he reobtained

$$\text{Area } BEHC = \frac{BE}{1} - \frac{BE^2}{2} + \frac{BE^3}{3} - \frac{BE^4}{4} + \frac{BM^5}{5} - \frac{BE^6}{6} + \dots$$

(see p. 19).

[60]Of course, Leibniz and the other mathematicians I discuss in this book did not possess the modern notion of interval of convergence, mainly because of the lacking of $\mathbb{R}$ (see Chapter 7). Nevertheless, the term "interval of convergence" can be conveniently used, provided when referring to the interval over which the power series $\sum_{i=0}^{\infty} a_i x^i$ converges, one does not refer to a subset of $\mathbb{R}$, but simply that the series is convergent when the quantity x varies within appropriate limits.

The logical rule, according to which all the consequences of a theorem are subject to the same conditions of that theorem, does not seem to be applied. In my opinion, Leibniz did not break this law but merely thought that one could operate on a series independently of the conditions of convergence. What was important was that the results had a numerical or geometrical meaning, whereas the steps of the formal procedure that yielded a certain result might lack a numerical or geometrical meaning.[61]

The conditions of convergence were viewed as extrinsic to the question of finding the rule of the development of a quantity, whereas they actually concerned the application of a rule to the given problem. If one had to square a fixed part of the hyperbola, then one fitted the series to a specific situation: This was possible because the question of convergence could easily be circumvented by an appropriate choice of the coordinate system.

It was precisely the geometrical context of *De quadratura* that made this approach possible. Not only was Leibniz's main objective the computation of areas and other geometric quantities, but the theory of series also used figural representations in a crucial way. Here is another example. After having introduced logarithms,[62] Leibniz showed that

$$\log_c \frac{AX}{AB} = \log_c \frac{AB + BX}{AB} = \log_c \frac{b+n}{b}$$

and

$$\log_c \frac{AZ}{AB} = \log_c \frac{AB - BZ}{AB} = \log_c \frac{b-m}{b}$$

(where $AB = b$; $BX = n$; $BZ = m$) were proportional to the decreasing series

$$n - \frac{n^2}{2b} + \frac{n^3}{3b^2} - \frac{n^4}{4b^3} + \frac{n^5}{5b^4} - \frac{n^6}{6b^5} + \dots \tag{23}$$

and

$$m + \frac{m^2}{2b} + \frac{m^3}{3b^2} + \frac{m^4}{4b^3} + \frac{m^5}{5b^4} + \frac{m^6}{6b^5} + \dots, \tag{24}$$

respectively (see Fig. 9).

On p. 121 of *De quadratura*, Leibniz observed that if $n > b$, then the series (23) was not decreasing. However, in order to compute $\log_c \frac{b+n}{b}$, namely to calculate the area $BXLK$, one could take the figure $BKHZ$ such that the area $BXLK$ was equal to $BKHZ$ and calculate the area $BKMZ$ (which is equal to $\log_c \frac{b-m}{b}$). Indeed, if $n > b$, then $m < b$, and (24) was decreasing.

---

[61] Leibniz referred to this kind of derivation during the debate about the series $1 - 1 + 1 - 1 + \dots$ (see Chapter 9).
[62] See Leibniz [KQA, 94–101].

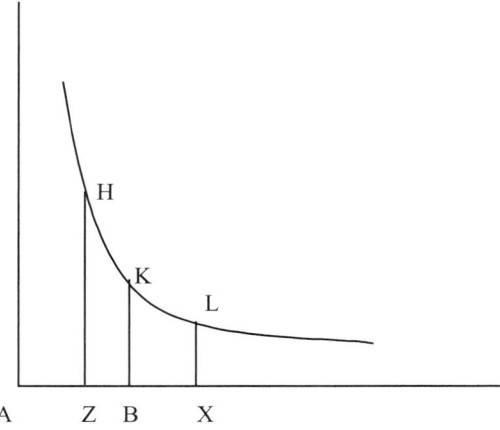

Fig. 9

In *Quadratura arithmetica communis sectionum conicarum*, a short paper written in 1691, Leibniz published his main results on the series expansions of geometrical quantities. In this article, apart from the expansion of logarithm (15), he provided the expansions of arctangent, sine, cosine, and exponential quantities:[63]

$$\arctan x = x - \frac{1}{3}x^3 + \frac{1}{5}x^5 - \ldots, \tag{25}$$

$$\sin x = x - \frac{1}{3!}x^3 + \frac{1}{5!}x^5 - \ldots,$$

$$1 - \cos x = \frac{1}{2!}x^2 - \frac{1}{4!}x^4 + \ldots,$$

$$e^x - 1 = x + \frac{1}{2!}x^2 + \frac{1}{3!}x^3 + \ldots.$$

Later in *Supplementum geometriae praticae* [1693], Leibniz presented a method that made it possible to solve differential equations using series. This method, which is known today as the *method of indeterminate coefficients*, is based upon the following principle:

---

[63] Leibniz derived these expansions by methods similar to the ones we have just seen for logarithms (see [KQA]). For instance, the expansion of $\arctan x$ can be briefly summarized in modern terms as follows. Since $\arctan x = \int \frac{1}{1+x^2}\,dx$, the term-by-term integration of the expansion of $\frac{1}{1+x^2}$ provides (25). From this series one can obtain $\pi/4$.

(**PIC**) *A series* $\sum\limits_{k=0}^{\infty} b_k x^{\alpha_k}$ *is equal to 0 for every x on an interval I if and only if all the coefficients b_k ($k = 0, 1, \ldots$) are separately equal to zero.*

Leibniz illustrated the method by means of several examples. I will give two of them here. The first again concerns the expansion of logarithm. In *Supplementum geometriae practicae* [1693, 286], Leibniz stated that if $y = a \log \frac{a+x}{a}$, then

$$dy = \frac{adx}{a+x}.$$

Hence,

$$a\frac{dy}{dx} + x\frac{dy}{dx} - a = 0. \tag{26}$$

He set

$$y = Bx + Cx^2 + Dx^3 + Ex^4 + \ldots$$

(he assumed the arbitrary constant to be equal 0). He differentiated term by term and obtained

$$\frac{dy}{dx} = B + 2Cx + 3Dx^3 + 4Ex^3 + \ldots \tag{27}$$

By replacing (27) into (26), he derived

$$(aB - a) + (2aC + B)x + (3aD + 2C)x^2 + (4aE + 2D)x^3 + \ldots = 0.$$

Equating the coefficients to zero, he obtained

$$aB - a = 0$$
$$2aC + B = 0$$
$$3aD + 2C = 0$$
$$4aE + 2D = 0$$
$$\ldots\ldots$$

Hence,

$$B = 1, \quad C = -\frac{1}{2a}, \quad D = \frac{1}{3a^2}, \quad E = -\frac{1}{4a^3}, \quad \ldots,$$

and

$$y = x - \frac{x^2}{2a} + \frac{x^3}{3a^2} - \frac{x^4}{4a^3} + \ldots . \tag{28}$$

The second example concerns the equation

$$a^2 dx^2 = a^2 dy^2 + y^2 dx^2. \tag{29}$$

This equation has an easily understood geometric meaning. Indeed, given any circle, if the sine, the arc and the radius are denoted by y, x, and a, then (29) represents the sine in terms of the arc, and radius. In *Supplementum geometriae practicae* [1693, 287], Leibniz differentiated (29) and derived

$$0 = 2a^2 dy d^2 y + 2y dy d^2 x$$

(he supposed that dx was a constant, namely, in more modern terms, that y was the independent variable). Hence,

$$a^2 \frac{d^2 y}{d^2 x} + y = 0. \tag{30}$$

Leibniz set

$$y = Bx + Cx^3 + Dx^5 + Ex^7 + \dots$$

and obtained

$$\frac{d^2 y}{d^2 x} = 2 \cdot 3Cx + 4 \cdot 5Dx^3 + 6 \cdot 7Ex^5 + \dots \ . \tag{31}$$

Substituting (31) into (30), he had

$$(2 \cdot 3a^2 C + B)x + (4 \cdot 5a^2 D + C)x^3 + (6 \cdot 7a^2 E + D)x^5 + \dots = 0.$$

By applying the principle of indeterminate coefficients and assuming $B = 1$, he obtained

$$B = 1, \quad C = -\frac{1}{2 \cdot 3a^2}, \quad D = -\frac{C}{4 \cdot 5a^2}, \quad E = -\frac{D}{6 \cdot 7a^2}, \quad \dots,$$

and

$$y = x - \frac{x^3}{2 \cdot 3a^2} + \frac{x^3}{2 \cdot 3 \cdot 4 \cdot 5a^2} - \frac{x^7}{2 \cdot 3 \cdot 4 \cdot 5 \cdot 6 \cdot 7a^2} + \dots . \tag{32}$$

In *Supplementum geometriae praticae*, Leibniz again stated that a series provided the exact solution to a differential equation only if it was considered as a whole; however, if one considered the partial sums of the series, then one obtained a solution that could be approximated as desired (Leibniz [1693, 286]). Leibniz made no further mention of convergence in this article, unlike in *De quadratura* and also *Quadratura arithmetica communis sectionum conicarum* [1691]. This stemmed from the fact that his aim in *Supplementum geometriae praticae* was to determine the form of the development, and questions concerning convergence therefore could be neglected.

It is worth noting that Leibniz regarded the variables x and y in Equation (26) as geometric quantities connected to the quadrature of hyperbola and the logarithmic curve. Similarly, the variables x and y in Equation (29) are geometric quantities connected to the circle. However, in *Supplementum geometriae praticae*, the immediate reference to geometrical figures is missing, and the reasoning regarded the way symbols were combined. Geometric applications were still the final aim of his study of series, but these applications were not the heart of the *Supplementum geometriae praticae:* While the objective in *De quadratura* was to find the area of the given figure, the aim in *Supplementum geometriae praticae* was restricted to finding the form of the development of the given quantity.

However much Leibniz still intended to solve geometrical questions, the problem of the quadrature assumed a more analytical aspect; in other words, it became a question of manipulating symbolic expressions, which began to be isolated from the geometrical aspect. In this way, the search for the development of a quantity and the application of this development began to be viewed as two separate problems. Leibniz only dealt with the first question in *Supplementum geometriae praticae*, whereas he examined both in *De quadratura*. Despite the difference in perspective, conditions of convergence were always considered as *a posteriori* conditions for the applicability of a development and not as *a priori* conditions for the development of quantities. Since convergence concerned the use of a specific series, conditions of convergence were not treated when the expansion of a quantity was sought, as is the case in *Supplementum geometriae praticae*.

It can therefore be stated that the Leibnizian theory of series always contained a certain degree of formalism though it was hidden from the immediacy of geometrical reference. However, in some cases, the phase of application of the results is removed from the search for expansion and the formal character of series theory became more evident. In the years that followed, this trend toward a stronger type of formalism, which was still in an initial stage in Leibniz, was progressively emphasized, as we shall in Part II of this book.

3 The Bernoulli series and Leibniz's analogy

In this chapter I follow the debate between Leibniz and Johann Bernoulli about the Bernoulli series and Leibniz's analogy. This provides the possibility of highlighting the role of Johann Bernoulli, who, together with his brother Jacob, were the first scholars who developed the work of Leibniz, and also of illustrating some other aspects of Leibniz's work. In particular, I show the kind of results that were obtained when analysis did not concern concrete geometrical figures directly but concentrated on the manipulation of symbols.

In 1694, following and developing Leibniz's search of *Supplementum geometriae praticae*, Johann Bernoulli formulated the theorem:[64]

> If n and z are two analytically expressed quantities,[65] then

$$\int ndz = nz - \frac{z^2 dn}{2!dz} + \frac{z^3 d^2 n}{3!dz} + \dots . \tag{33}$$

This series, which was later known as the Bernoulli series, is equivalent to the Taylor theorem. According to Bernoulli, (33) was an improvement over the method of indeterminate coefficients that Leibniz had used in *Supplementum geometriae praticae* [1693]. Bernoulli asserted that Leibniz's method was "rather ingenious"; however for any application an individual calculation had to be undertaken by which only a particular series was produced. Instead, Bernoulli series was "a universal series which expressed generally all quadratures, rectifications, and integrals of other differentials" (Johann Bernoulli [1694, 437]).

To derive (33) Bernoulli considered a quantity n and wrote

$$ndz = ndz + (zdn - zdn) + \left(-\frac{z^2 dn}{2!dz} + \frac{z^2 dn}{2!dz} \right) + \left(\frac{z^3 d^2 n}{3!dz} - \frac{z^3 d^2 n}{3!dz} \right) - \dots . \tag{34}$$

---

[64]On the Bernoulli series, see Feigenbaum [1985] and Panza [1992, 394–406].

[65]Bernoulli stated that n is "a quantity formed in any way of indeterminates and constants". This is a clear anticipation of the 18th-century notion of a function. It was precisely Johann Bernoulli who introduced this notion in his [1718], by using very similar words: "I call a function of a variable quantity, a quantity composed in whatever way of that variable quantity and constants" (Bernoulli [1718, 241]). It is appropriate to note that Bernoulli assumed dz to be a constant (z is an independent variable) and that he did not use the symbol of integral yet, but wrote *integr. ndz*.

By rearranging and integrating appropriately, he had

$$
\begin{aligned}
\int n\,dz &= \int (n\,dz + z\,dn) - \int \left(z\,dn + \frac{z^2\,dn}{2!\,dz} \right) + \int \left(\frac{z^2\,dn}{2!\,dz} + \frac{z^3\,d^2n}{3!\,dz} \right) - \cdots \\
&= \int d(nz) - \int d\left(\frac{z^2\,dn}{2!\,dz} \right) + \int d\left(\frac{z^3\,d^2n}{3!\,dz} \right) + \cdots \\
&= nz - \frac{z^2\,dn}{2!\,dz} + \frac{z^3\,d^2n}{3!\,dz} + \cdots.
\end{aligned}
$$

The application of formula (33) allowed Bernoulli to obtain the expansions that had already been derived by Leibniz in *Supplementum geometriae practicae* [1693]. For example, in order to derive the expansion of the logarithm

$$
y = a \log \frac{a + x}{a},
$$

Bernoulli observed that

$$
dy = \frac{a\,dx}{a + x}
$$

and set $n = \frac{1}{a+x}$ and $dz = dx$. He obtained

$$
\frac{dn}{dz} = -\frac{1}{(a + x)^2},
$$

$$
\frac{d^2n}{dz^2} = \frac{1}{2(a + x)^3},
$$
$$
\cdots
$$

By applying (33), he had

$$
\frac{y}{a} = \int \frac{dx}{a + x} = \frac{x}{a + x} + \frac{x^2}{2(a + x)^2} + \frac{x^3}{3(a + x)^3} + \cdots . \tag{35}
$$

Bernoulli observed that this formula was different from that of Leibniz [he referred to (28)], "nevertheless it has the same value"[66] (Bernoulli [1694, 438]).

Bernoulli also dealt with the differential equation for the sine. Indeed from (29) one had

$$
dy = \frac{\sqrt{a^2 - y^2}\,dx}{a}
$$

and, by setting $dz = dx$ and

$$
n = \frac{\sqrt{a^2 - y^2}}{a},
$$

---

[66]See Feigenbaum [1985, 83–84].

it is easy to obtain

$$\frac{dn}{dz} = -\frac{y\frac{dy}{dx}}{a\sqrt{a^2-y^2}} = -\frac{y}{a^2}$$

$$\frac{d^2n}{dz^2} = -\frac{1}{a^2}\frac{dy}{dx} = -\frac{\sqrt{a^2-y^2}}{a^3}$$

$$\frac{d^3n}{dz^3} = \frac{y\frac{dy}{dx}}{a^3\sqrt{a^2-y^2}} = \frac{y}{a^4}$$

$$\cdots$$

Hence,

$$y = \int\frac{\sqrt{a^2-y^2}}{a}dx = \frac{\sqrt{a^2-y^2}}{a}x + \frac{y}{2a^2}x^2 - \frac{\sqrt{a^2-y^2}}{3!a^3}x^3 - \frac{y}{4!a^3}x^4 + \cdots.$$

Bernoulli divided this equation by $\sqrt{a^2-y^2}$ and obtained

$$\frac{y}{\sqrt{a^2-y^2}} = \frac{x}{a} + \frac{yx^2}{2a^2\sqrt{a^2-y^2}} - \frac{x^3}{3!a^3} - \frac{yx^4}{4!a^3\sqrt{a^2-y^2}} + \cdots.$$

Hence,

$$\frac{y}{\sqrt{a^2-y^2}}\left(1 - \frac{x^2}{2a^2} + \frac{x^4}{4!a^3} + \cdots\right) = \frac{x}{a} - \frac{x^3}{3!a^3} + \cdots$$

and

$$\frac{y}{\sqrt{a^2-y^2}} = \frac{x - \frac{x^3}{3!a^2} + \frac{x^5}{5!a^4} - \cdots}{a - \frac{x^2}{2a} + \frac{x^4}{4!a^3} + \cdots}.$$

Bernoulli observed that the numerator of

$$\frac{x - \frac{x^3}{3!a^2} + \frac{x^5}{5!a^4} - \cdots}{a - \frac{x^2}{2a} + \frac{x^4}{4!a^3} + \cdots}$$

is the sine series. This example shows that the Bernoulli series did not always succeed in isolating the sought-after quantity. Bernoulli himself noted this disadvantage of his method with respect to Leibniz's:

> This is to remarked to the credit of the Incomparable Leibniz, that by his method what is sought results at once alone and not involved with other quantities. However, he will likewise not deny that mine is highly praiseworthy for its universality. (Johann Bernoulli [1694, 440–441])

Johann Bernoulli communicated this theorem to Leibniz on September 2, 1694 (see Leibniz [GMS, 3:150–152]). In response, in December 1694 (see Leibniz [GMS, 3:150–152]), Leibniz wrote that he had already found the expansion $\int n\,dz = nz - \frac{z^2\,dn}{2!\,dz} + \frac{z^3\,d^2n}{3!\,dz} + \dots$ by means of techniques connected to his first studies of Pascal's triangle (see Leibniz [GMS 5, 396–397]). Leibniz's proof goes as follows. Given a decreasing sequence a_n, one can write

$$a_1 = a_1 - a_2 + a_2 - a_3 + a_3 + \dots = \sum_{k=1}^{\infty} \Delta a_k,$$

where $\Delta a_k = a_{k+1} - a_k$. Since

$$\Delta a_k = \sum_{n=0}^{k-1} (-1)^n \binom{k-1}{n} \Delta^{n+1} a_1,$$

where $\Delta^n a_k = \Delta^{n-1} a_{k+1} - \Delta^{n-1} a_k$, one has

$$a_1 = \sum_{k=1}^{\infty} \Delta a_1 - \sum_{k=1}^{\infty} \binom{k}{1} \Delta^2 a_1 + \sum_{k=1}^{\infty} \binom{k}{2} \Delta^3 a_1 - \dots. \tag{36}$$

According to Leibniz,

$$\sum_{k=1}^{\infty} 1 = x,$$

$$\sum_{k=1}^{\infty} k = \int x,$$

$$\sum_{k=1}^{\infty} \frac{k(k+1)}{2} = \int\int x,$$

$$\dots$$

Leibniz does not explain how these sums, which are rather similar to the sums (5) of Wallis and the sums (16) of Mercator, are derived. However, one might observe that, by considering $x = 1 + 1 + 1 + \dots$ as an infinite number, one has

$$\sum_{k=1}^{\infty} k = \sum_{k=1}^{\infty} \left(\sum_{j=1}^{k} 1 \right) = \sum_{j=1}^{\infty} \left(\sum_{k=j}^{\infty} 1 \right) = \sum_{j=1}^{\infty} x = \int x, \quad \text{etc.}$$

Leibniz sets $y = a_1$ and $dy = \Delta a_1$, where dy is an infinitesimal quantity, and assumes that dx is an infinitely small quantity taken as a unity. He rewrote Equation (36) in the form

$$y = x\,dy - d^2y \int x + d^3y \int\int x - \dots.$$

Since $\int x = \frac{x^2}{2}$, $\int \int x = \frac{x^3}{6}$, ..., Leibniz obtains

$$y = x\,dy - \frac{x^2}{2}d^2y + \frac{x^3}{6}d^3y - \dots ,$$

By taking $dx = 1$, one has

$$y = x\frac{dy}{dx} - \frac{x^2}{2}\frac{d^2y}{dx^2} + \frac{x^3}{6}\frac{d^3y}{dx^3} - \dots \qquad (37)$$

and, "by moving ahead the y, dy, ddy, etc. into $\int y$, y, dy, etc., respectively",

$$\int y\,dx = xy - \frac{x^2}{2}\frac{dy}{dx} + \frac{x^3}{6}\frac{d^2y}{dx^2} - \dots . \qquad (38)$$

Leibniz started from a decreasing sequence a_n, which was later identified with a continuous variable y. The proposition would have to be valid for $y(x)$ decreasing. However, it is clear from the context that the conditions under which this series was derived were of little interest to Leibniz and Bernoulli. The result was acknowledged to be general: The condition regarding the decrease of the series did not limit the result.

Bernoulli's and Leibniz's derivations of the Bernoulli series highlighted the fact that the formal aspect of the manipulation of series became much accentuated when the object of the study was not a specific series derived from specific geometrical quantities with reference to a concrete figure, but a generic series derived from a generic quantity. In this case, mathematicians could not refer to the properties of the specific geometrical quantities; rather they investigated what could be achieved through the mere combination of symbols that were not subjected to *a priori* conditions. This approach led Johann Bernoulli and Leibniz to obtain several interesting results. Indeed, in a letter written on February 28, 1695 (see Leibniz [GMS, 3:164–169]), Leibniz derived an expansion of $\int x^n d^m y$ by reiterative applications of the formula

$$\int x^n d^m y = x^n d^{m-1}y - \int n x^{n-1}\,dx\,d^{m-1}y.$$

In his proof there was an error of calculation; Bernoulli readily corrected it and gave the precise expansion:

$$\int x^n d^m y = x^n d^{m-1}y - n x^{n-1}d^{m-2}y\,dx$$
$$+ n(n-1)x^{n-2}d^{m-3}y\,dx^2 - \dots \quad (dx \text{ constant})$$

(see Leibniz [GMS, 3:169–174]). According to Leibniz, this formula implied differentials with negative exponents to be regarded as equivalent to integrations, namely

$$d^0 y = \int {}^0 y = y, \quad d^{-1}y = \int {}^1 y, \quad d^{-2}y = \int \int y = \int {}^2 y, \quad \dots$$

(see Leibniz [GMS, 3:167]). This led Leibniz to formulate the famous analogy between the powers of a binomial and differentials of a product by observing that

$$(x + y)^1 = 1x + 1y = 1x^1y^0 + 1x^0y^1,$$
$$(x + y)^2 = 1x^2 + 2xy + 1y^2,$$
$$(x + y)^2 = 1x^3 + 3x^2y + 3xy^2 + 1y^3,$$
etc.

and

$$d^1(xy) = 1ydx + 1xdy = 1d^1xd^0y + 1d^0xd^1y,$$
$$d^2(xy) = 1d^2x + 2dxdy + 1d^2y,$$
$$d^3(xy) = 1d^3x + 3d^2xdy + 3dxd^2y + 1d^3y,$$
etc.

obey the same rule[67] (see Leibniz [GMS, 3:174–177]). The combinatorial aspect, previously incorporated to diagrammatic representation, is now freely shown.

In response, on June 1695, Bernoulli observed that d, d^2, d^3, ... could be considered "as algebraic quantities and not just as characteristic letters" (see Leibniz [GMS, 3:180]). For instance, he considered the proportions

$$d^3 : d^2 = d^2 : d, \ \ d^4 : d^3 = d^3 : d^2, \ \ldots$$

Later, in a letter on July 17, 1695 (see Leibniz [GMS, 3:197–205]), Bernoulli developed this idea. He stated that the third proportional between

$$\int ndz = d^{-1}(ndz) \quad \text{and} \quad ndz = d^0(ndz)$$

was

$$d(ndz).$$

Indeed, if one operated in the same way as in Leibniz's analogy, one had

$$d^{-1}(ndz) : d^0(ndz) = d^0(ndz) : d(ndz).$$

By differentiating $d(ndz)$, he obtained $d(ndz) = d^0nd^2z + dndz$ and, therefore,

$$\int ndz = d^{-1}(ndz) = \frac{(d^0(ndz))^2}{d(ndz)} = \frac{d^0nd^2z}{d^0nd^2z + dndz} = \frac{d^0ndz}{d^0ndz + dn}.$$

---

[67]In 1710 Leibniz published this analogy in a paper entitled *Symbolismus memorabilis calculi algebraici et infinitesimalis in comparatione potentiarum et differentiarum, et de lege homogeneorum transendentali* [1710]. Here he used the symbols $p^e(x+y)$ and $d^e(x+y)$, as in a letter to de l'Hôpital on September 30, 1695 [GMS, 1:301–302].

Since Mercator's rule led one to find two different expansions of a quantity $\frac{a}{b+c}$, Bernoulli obtained two different expansions of $\int n\,dz = \frac{d^0 n dz}{d^0 ndz + dn}$, namely

$$
\begin{aligned}
\int n\,dz &= \frac{d^0 ndz}{d^0 ndz + dn} = d^0 n d^0 z - d^1 n d^{-1} z + d^2 n d^{-2} z - d^3 n d^{-3} z + \ldots \\
&= nz - dn \int z + d^2 n \int{}^2 z - d^3 n \int{}^3 z + \ldots \\
&= nz - \frac{z^2}{2dz} dn + \frac{z^3}{6dz^2} d^2 n - \frac{z^4}{24dz^3} d^3 n + \ldots \quad (dz \text{ constant})
\end{aligned}
$$

and

$$
\begin{aligned}
\int n\,dz &= \frac{d^0 ndz}{dn + d^0 ndz} = d^1 z d^{-1} n - d^2 z d^{-2} n + d^3 z d^{-3} n - \ldots \\
&= d^1 z \int n - d^2 z \int{}^2 n + d^3 z \int{}^3 n - \ldots \\
&= \frac{n^2}{2dn} dz + \frac{n^3}{6dn^2} d^2 z - \frac{n^4}{24dn^3} d^3 z + \ldots \quad (dn \text{ constant})
\end{aligned}
$$

Thus, starting from Leibniz's analogy, the Bernoulli series was once again obtained.[68] Bernoulli stated:

> [W]hen I began to write these things, I did not indeed expect this result, thinking that I would arrive at a far different series by this method. This elegant agreement wonderfully confirms the probity of the methods, especially of this last, where so remarkably and contrary to all practice it is advanced with the letters d. (see Leibniz [GMS, 3:200], translation in Feigenbaum [1985, 89])

It is worthwhile noting that Leibniz also obtained a generalization of the Bernoulli series. In October 1695 he wrote to Johann Bernoulli that

$$
\int{}^n (xy) = y \int{}^{n-1} z - n\,dy \int{}^n z + \frac{n(n+1)}{2!} d^2 y \int{}^{n+1} z + \ldots
$$

followed from

$$
d^m (xy) = d^m x d^0 y + m d^{m-1} x d^1 y + \frac{m(m-1)}{2!} d^{m-2} x d^2 y + \ldots
$$

by letting $m = -n$ and $x = dz$ (see Leibniz [GMS, 3:221]).

---

[68]Such an entirely formal procedure, which anticipated the typical aspects of the calculus of operations (see Chapter 21), was, however, a borderline case around 1700.

4 Newton's method of series

In this chapter I investigate Newton's contribution to the series theory. Newton obtained a large number of results concerning series, which he expounded upon in several papers but only partially published.[69] Of these papers, the most interesting for my purpose are:

- *De analysi per æquationes numero terminorum infinitas* (see Newton [MP, 2:206–247]), which was probably written 1669, on the basis of previous results. It was published in 1711 in I. Newton, *Analysis per quantitatum, series ac differentias cum enumeration linearum tertii ordinis*, edited by W. Jones, London: ex officina Pærsoniana.

- *De Methodis serierum et fluxionum* (see Newton [MP, 3:32–329]). It was probably written in 1671 (based on a manuscript draft of 1666, see Newton [MP, 1:400-448]) but was published only nine years after his death by J. Colson in English translation with the title *The Method of Fluxions and Infinite Series; with Its Application to the Geometry* [1736].

- *Tractatus de quadratura curvarum* [MP, 8:92–167]. It is a revision of *De quadratura* [MP, 7:48-129], which he wrote between December 1691 and the beginning of 1692. A wide-ranging abstract was also inserted by Wallis in his work *Algebra* (see Wallis [1693, 390–396]). It was published as one of the appendices of the *Opticks: Or a Treatise of the Reflections, Refractions, Inflections & Colours of Light. Also Two Treatises of the Species and Magnitude of Curvilinear Figures*, London: Smith and Walford (the other treatise being *Enumeratio linearum tertii ordinis*).

- some pieces of his correspondence in *Commercium epistolicum d. Johannis Collins et aliorium de analysi promota* [CE]. This paper originated by the polemics with Leibniz, which Collins edited in 1712. In particular, it contains two very famous letters that are addressed to Oldenburg though they were actually written for Leibniz. They are usually known as *epistola prior* (the first letter is dated June 13, 1676) and *epistola posterior* (the second is dated October 24, 1676).

- some manuscripts of Newton published by Whiteside in [MP], especially the manuscripts that Whiteside entitled A*nnotations from Wallis* [MP, 1:89–142].

Moreover, in discussing Newton's concept of limit, I mainly refer to

---

[69]For biographical information on Newton, see Westfall [1980] and Panza [2003].

- *Philosophia naturalis principia mathematica,* whose first edition dates back to 1687 and which was later republished in 1713 and 1726. It contains several mathematical propositions, in particular the much-debated 11 mathematical lemmas of Book 1, Section 1, concerning the method of first and last ratios.

Finally, in Chapter 6, while investigating the notion of quantity during the 17th and 18th centuries, I shall also deal with

- *Arithmetica universalis, sive de compositione et de resolutione arithmetica liber.* It was based on lectures that Newton had left with the university library. It was published in 1707 by William Whiston, Newton's successor as Lucasian Professor at Cambridge, without the name of the author. A second Latin edition was published in 1722 and an English version in 1720.

My investigation of Newton's method of series is subdivided into two sections. In Section 4.1, I first deal with Newton's procedures for expanding quantities —also expressed in the form of algebraic or differential equations— and stress the important role played by convergence, after which I examine the notion of limit and convergence. In Section 4.2, I discuss the manipulative aspects of the Newtonian theory of series and the relationship between manipulation and convergence. Newton's contribution to the Taylor series is illustrated in Chapter 6.

4.1 The expansion of quantities into convergent series

The reading of Wallis's *Arithmetica infinitorum*[70] in 1664 had a great influence on Newton's first mathematical investigations. In the winter of 1664–1665, he was able to derive his first mathematical discovery, the binomial expansion:

$$(p+q)^r = p^r + rp^{r-1}q + \frac{r\,(r-1)}{2!}p^{r-2}q^2 + \frac{r\,(r-1)\,(r-2)}{3!}p^{r-3}q^3 + \dots \quad (39)$$

(r is any rational exponent) by applying Wallis's method of interpolation[71].

Newton's procedure can be summarized as follows.[72] At that time, mathematicians knew how to square the curves $(1-x^2)$, $(1-x^2)^2$, $(1-x^2)^3$, ..., but they did not possess a general method for squaring the irrational curves

$$(1-x^2)^{\frac{1}{2}}, \quad (1-x^2)^{\frac{3}{2}}, \quad (1-x^2)^{\frac{5}{2}}, \quad \dots.$$

---

[70] On Wallis's influence on Newton, see Panza [1995].
[71] See Chapter 1.
[72] See Newton [CE, 67–86] and [MP, 1:104–111].

Newton regarded the problem of the quadrature of the curves $(1 - x^2)^{\frac{1}{2}}$, $(1 - x^2)^{\frac{3}{2}}$, $(1 - x^2)^{\frac{5}{2}}, \ldots$ as that of interpolating the following table:

Curves	Quadratures
$y = (1 - x^2)^{\frac{0}{2}} = 1$	$z = x$
$y = (1 - x^2)^{\frac{1}{2}}$	
$y = (1 - x^2)^{\frac{2}{2}} = (1 - x^2)$	$z = x - \frac{1}{3}x^3$
$y = (1 - x^2)^{\frac{3}{2}}$	
$y = (1 - x^2)^{\frac{4}{2}} = (1 - x^2)^2$	$z = x - \frac{2}{3}x^3 + \frac{1}{5}x^5$
$y = (1 - x^2)^{\frac{5}{2}}$	
$y = (1 - x^2)^{\frac{6}{2}} = (1 - x^2)^3$	$z = x - \frac{3}{3}x^3 + \frac{3}{5}x^5 - \frac{1}{7}x^7$

In other terms, the problem was to find an analytical expression $Q_m(x)$ such that, by giving the values $0, 1, 2, 3, \ldots$ to m, the analytical expressions $Q_0(x), Q_1(x), Q_2(x), Q_3(x), \ldots$ provide the quadratures of

$$1, \quad (1 - x^2), \quad (1 - x^2)^2, \quad (1 - x^2)^3, \quad \ldots .$$

Then the principle of Wallis's interpolation made it possible to state that by giving the values $\frac{1}{2}, \frac{3}{2}, \frac{5}{2}, \ldots$ to m, the analytical expressions

$$Q_{1/2}(x), \quad Q_{3/2}(x), \quad Q_{5/2}(x), \quad \ldots$$

provide the quadratures of the curves

$$(1 - x^2)^{\frac{1}{2}}, \quad (1 - x^2)^{\frac{3}{2}}, \quad (1 - x^2)^{\frac{5}{2}}, \quad \ldots .$$

Newton noted that the finite polynomials expressing the area under 1, $(1 - x^2)$, $(1 - x^2)^2$, $(1 - x^2)^3 \ldots$, were of the form

$$Q_m(x) = x - \frac{m}{3}x^3 + \frac{m\frac{m-1}{2}}{5}x^5 - \frac{m\frac{m-1}{2}\frac{m-2}{3}}{7}x^7 + \ldots, \tag{40}$$

where m is an integer. He considered (40) valid even when m is not an integer; by setting $m = \frac{1}{2}$, he obtained the infinite series

$$Q_{1/2}(x) = x - \frac{\frac{1}{2}}{3}x^3 - \frac{\frac{1}{8}}{5}x^5 - \frac{\frac{1}{16}}{7}x^7 + \ldots, \tag{41}$$

by which the semicircle $\sqrt{1 - x^2}$ was squared.

Newton went on to observe that

$$y = (1 - x^2)^{\frac{0}{2}} = 1,$$
$$y = (1 - x^2)^{\frac{2}{2}} = 1 + x^2,$$
$$y = (1 - x^2)^{\frac{4}{2}} = 1 - 2x^2 + x^4,$$
$$y = (1 - x^2)^{\frac{6}{2}} = 1 - 3x^2 + 3x^4 - x^6.$$

These expansions obeyed the rule

$$(1 - x^2)^m = 1 - mx^2 + m\frac{m-1}{2}x^4 - m\frac{m-1}{2}\frac{m-3}{3}x^6 + \dots \qquad (42)$$

By considering (42) valid for a noninteger m, he obtained the binomial expansion. In particular,[73] by setting $m = \frac{1}{2}$, he had

$$\sqrt{1 - x^2} = 1 - \frac{1}{2}x^2 - \frac{1}{8}x^4 - \frac{1}{16}x^6 + \dots \qquad (43)$$

Newton also obtained the quadrature of the hyperbola (and therefore the expansion of the logarithm[74]) using the same method. Indeed, in his [MP, 1:112-115] he considered the following scheme

Curve	Quadrature
$y = (1 + x)^0 = 1$	$z = x$
$y = (1 + x)^1 = 1 + x$	$z = x + \frac{1}{2}x^2$
$y = (1 + x)^2 = 1 + 2x + x^2$	$z = x + x^2 + \frac{1}{3}x^3$
$y = (1 + x)^3 = 1 + 3x + 3x^2 + x^3$	$z = x + \frac{3}{2}x^2 + x^3 + \frac{1}{4}x^4$

and noted that the general rule was

Curve
$y = (1 + x)^n = 1 + nx + \frac{n(n-1)}{1\cdot 2}x^2 + \frac{n(n-1)(n-2)}{1\cdot 2\cdot 3}x^3 + \dots$

Quadrature
$z = x + \frac{n}{2}x^2 + \frac{n(n-1)}{1\cdot 2\cdot 3}x^3 + \frac{n(n-1)(n-2)}{1\cdot 2\cdot 3\cdot 4}x^4 + \dots$

Then the quadrature of the hyperbola could be obtained by setting $n = -1$

Curve	Quadrature
$y = \frac{1}{1+x} = 1 - x + x^2 - x^3 + \dots$	$z = x - \frac{1}{2}x^2 + \frac{1}{3}x^3 - \frac{1}{4}x^4 + \dots$

Three comments should be made at this point. First, making the step from an integer n to a fractional or negative n involved the step from a finite series to an infinite series. Newton regarded this as unproblematic: He made no distinction between finite series and infinite series. For instance, in *De analysi* he wrote:

> Whatever common analysis performs by means of an equation with a finite number of terms (whenever it may be possible) can be performed by means of infinite equations. (Newton [1711, 1:280])

---

[73] Newton also understood the relationship between the coefficients of the series (41) and (43) and so he poses the basis for the algorithm of the calculus (on this, see Panza [1989]).

[74] It should be noted that while in his letter to Oldenburg on October 24, 1676, Newton explicitly identified the area of hyperbola with logarithms (cf. Newton [CE, 67–86]), in his young paper he made no mention of this fact (cf. Newton [MP, 1:104–111]).

Second, even if Newton's derivation of the binomial expansion (39) is formal, he payed attention to convergence. For instance, in writing to Leibniz, Newton considered several applications of the binomial expansion, which he wrote in the form

$$(P + PQ)^{\frac{m}{n}} = P^{\frac{m}{n}} + \frac{m}{n} P^{\frac{m}{n}} Q + \frac{m(m-n)}{2n^2} P^{\frac{m}{n}} Q^2 + \ldots \qquad (44)$$

In the case of the function

$$\sqrt[5]{c^5 + c^4 x - x^5},$$

he first set

$$P = c^5 \quad \text{and} \quad Q = \frac{c^4 x - x^5}{c^5}$$

and obtained

$$\sqrt[5]{c^5 + c^4 x - x^5} = c + \frac{c^4 x - x^5}{5c^4} - \frac{2c^8 x^2 - 4c^4 x^6 + 2x^{10}}{25c^9} + \ldots$$

He then set

$$P = -x^5 \quad \text{and} \quad Q = -\frac{c^4 x + c^5}{x^5}$$

and obtained

$$\sqrt[5]{c^5 + c^4 x - x^5} = -x + \frac{c^4 x + c^5}{5x^4} + \frac{2c^8 x^2 + 4c^9 x + 2c^{10}}{25x^9} + \ldots.$$

Finally, he observed that the first procedure is preferable when x is very small, the second when it is very large (see Newton [OO, 1: 285-289]).

Third, Newton used Wallis's interpolation in an improved and simplified form, which can be schematized as follows:

1. Some specific cases $P(1)$, $P(2)$, ... are given;

2. an appropriate formula $P(k)$ is determined such that $P(1)$, $P(2)$, ... are specific cases of $P(n)$;

3. if the formula $P(n)$ has an appropriate quantitative meaning for nonnatural numbers (usually a rational number, but in principle any number), then $P(n)$ is considered valid for nonnatural numbers.

The complicated argumentation used by Wallis, which was often based on geometrical reasoning and consisted of the investigation of specific tables of numerical values, had now disappeared. They are replaced by the search for a general formula. This interpretation of Wallis's interpolation paved the way for the research of later mathematicians.

Newton invented other procedures for expanding analytically expressed quantities. Two of them — the long division and extraction of root— are known as Mercator's rules, even if Mercator published only the long division in his *Logarithmotechnia*. I have already dealt with long division in Chapter 1, p. 20 and Section 2.2, p. 37. Concerning the extraction of roots, Newton noted that the common procedure of extraction of roots made it possible to calculate the terms of the development of an analytical expression of the type $\sqrt{f(x)}$. For instance, to obtain the development $\sum_{i=0}^{\infty} a_i x^i$ of $\sqrt{a^2 + x^2}$ (see Newton [MP, 3:45-50]), one can proceed as follows:[75]

i. The first term a_0 of the expansion was the square root of a^2, namely

$$a_0 = \sqrt{a^2} = a;$$

ii. one calculated the first remainder

$$R_0 = (a^2 + x^2) - a^2 = x^2;$$

iii. one looked for the second term $a_1 x$ such that $(a_0 + a_1 x)^2 = a^2 + R_0 + (a_1 x)^2$, and got

$$a_1 x = \frac{R_0}{2a} = \frac{x^2}{2a};$$

iv. one multiplied $\frac{x^2}{2a}$ by $\left(2a + \frac{x^2}{2a}\right)$ and obtained

$$\Gamma_1 = x^2 + \frac{x^4}{4a^2};$$

v. one calculated the second remainder R_1

$$R_1 = R_0 - \Gamma_1 = -\frac{x^4}{4a^2};$$

vi one repeated the step (*iii*) on R_1 in order to find $a_2 x^2$, that is, one looked for a_2 such that $\left(a_0 + a_2 x^2\right)^2 = a^2 + R_1 + x^2$, and got

$$a_2 x^2 = \frac{R_1}{2a} = -\frac{x^4}{8a^3};$$

vii. one considered

$$\Gamma_2 = -\frac{x^4}{8a^3}\left(2\left(a + \frac{x^2}{2a}\right) - \frac{x^4}{8a^3}\right) = -\frac{x^4}{4a^2} - \frac{x^2}{8a^4} + \frac{x^8}{16a^6};$$

viii. one calculated the third remainder

$$R_2 = R_1 - \Gamma_2 = -\frac{x^2}{8a^4} + \frac{x^8}{16a^6};$$

---

[75]See Ferraro and Panza [2003, 24–25].

and so on.

One thus obtained[76]

$$\sqrt{a^2 + x^2} = a + \frac{x^2}{2a} - \frac{x^4}{8a^3} + \frac{x^6}{16a^5} + \dots. \tag{45}$$

The following scheme clarified the different steps of the procedure:

$$a^2 + x^2 \qquad\qquad\qquad\qquad a + \frac{x^2}{2a} - \frac{x^4}{8a^3} + \dots$$

$$\underline{a^2}$$

$$+ x^2$$

$$\dots \quad \underline{x^2 + \frac{x^4}{4a^2}}$$

$$-\frac{x^4}{4a^2}$$

$$\underline{-\frac{x^4}{4a^2} - \frac{x^2}{8a^4} + \frac{x^8}{16a^6}}$$

$$\frac{x^2}{8a^4} - \frac{x^8}{16a^6}$$

$$\dots$$

Another of Newton's procedures is today known as Newton's method of parallelogram. It makes it possible to solve a given algebraic equation

$$P(x, y) = \sum_{i,j} a_{i,j} x^j y^i = 0$$

by means of a power series $\sum_{k=0}^{\infty} b_k x^k$ or, in certain cases, by means of a series of the kind $\sum_{k=0}^{\infty} b_k x^{\alpha_k}$, where α_k is a rational number.[77] Newton considers tables of the type

j				
x^4	$x^4 y$	$x^4 y^2$	$x^4 y^3$	$x^4 y^4$
x^3	$x^3 y$	$x^3 y^2$	$x^3 y^3$	$x^3 y^4$
x^2	$x^2 y$	$x^2 y^2$	$x^2 y^2$	$x^2 y^4$
x	xy	xy^2	xy^3	xy^4
0	y	y^2	y^3	y^4	i

and marks all the squares of coordinates (i, j) corresponding to the terms $a_{i,j} x^j y^i$ of the given equation $\sum_{i,j} a_{i,j} x^j y^i = 0$. For instance, if the equation

$$y^3 + a^2 y + axy - 2a^3 - x^3 = 0$$

---

[76]The same result could be derived by applying (44).

[77]See Newton [MP, 3:51–57] and [MP, 2:218–233].

is given, one marks the squares $(3,0)$, $(1,0)$, $(1,1)$, $(0,3)$, $(0,0)$ corresponding to the terms y^3, y, xy, x^3, 0.

One chooses the lowest marked square in the column farthest to the left, namely, one chooses the square (i_1, j_1), where

- $i_1 = \min\{i : (i, j) \in S\}$,

- $j_1 = \min\{j : (i_1, j) \in S\}$,

- $S = \{(i, j)$ such that (i, j) are the coordinates of marked squares$\}$.

One then draws a line from the square $A \equiv (i_1, j_1)$ to another marked square, say $B \equiv (i_2, j_2)$, such that all the rest of the marked squares either are in contact with the line AB or lie above it. Then one considers all the terms that are found to be in contact with the line AB and forms a "fictitious" equation by making terms equal to zero. An appropriate root of this equation is the first term, say y_1, of the sought-after expansion of $P(x, y) = 0$.

In the example the line AB is the horizontal line joining $(0, 0)$ and $(3, 0)$ and the considered terms are $-2a^3$, a^2y, y^3. He thus obtains the equation

$$y^3 + a^2y - 2a^3 = 0.$$

One of the roots of this equation is $y_1 = a$; then

$$y = a$$

is the first term of the sought-after expansion $\sum_{k=0}^{\infty} b_k x^k$.

By replacing

$$y = y_1 + p$$

into the given equation $P(x, y) = 0$, Newton obtains a new equation $Q(x, p) = 0$ and, by repeating the procedure, he finds the other terms of the expansion. In the example he replaces $y = a + p$ in the given equation and obtains the equation

$$p^3 + 3ap^2 + axp + 4a^2p + a^2x - x^3 = 0. \qquad (46)$$

The corresponding diagram is

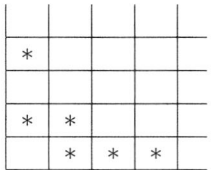

By repeating the reasoning, it is possible to find a new fictitious equation

$$a^2 x + 4a^2 p = 0.$$

The root is $p = -\frac{1}{4}x$. This is the second term of the sought-after expansion. Then he sets

$$p = -\frac{1}{4}x + q$$

and substitutes it into (46), etc. In this way Newton obtains

$$y = a - \frac{x}{4} + \frac{x^2}{64a} + \frac{131x^3}{512a^2} + \ldots.$$

The crucial idea of this procedure is the following. By substituting the indeterminate series $\sum_{k=0}^{\infty} b_k x^k$ for y in $P(x,y) = \sum_{i,j} a_{i,j} x^j y^i$, one should obtain a new polynomial

$$Q(x) = \sum_{i,j} a_{i,j} \left(\sum_k b_k x^{j+ki} \right) = \sum_{i,j,k} A_{i,j} x^{j+ki} = \sum_r B_r x^r.$$

If the series $\sum_{k=0}^{\infty} b_k x^k$ is the sought-after solution of $P(x,y) = 0$, then

$$\sum_r B_r x^r = 0$$

and all the coefficients B_r of the powers x^r have to be separately equal to zero, namely the terms x^{j+ki} should cancel each other out. Therefore, Newton *de facto* uses the principle of indeterminate coefficients.[78] In order to justify how Newton draws the lines, note that if x^s is the term of lower degree in $\sum_r B_r x^r$ and if B_s is equal to 0, then the exponent s is obtained in at least two different ways as a sum of the type $j + ki$, say $j_1 + ki_1$ and $j_2 + ki_2$. Moreover,

$$x^s = x^{j_1 + ki_1} = x^{j_2 + ki_2}$$

---

[78] See Chapter 2.

is effectively the term of lower degree only if there is no point (i, j) below the straight line joining (i_1, j_1) and (i_2, j_2).[79]

Newton observed that if one did not consider small values of[80] x, but values that differed by a small amount from a number c, one could replace x by $z = x - c$ into the equation. Further if x was large, one could replace x by $z = \frac{1}{x}$. In this way, he was sure to obtain a good approximation of y as a function of x since the series was convergent over at least an appropriate certain interval.

In the 1692 version of *De quadratura curvarum* [MP, 7:93-96] Newton also illustrated a method to expand the solution of a differential equation into series that can be considered a generalization of the previous method. Indeed Proposition 12 of this treatise contains the following problem:

> *Proposition 12. Given an equation involving two fluent quantities (if any, together with their fluxions), express the relation between these quantities by means of a convergent series.* (Newton [MP, 7:93])

To solve this problem, Newton (see [MP, 7:93-100]) considered the differential equation $F(z, y, y', y'', \ldots) = 0$, where F is an expression of the type

$$F(z, y, y', y'', \ldots, y^{(k)}) = \sum_i p_i z^{\mu_i} y^{\alpha_i} (y')^{\beta_i} (y'')^{\gamma_i} \ldots (y^{(k)})^{\omega_i}. \qquad (47)$$

If one substitutes $y = az^n$ (a and n to be determined) into the equation $F(z, y, y', y'', \ldots) = 0$, then one has

$$F[z, az^n, anz^{n-1}, an(n-1)z^{n-2}, \ldots] = 0.$$

This equation can be written as

$$F[z, az^n, anz^{n-1}, an(n-1)z^{n-2}, \ldots] = Az^\alpha + Bz^\beta + Cz^\gamma + \ldots + Sz^\sigma = 0,$$

for $\alpha < \beta < \gamma < \ldots < \sigma$. The coefficients A, B, C, ..., S depend on a, while the exponents α, β, γ, ..., σ depend on n and have the form

$$\mu_i + (\alpha_i + \beta_i + \gamma_i + \delta_i + \ldots)n - (\beta_i + 2\gamma_i + 3\delta_i + \ldots).$$

The number n is taken by considering the summands of $F(x, y, y', y'', \ldots)$ of the type Kz^{μ_i} (without y, y', y'', ...) and setting

$$n = \max_i \frac{\lambda - \mu_i + \beta_i + 2\gamma_i + 3\delta_i + \ldots}{\alpha_i + \beta_i + \gamma_i + \delta_i + \ldots},$$

---

[79]Newton gave examples where the points (i_1, j_1) and (i_2, j_2) are of the types $(0, j_1)$ and $(i_2, 0)$. It can happen, however, that the distribution of the marked rectangles (i, j) gives rise to more complicated situations (see Chabert [1999, 195–196]).

[80]The series $\sum_{k=0} b_k x^k$ is convergent if x is small.

where λ is the minimum exponent μ_i. Thus, the lowest exponent α in $Az^\alpha + Bz^\beta + Cz^\gamma + \ldots + Sz^\sigma$ is equal to λ. Since

$$
\begin{aligned}
& Az^\alpha + Bz^\beta + Cz^\gamma + \ldots + Sz^\sigma \qquad\qquad (48) \\
= & \; z^\alpha(A + Bz^{\beta-\alpha} + Cz^{\gamma-\alpha} + \ldots + Sz^{\sigma-\alpha}) \\
= & \; 0,
\end{aligned}
$$

one can find a by solving the equation $A = 0$.[81]

As the next step in the procedure, one sets

$$ y = az^n + p $$

and replaces $y, y', \ldots$ into $F(x, y, y', \ldots)$: Thus one obtains a new function $G(z, p, p', p'', \ldots)$ and can repeat the same reasoning. In this way, step by step, one finds a series that is a solution of $F(z, y, y', y'', \ldots) = 0$.

Newton observed that this method worked well when the quantity x is small enough. This is equivalent to recognizing that the series is not necessarily always convergent. However, if one desires the development at a point x that is not small enough, it is sufficient to set $x = w + v$, where w is an appropriate constant, and to proceed as earlier with respect to the variable v. It is clear that in this case Newton found an expansion of the type

$$ \sum_{i=0}^{\infty} A_i (x - w)^i, $$

convergent over a neighborhood of w.[82]

In a letter to Wallis (see Newton [OO, 1: 294]), the method was applied to

$$ y^2 - z^2 y' - d^2 + dz = 0. $$

In the first step, the summand of the lowest degree (excepting the terms with y and y') is d^2, and therefore $\lambda = 0$. Since

$$ n = \max_{i=1,2} \frac{\lambda - \mu_i + \beta_i}{\alpha_i + \beta_i} = 0, $$

---

[81] Higher-order infinitesimals are neglected near zero and so Equation (81) is reduced to $Az^\alpha = 0$, which is constantly equal to zero when $A = 0$. In this way an approximate solution is established.

[82] Newton also considered the case when x is large (*quantitas permagna*). In this case, in order to determine the exponent n of $y = az^n$ one always considers the summands of $F(x, y, y', y'', \ldots)$ of the type Kz^{μ_i} (without $y, y', y'', \ldots$) but sets

$$ n = \max_i \frac{\xi - \mu_i + \beta_i + 2\gamma_i + 3\delta_i + \ldots}{\alpha_i + \beta_i + \gamma_i + \delta_i + \ldots}, $$

where ξ is the maximum exponent μ_i.

he had $y = a$ and $y' = 0$. By substituting into the initial equation, he obtained

$$a^2 - d^2 + dz = 0.$$

The equation $a^2 - d^2 = 0$ allowed him to determine $a = d$. In the second step, he set $y = d + p$ and obtained

$$2dp + p^2 - z^2 p' + dz = 0.$$

It is easy to verify that $\lambda = 1$ and $n = 1$; hence, he set $p = az$ and, substituting it into the last equation, found $a = \frac{1}{2}$. Consequently, the first two terms of the expansion of y are $d - \frac{z}{2}$. In the third step, he set $p = -\frac{z}{2} + q$, repeated the reasoning, and found $q = -\frac{3z^2}{8d}$. Proceeding in this way, he obtained the expansion

$$y = d - \frac{1}{2}z - \frac{3}{8}\frac{z^2}{d} - \frac{9}{16}\frac{z^3}{d^2} + \ldots.$$

At this juncture, I explicitly note that Newton's procedures cannot be applied to a generic function, if one gives to this term the modern meaning (namely, if function is taken to mean a relation that associates an element y of a given set B to an element of another given set A). Newton's procedures were applied to the analytical expressions of geometrical quantities; more precisely, they were applied to a very special kind of analytical expression: the finite analytical expressions obtained by the composition of elementary operations (addition, subtraction, multiplication, division, extraction of root, raising to power).[83] Consequently, when Newton used the term "equation", he was referring to the equality between the analytical expressions composed by means of elementary operations[84]. For instance, the method of Proposition 12 is the general solution of the problem of finding the solution to a fluxional equation since, in Newton's opinion, *any equation* involving two fluent quantities and their fluxions had the form $F(x, y, y', y'', \ldots) = 0$, where F is an analytical expression of type (47).[85]

I would also like to emphasize the fact that Newton thought that *the power series that resulted from the expansion of any quantity was convergent*

---

[83]The expansion of quantities, such as trigonometric quantities, that lacked an analytical expression at the time, was derived by using their connection with certain analytical expressions. For instance, the expansion of the sine was derived by using the fact that the fluxion of arcsine is an analytical expression (see Section 4.2).

[84]One might distinguish between algebraic and fluxional equations (but the distinction is not explicit in Newton). If the variables are fluent and fluxions, we can refer to them as fluxional equations.

[85]For this reason, if one might for the sake of discussion use the term "function" in reference to Newton (Newton, as is known, never used it), provided the word "function" is taken to mean an algebraic function. The 18th-century notion of a function is examined in Chapter 18.

at least when the variable varied for an appropriate interval of values of x.
This opinion, for instance, is the basis of the following proposition, which is
found in the 1692 version of *De quadratura* (it is a corollary of Proposition
12):

> *All curves can be squared by means of indeterminate convergent*
> *series.* (Newton [MP, 7:96])

Newton explained that, given a curve $F(x, y) = 0$, it is sufficient to
express the ordinate $y = y(x)$ by a series using the above method and
integrating it term by term.

At this point, it is appropriate to try to clarify the meaning of the expression "convergent series" in Newton's work. In order to do this, I investigate
the closely connected concept of limit as formulated in Book 1, Section 1 of
Principia mathematica, where Newton expounded the basic notions of the
calculus in geometrical forms. In the final scholium Newton stated:

> For those ultimate ratios with which quantities vanish, are
> not truly the ratios of ultimate quantities, but limits towards
> which the ratios of quantities, decreasing without limit, do al-
> ways converge; and to which they approach nearer than by any
> given difference, but never go beyond, nor in effect attain to, till
> the quantities are diminished *in infinitum*. (Newton [PN, 87])[86]

One might be tempted to consider this sentences as a definition of limit,
and in this case the expression "ultimate ratio" would denote something
very similar to the modern limit concept (the point of view, e.g., of Pourciau [1998] and [2001]). However, I believe that a different interpretation is
preferable. For instance, consider the first of the 11 mathematical lemmas
of Book 1 of *Principia*:

> *Lemma 1. Quantities, and the ratios of quantities, which in any*
> *finite time converge continually to equality, and before the end of*
> *that time approach nearer the one to the other than by any given*
> *difference, become ultimately equal.* (Newton [PN, 73])

This lemma contains the following explicit hypotheses:

(H1) two quantities, say $A(t)$ and $B(t)$, approach closer and closer to one
     other, when t varies over a finite interval I, whose endpoints are a and
     c, and approach c.

(H2) $A(t)$ and $B(t)$ approach so close to one other that their difference is
     less than any given quantity, namely it is $|A(t) - B(t)| < \varepsilon$, when $t < c$
     but near enough to c.

---

[86]The translations from *Principia* are by Motte (see Newton [M]).

Hypothesis (H1) implies that A and B approach each other, but this does not necessarily mean that the distance between A and B becomes smaller than any quantity [for instance, $A(t) = -t^2$ and $B(t) = t^2 + 1$ $(t \to 0)$] satisfy hypothesis (H1)). Hypothesis (H2) guarantees that the distance actually becomes smaller than any given quantity. The thesis is

(T) $A(t)$ and $B(t)$ are ultimately equal.

The thesis states that the two quantities effectively reach each other when $t = c$. If we use the term "limit" in the same way as Newton does in the scholium, the thesis states that the quantities $A(t)$ and $B(t)$ have the same limit or that the limit of their difference is 0.

In the proof, Newton[87] assumes that $D > 0$ is the ultimate difference, namely, $|A(c) - B(c)| = D$; then $|A(t) - B(t)|$ does not become less than D, contrary to hypothesis (H2).

It is clear that if the proof is to be taken seriously, (H2) and (T) are not the same thing, and this implies that Newton did not think of (H2) as the definition of limit or ultimately equal.[88] (H2) is an essential property of limit but not the definition. In effect, Newton did not define the terms "limit" and "ultimate ratio": These terms had a clear intuitive meaning to him.[89] In the *Principia* he illustrated this intuitive meaning by referring to the "limit" as the last place or the last velocity of a motion.[90] For Newton, the notion of limit or ultimate value was an idea borrowed from nature; it was not a mathematical notion determined by its definition. A translation into modern terminology would strain Newton's concept.[91]

---

[87] "If you deny it; suppose them to be ultimately unequal, and let D be their ultimate difference. Therefore they cannot approach nearer to equality than by that given difference D; which is against the supposition" (Newton [PN, 73]).

[88] See also Panza [2003, 194–195].

[89] This behavior is similar to Leibniz's, who did not define the "ultimate value" (see Chapter 2).

[90] "Perhaps it may be objected, that there is no ultimate proportion of evanescent quantities; because the proportion, before the quantities have vanished, is not the ultimate, and when they are vanished, is none. But by the same argument it may be alleged, that a body arriving at a certain place, and there stopping, has no ultimate velocity; because the velocity, before the body comes to the place, is not its ultimate velocity; when it has arrived, is none. But the answer is easy; for by the ultimate velocity is meant that with which the body is moved, neither before it arrives at its last place and the motion ceases, nor after, but at the very instant it arrives; that is, that velocity with which the body arrives at its last place, and with which the motion ceases. And in like manner, by the ultimate ratio of evanescent quantities is to be understood the ratio of the quantities, not before they vanish, nor afterwards, but with which they vanish. In like manner the first ratio of nascent quantities is that with which they begin to be. And the first or last sum is that with which they begin and cease to be (or to be augmented or diminished). There is a limit which the velocity at the end of the motion may attain, but not exceed. This is the ultimate velocity. And there is the like limit in all quantities and proportions that begin and cease to be" (Newton [PN, 87]).

[91] For a more general discussion of the differences between the 17th- and 18th-century

Note that the absence of appropriate definitions was not restricted to this case. For instance, in Lemma 2 of Book 1 of *Principia* [PN, 73–74], Newton proved that the area under a curve was the limit of the circumscribed and inscribed rectangles. However, he did not define the area (nor the integral[92]). The area is a geometrical and physical entity that existed *per se* prior to any possible definition.[93]

Of course, this concept of limit has remarkable consequences for the notion of the sum. For Newton, summing a series $\sum_{i=0}^{\infty} a_i$ meant gradually combining all its terms in order to obtain the last term

$$a_1 + a_2,$$
$$(a_1 + a_2) + a_3,$$
$$(a_1 + a_2 + a_3) + a_4,$$
$$\ldots,$$
$$(a_1 + a_2 + \ldots + a_{n-1}) + a_n,$$
$$\ldots.$$

It made no difference whether the last term n was finite or infinite. It was intuitively clear that if S was the sum of the series $\sum_{i=0}^{\infty} a_i$, the property

$$(*) \qquad \left| S - \sum_{i=0}^{k} a_i \right| < \varepsilon, \qquad \text{for every } \varepsilon \text{ and large } k$$

held, and *vice versa* if property (∗) held, S was the sum of the series. However, Newton did not regard (∗) as the definition of the sum. According to him, the finiteness of human beings precluded them from naming (designating) and conceiving all the terms of a series and therefore determining the precise value of the quantity expressed by the series[94] (Newton [1711, 280]). As we shall see below, this conception prevented the development of the notion of limit in a modern sense and favoured the growth of a more formal concept of the sum of a series.

4.2   On Newton's manipulations of power series

Power series played a fundamental role in Newtonian calculus. For instance, in *De analysi* [MP, 2:206–207], Newton based his method of squaring a curve upon the following three rules:

---

notion of limit and the modern one, see Chapter 7

[92] For a different opinion, see Pourciau [1998].

[93] On the role of definitions in the 17th- and 18th-century mathematics, see Chapter 18.

[94] However, he stated that reasonings about series were certain and equations involving series were definitely correct (Newton [1711, 280]). In other words, a series could adequately represent a quantity in analytical calculations, but only the knowledge of the whole series provided the exact values of a quantity.

1. The area[95] under the curve of equation

$$y = ax^{m/n}$$

is

$$\frac{na}{m+n}ax^{\frac{m+n}{n}}.$$

2. If the equation $y = y(x)$ of a curve is given by the sum of a finite or infinite number of terms $y_1 + y_2 + y_3 + \ldots$, the area under the curve y is equal to the sum of the areas of all terms $y_1, y_2, y_3, \ldots$.

3. If the curve has a more complicated form, then one must expand the equation of the curve into a series of the type $\sum a_k x^{\alpha_k}$, where α_k is a rational number, and apply rules 1 and 2.

In the third point, Newton assumed that any analytically expressed quantity could be expanded into a series of the type $\sum a_k x^{\alpha_k}$, where α_k is a rational number. This series can be squared by integrating term by term using rule 1.

Similarly, in *De Methodis* [MP, 3:70–71], Newton reduced many geometrical problems (determination of tangents, maxima and minima, areas, surfaces, curvatures, arc lengths, centers of gravity, etc.) to two problems, which he formulated as follows:

(1a) Given the length of the space continuously (that is, at every time), find the velocity of motion at any time proposed.

(2a) Given the speed of the motion continuously, find the length of the space described at any time proposed.

When Newton introduced the terms "fluent" and "fluxion", these problems became

(1b) Given the relation between the fluents, find the relation between the fluxions.

(2b) Given the relation between the fluxions, find the relation between the fluents.

In modern terms, problem (2b) consists of seeking the solution to a differential equation. Newton suggested two ways of solving it.

---

[95]This rule was already known (see Chapter 1). On the rise of Newton's calculus, see Panza [1995].

First, one tried to transform a fluxional equation $f(x, y, \dot{x}, \dot{y}) = 0$ so that the problem could be reduced to seek the fluent in a known catalogue of fluents (in more modern terms, a table of integrals).

Second, one sought a solution expressed using series. We have seen that in the 1692 version of *De quadratura*, Newton proposed a general method for solving equations by series. Instead, in *de Methodis* to solve an equation $f(x, y, \dot{x}, \dot{y}) = 0$, Newton transformed it in the form

$$\frac{\dot{y}}{\dot{x}} = g(x, y);$$

then he expanded $g(x, y)$ into series and integrated term by term.

It is worthwhile noting that, despite the huge importance of series in the calculus, *solutions in closed form were preferred with respect to series solution*. For instance, in Proposition 11 of *De quadratura*, Newton formulated the following proposition:

> *Proposition 11. Given an equation involving the fluxions of two fluent quantities, find the relation between these quantities.*

This problem is similar to the problem of the subsequent Proposition[96] 12, the difference is that, in Proposition 11, Newton sought a solution in finite form (see Newton [MP, 1:72–92]). Only when the equation could not be solved by means of a finite expression did Newton suggest solving this by means of convergent series, according to the method of Proposition 12, and then by trying to reduce the series to a finite equation. The series solution offered a useful representation of a quantity, which could be handled and made it possible to compute the value of the quantity by approximation, but, where possible, one had to seek solutions in closed form.

In the previous Section 4.1, I underlined the importance of convergence for Newton, although we also saw that Newton's procedures were formal ones based upon the infinite extension of finite procedures. I shall now move on to investigate the relationship between convergence and manipulation of series in more detail, and I shall show that the Newtonian conception was not substantially different to Leibniz's with regard to this topic (in particular, convergence was not a preliminary condition to manipulation of power series).

For instance, to find the area enclosed by the curve

$$y = \frac{1}{1 + x^2}$$

and the axis of abscissa, Newton (see [OO, 1:264]) observed that

$$1 : (1 + x^2)$$

---

[96]See Proposition 12 on p. 62.

is equal to $1 - x^2 + x^4 + \ldots$ by continued division. It is then easy to express the area by the series

$$y = x - \frac{1}{3}x^3 + \frac{1}{5}x^5 + \ldots$$

by integrating term by term. According to Newton, however, one can also divide

$$1 : (x^2 + 1).$$

In this case the result is $y = x^{-2} - x^{-4} + x^{-6} + \ldots$ and the area is given from

$$y = -x^{-1} - \frac{1}{3}x^{-3} + \frac{1}{5}x^{-5} + \ldots.$$

The former expansion, Newton said, was to be used when x was small (*satis parva*) enough whereas the latter was to be used when x was large enough (*satis magna*). The difference of the developments $1 : (1 + x^2)$ and $1 : (x^2 + 1)$ was not viewed as a defect but as a useful tool for facilitating the computation of areas. A preliminary analysis of convergence seemed not only superfluous but even counterproductive. It was preferable to expand a series formally: After determining the coefficients and form of development, one verified if the interval of convergence was suitable for the specific problem and, if there were any, chose the series that fitted it best.[97] Convergence concerned the moment of the application, namely when one computed the values of the given series in order to find the values of a certain geometrical (or physical) quantity.

As we already saw in Leibniz,[98] such an approach was encouraged by the facts that, for practical aims, the lack of convergence in a certain interval appeared fairly easy to avoid by employing appropriate strategies. For instance, consider the fluxional equation

$$\frac{\dot{y}}{x} = \frac{2}{x} + 3 - x^2 \ldots.$$

Newton [OO, 1:418] solved it by replacing x by $z + 1$ and expanding the right side into power series. He obtained

$$\frac{\dot{y}}{\dot{z}} = \frac{2}{1+z} + 2 - 2z - z^2 = 4 - 4z + z^2 - 2z^3 + 2z^4 + \ldots$$

and

$$y = 4z - 2z^2 - \frac{z^3}{3} + \ldots.$$

---

[97] See also the example on p. 57.
[98] See Section 2.2.

Of course, this series converges only for $|z| < 1$ $(0 < x < 2)$. However, this difficulty can be overcome by using a different substitution $b + z$ and computing $y(x)$, for every $|x - b| = |z| < b$.

This method of tackling convergence was based on the assumption, which I mentioned on p. 64, that power series were always convergent for appropriate values of variables. It went tacitly that a convergent series at a specific point x could be easily determined by means of appropriate substitutions, and this was sufficient to obtain numerical approximations.

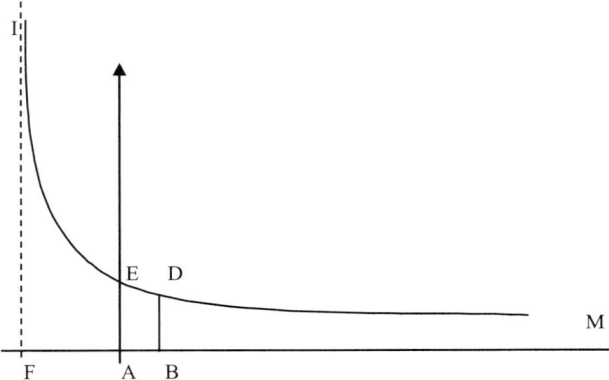

Fig. 10

This approach was closely connected with geometric aims of the early calculus. For instance, to calculate the area $ABDE$ under the hyperbola

$$y = \frac{a^2}{b + x}$$

(see Fig. 10), Newton expanded this analytical expression so to obtain series

$$\frac{a^2}{b + x} = \frac{a^2}{b} - \frac{a^2}{b^2}x + \frac{a^2}{b^3}x^2 - \ldots \tag{49}$$

By integrating this series, he obtained[99]

$$\text{area } (ABDE) = \frac{a^2 x}{b} - \frac{a^2}{2b^2}x^2 + \frac{a^2}{3b^3}x^3 - \frac{a^2}{4b^4}x^4 + \dots.$$

Of course, this derivation is valid under the condition $(0 <)x < b$; however, Newton did not point it out expressly. In a footnote of *Mathematical Papers*, Whiteside observed that it is possible that Newton is aware of the last restriction and seeks to convey it in his accompanying figure by setting[100] $AB < FA$ (see [MP, 1:129]).

Newton had in effect no need to specify the condition of convergence since *the geometrical context made it clear*.[101] As long as series were applied to an immediately geometric situation, the formalism inherent in such an approach did not disclose and, in a sense, was hidden by the geometrical situation. During the 18th century, the calculus grew and developed into a separate branch of mathematics, so the phase of the applications became further removed from the search for the development. Since convergence concerned only the application, the formal characteristics were no longer tempered by the reference to any geometrical figure or a concrete situation and became progressively more evident, giving rise to the typical 18th-century formalism (see Part II).

Moreover, I must highlight another important of series theory: the relation between a series and an analytically expressed quantity was conceived as valid for any value of variables, even if the series was convergent only for a certain interval of variation of the variable. For instance, Newton considered the curve

$$y = \frac{a^2}{(b+x)^2} = \frac{a^2}{b^2} - 2\frac{a^2}{b^3}x + 3\frac{a^2}{b^4}x^2 - 4\frac{a^2}{b^5}x^3 + \dots$$

---

[99] In modern symbols,

$$\int_0^x \frac{a^2}{b+x}dx = \int_0^x \left(\frac{a^2}{b} - \frac{a^2}{b^2}x + \frac{a^2}{b^3}x^2 - \dots \right) dx$$

$$= \frac{a^2 x}{b} - \frac{a^2}{2b^2}x^2 + \frac{a^2}{3b^3}x^3 + \dots$$

[100] I modernize the figure that is found in [MP, 1:129]. In this figure the segment FA is taken less than AB. Newton used the same figure to illustrate other results. In the present case the equation of EDM is $y = \frac{a^2}{b+x}$ and $FA = b$, $AE = \frac{a^2}{b}$.

[101] Newton observed that this series provided the area $ABCD$ by means of an infinite series; however, in practice, a few initial terms are as exact as desired (*quamvis satis exact*) if x is several times less than (*aliquoties minor quam*) b. This clarification concerns the speed of convergence rather convergence itself (with reference to a modern branch of mathematics, they concern numerical analysis) and has the practical aims of effectively computing the given series rather than the theoretical objective of fixing the interval of convergence (see Newton [1711, 263]).

and found that the area[102] $ABDE$ enclosed by the curve and the axis of abscissa is

$$\frac{a^2}{b} - \frac{a^2}{b+x},$$

since

$$
\begin{aligned}
area\ (ABDE) &= \int_0^x \frac{a^2}{(b+x)^2}dx = \frac{a^2}{b^2}x - \frac{a^2}{b^3}x^2 + \frac{a^2}{b^4}x^3 - \frac{a^2}{b^5}x^4 + \ldots \\
&= \frac{a^2}{b} - \left(\frac{a^2}{b} - \frac{a^2}{b^2}x + \frac{a^2}{b^3}x^2 - \frac{a^2}{b^4}x^3 + \frac{a^2}{b^5}x^4 - \ldots\right) \\
&= \frac{a^2}{b} - \frac{a^2}{b+x}.
\end{aligned}
$$

Newton [MP, 1:129–30] stated that this result "may also thus appeare": if $z = b + x$, then

$$\frac{a^2}{(b+x)^2} = \frac{a^2}{z^2};$$

therefore

area $(DBM) = \frac{a^2}{b+x}$,
area $(EAM) = \frac{a^2}{b}$,
area $(ABDE)$ = area (EAM)-area $(DBM) = \frac{a^2}{b} - \frac{a^2}{b+x}$.

In modern symbols,

$$
\begin{aligned}
\int_0^x \frac{a^2}{(b+x)^2}dx &= \int_0^\infty \frac{a^2}{(b+x)^2}dx - \int_x^\infty \frac{a^2}{(b+x)^2}dx \\
&= \int_0^\infty \frac{a^2}{z^2}dz - \int_x^\infty \frac{a^2}{z^2}dz = \frac{a^2}{b} - \frac{a^2}{b+x}.
\end{aligned}
$$

Thus, Newton obtained

$$area\ (ABDE) = \int_0^x \frac{a^2}{(b+x)^2}dx = \frac{a^2}{b} - \frac{a^2}{b+x} \tag{50}$$

in two different ways. However, the first proof uses series and so Equation (50) is demonstrated only for $|x| < b$. In the second proof, Equation (50) is not subjected to this restriction. From a modern point of view, the two proofs provide different results. Newton instead did not mention this fact and seems to consider the proofs as equivalent: He approached the matter as if results obtained handling a series convergent over an interval were valid for any value of x, independently of the domain of validity of series.

---

[102]Newton referred to Fig. 10 again. Now the equation of the curve EDM is $\frac{a^2}{(b+x)^2}$.

$$* \quad * \quad *$$

In the letter to Oldenburg on October 24, 1676, Newton considered

$$z = x + \frac{1}{2}x^2 + \frac{1}{3}x^3 + \frac{1}{4}x^4 + \ldots, \tag{51}$$

namely $z = \log \frac{1}{1-x}$, and tried to invert this series, that is, he tried to express x by means of a power series of the type

$$x = b_1 z + b_2 z^2 + b_3 z^3 + b_4 z^4 + \ldots.$$

He first calculated the powers

$$z^2 = x^2 + x^3 + \frac{11}{12}x^4 + \frac{5}{6}x^5 + \ldots,$$

$$z^3 = x^3 + \frac{3}{2}x^4 + \frac{7}{4}x^4 + \ldots,$$

$$\ldots$$

By multiplying the expansion of z^2 by b_2 and subtracting the result from (51), he obtained

$$z - b_2 z^2 = x + \left(\frac{1}{2} - b_2\right)x^2 + \left(\frac{1}{3} - b_2\right)x^3 + \left(\frac{1}{4} - b_2\right)x^4 + \ldots.$$

Since an expression of the type $x = b_1 z + b_2 z^2 + b_3 z^3 + b_4 z^4 + \ldots$ is desired,

$$b_2 = \frac{1}{2}$$

is required. Hence,

$$z - \frac{1}{2}z^2 = x - \frac{1}{6}x^3 - \frac{5}{24}x^4 + \ldots.$$

Similarly, Newton had

$$z - \frac{1}{2}z^2 + b_3 z^3 = x + \left(b_3 - \frac{1}{6}\right)x^3 + \left(b_3 - \frac{5}{24}\right)x^4 + \ldots.$$

Hence,

$$b_3 = \frac{1}{6}$$

and

$$z - \frac{1}{2}z^2 + \frac{1}{6}z^3 = x + \frac{1}{24}x^4 + \ldots,$$

and so on. The series of the logarithm converges for $-1 \leq x < 1$ and provided the variable z is greater than $\log \frac{1}{2}$. After inverting the series, Newton obtained

$$x = z - \frac{1}{2}z^2 + \frac{1}{6}z^3 - \frac{1}{24}z^4 + \ldots = 1 - e^{-z},$$

which instead converges everywhere and provides a relation that is valid for any z and for $x < 1$.[103]

Newton acted as if, beginning with a series $\sum a_n x^n$ convergent when $|x|$ is less than a certain number b, it could be used to derive a new result, which was considered legitimate independently of the values of x for which $\sum a_n x^n$ is convergent (of course, this had to be numerically valid, but this could be verified *a posteriori*). In modern terms, the domain of validity of $f(x) = \sum a_n x^n$ was enlarged to the whole set of real numbers.

This approach seems to me to be a natural consequence of the fact that convergence was not a preliminary condition to the manipulation of series and that it was considered only at the moment of the application of results. However, Newton did not discuss or refer even implicitly to this point, therefore, further remarks are appropriate. It is not, indeed, difficult to imagine different and more direct ways to derive $x = z - \frac{1}{2}z^2 + \frac{1}{6}z^3 - \frac{1}{24}z^4 + \ldots$: One could argue that the enlargement of the domain of validity was established on other grounds and that Newton did not intend actually to rely on the unrestricted manipulation of series. In my opinion, this interpretation is unlikely. In the letter to Oldenburg, Newton used this example as a special case of the general problem of the reversion of series:

Given the series

$$z = \sum_{n=1}^{\infty} a_n x^n,$$

find

$$x = \sum_{n=1}^{\infty} b_n z^n.$$

He gave the following general solution to this problem:[104]

$$x = \frac{1}{a_1}z - \frac{a_2}{a_1^3}z^2 + \frac{2a_2^2 - a_1 a_3}{a_1^5}z^3 - \frac{5a_1 a_2 a_3 - 5a_2^3 - a_1^2 a_4}{a_1^7}z^4 + \ldots. \qquad (52)$$

---

[103] See Newton [CE, 84–85].
[104] See Newton [CE, 85].

Newton did not provide a demonstration of (52), but it is clear that it can be obtained by operating in the same manner as he did in the example.[105] In the general case, the derivation can be intended only as a purely formal derivation without reference to convergence of series. This means that, given a series convergent over an interval I_1, one could derive series (52), which had the interval of convergence I_2: the link I_2 with I_1 was not established *a priori* by the theorem (the interval of convergence I_2 was determined by taking a specific series $\sum_{n=1}^{\infty} b_n z^n$ into account).

In my opinion, the observation that the expansion

$$1 - e^{-z} = x = z - \frac{1}{2}z^2 + \frac{1}{6}z^3 - \frac{1}{24}z^4 + \dots$$

could be derived not only by the reversion of the series

$$z = x + \frac{1}{2}x^2 + \frac{1}{3}x^3 + \frac{1}{4}x^4 + \dots$$

but also differently strengthened the idea that the unrestricted manipulation of series was legitimate and useful (I, however, do not mean that Newton chose this example for this purpose).

The method of series reversion is at the basis of Newton's treatment of the expansion of the sine series. In his letter to Oldenburg on June 13, 1676, Newton gave the expansions of the sine and of versed sine[106] (see Newton [OO, 1:285–288]). He did not explain how he had derived them.[107] In his explanatory notes to this letter, S. Horsley[108] reconstructed Newton's derivation of the sine expansion as follows.[109]

---

[105]By reasoning as in the example, one has

$$\frac{z}{a_1} = x + \frac{a_2}{a_1}x^2 + \frac{a_3}{a_1}x^3 + \dots$$

$$\left(\frac{z}{a_1}\right)^2 = x^2 + 2\frac{a_2}{a_1}x^3 + \left[\left(\frac{a_2}{a_1}\right)^2 + 2\left(\frac{a_3}{a_1}\right)^2\right]x^4 + \dots$$

$$\left(\frac{z}{a_1}\right)^3 = x^3 + 3\frac{a_2}{a_1}x^4 + \left[3\left(\frac{a_2}{a_1}\right)^2 + 3\left(\frac{a_3}{a_1}\right)^2\right]x^5 + \dots\dots$$

$$\dots$$

and then

$$\frac{z}{a_1} - \frac{a_2}{a_1^3}z^2 = x + \frac{2a_2^2 - a_1 a_3}{a_1^5}z^3 - \dots$$

and so on (see Panza [1992, 559]).

[106]The versed sine or versine of an angle z is equal to $1 - \cos z$.

[107]These expansions was not published in his *De analysi* and *De methodis*, probably because he considered it a mere application of the procedures mentioned in the last section and of series reversion (see Panza [1989, 161]).

[108]See Newton [OO, 1:297–98].

[109]The reconstruction uses typical Newtonian procedures.

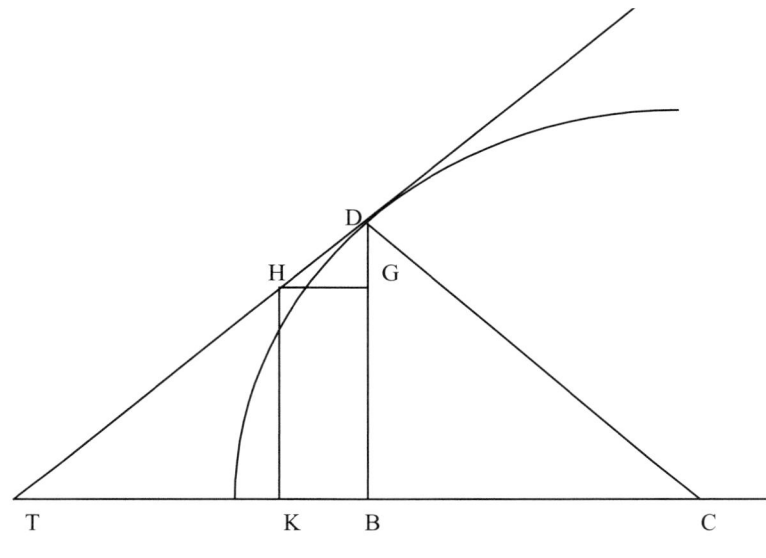

Fig. 11

With reference to Fig. 11, we set the radius $DC = 1$, the arc $DA = x$, and $DB = \sin x$. Since the triangles DHG (it is a characteristic triangle, in Leibniz's terms) and DBC are similar, $DH : DG = DC : BC$, and then

$$DH = \frac{DG \cdot DC}{BC} = \frac{DG}{\sqrt{1 - \sin^2 x}}$$

DH is the moment of the arc $x = dx$. If we take $DG = dz \ (= 1$, namely z is the independent variable),

$$DH = \frac{DG \cdot DC}{BC} = \frac{DG}{\sqrt{1 - DB^2}} = \frac{DG}{\sqrt{1 - \sin^2 x}},$$

$$DH = \frac{1}{\sqrt{1 - DB^2}} = 1 + \frac{1}{2} DB^2 + \frac{3}{8} DB^4 + \frac{5}{16} DB^6 + \dots.$$

By integrating

$$x = DB + \frac{1}{6} DB^3 + \frac{3}{40} DB^5 + \frac{5}{112} DB^7 + \dots,$$

we can write this series in the form

$$x = \sin x + \frac{1}{6} \sin^3 x + \frac{3}{40} \sin^5 x + \frac{5}{112} \sin^7 x + \dots$$

If we assume that $\sin x$ can be expanded into a power series:

$$\sin x = b_1 x + b_2 x^2 + b_3 x^3 + b_4 x^4 + \ldots,$$

then it is easy to obtain

$$\sin x = x - \frac{1}{6}x^3 + \frac{1}{120}x^5 - \frac{1}{5040}x^7 + \ldots$$

by using the procedure of reversion.

In this procedure the sine is not actually a function or analytical expression; it is a geometrical quantity, which is expanded by using the fact that the fluxion of another geometrical quantity (the arcsine) has an analytical expression.

5   Jacob Bernoulli's treatise on series

Jacob Bernoulli dealt with series at length. His work was not as innovative as that of Newton and Leibniz; nevertheless, he achieved some interesting results and also wrote a treatise, the *Positiones arithmeticae de seriebus infinitis, earumque summa finita*, in which he tried to systematize the results obtained during the 17th century and give an organic and coherent exposition of the theory of series. This treatise was first published between 1689 and 1704 in five separate papers, which were later collected in the appendix of the posthumously published *Ars conjectandi* (see Jacob Bernoulli [P]).

Bernoulli attempted to give a Euclidean form to the theory of series, namely to organize it as a sequence of propositions derived from certain axioms. Unlike modern axiomatic theory and similarly to Euclid's *Elements*, Bernoulli based his treatment on various implicit principles (for instance, the possibility of infinite extension of rules that were valid for finite processes) and undefined notions (e.g., he did not provide a definition of the sum and limit). He began his book with three axioms (see Bernoulli [P, 243]):

1. Each quantity can be divided into smaller parts.

2. Given a finite quantity, one can assume a quantity that is greater than this quantity.

3. Given a quantity, if one takes away any part from it and subtracts what is left from the original quantity, then one obtains that part (that is to say, given A, if $A - B = C$, then $A - C = B$).

The first two propositions are

a. A quantity less than any other quantity is non-quantum or nothing.

b. A quantity greater than any quantity is infinity (Bernoulli [P, 243]).

In order to prove theorem a, Bernoulli stated that if a quantity less than any other quantity was not nothing, then it could be divided into smaller parts and this is contrary to the hypothesis. Similarly, in theorem b, if a quantity was greater than any quantity and was finite, then it could not be increased, which is contrary to axiom 2.

After stating that every geometric progression can be continued *in infinitum*, because new terms can be added (none of which is zero or infinite), he proved that

c. given a geometric progression A, B, C, D, E and an arithmetic progression A, B, F, G, H, if the first two terms of both progressions are equal, then the terms C, D, E of a geometric progression are greater than the corresponding terms F, G, H of the arithmetic progression.

If we set $A = 1$ and $B = 1 + a$, this theorem provides the Bernoulli's inequality:[110]

$$(1 + a)^n > 1 + na.$$

Bernoulli's proof runs as follows. Since $A : B = B : C$, then $A + C > 2B = A + F$[111] and $C > F$. Moreover,

$$A + D > B + C > B + F = A + G.$$

Hence, $D > G$, so on.

Then Bernoulli proved the following two statements.

d. An increasing geometric progression[112] a_1, a_2, ... can be continued up to a term a_n greater than any given number Z.

e. A decreasing geometric progression a_1, a_2, ... can be continued up to a term a_n less than any given number Z.

To prove theorem e, Bernoulli considered an increasing geometric progression b_1, b_2, ... such that $b_1 = Z$ and the ratio $b_1 : b_2$ is the reciprocal of the ratio $a_1 : a_2$. This progression can be continued up to a term $b_n > a_1$ (for proposition d). Now continue the progression a_1, a_2, ... up to a_n. Since

$$\frac{a_n}{a_1} = \frac{b_1}{b_n} \quad \text{and} \quad a_1 < b_n,$$

one obtains

$$a_n < b_1 = Z$$

(see Bernoulli [P, 244]).

Bernoulli gave two corollaries to the theorems d and e:

f. The *ultimate* terms of an increasing geometric progression is ∞.

g. The *ultimate* terms of a decreasing geometric progression is 0 (see Bernoulli [P, 244]).

---

[110]This result had already been proved in a similar way by Barrow in his *Lectiones geometricae* [1670, 60].

[111]Bernoulli explicitly refers to Euclid's *Elements*, Book 5, Proposition 25: If four magnitudes are proportional, then the sum of the greatest and the least is greater than the sum of the remaining two.

[112]Bernoulli indicated the progression by A, B, C, D, E as in the previous theorem. I prefer to use a more modern symbolism.

Following Euclid, Bernoulli stated that if a_i, $i = 1, \ldots, n$, is a finite geometric progression then

$$a_1 : a_2 = \sum_{i=1}^{n-1} a_i : \sum_{i=2}^{n} a_i.$$

Therefore,

$$s_n = \frac{a_1(a_1 - a_n)}{a_1 - a_2} + a_n,$$

where $s_n = \sum_{i=1}^{n} a_i$. Then, following Viète (see p. 5), he showed that if the geometric progression is decreasing, then $\sum_{i=1}^{\infty} a_i$ is equal to

$$\frac{(a_1)^2}{a_1 - a_2},$$

since a_n becomes less than any given quantity and the last term of the sequence is 0 (for proposition g) (see Bernoulli [P, 244–245]).

Bernoulli then proved that the ultimate term had to be zero in order for the sum to be finite (see Bernoulli [P, 249]). He also showed that the harmonic series $\frac{1}{n}$ has an infinite sum. To prove this theorem he observed that

$$\frac{1}{k} + \frac{1}{k+1} + \frac{1}{k+2} + \ldots + \frac{1}{k^2} > \frac{1}{k} + \frac{1}{k^2} + \frac{1}{k^2} + \ldots + \frac{1}{k^2}$$
$$= \frac{1}{k} + (k^2 - k)\frac{1}{k^2} = 1.$$

Therefore, given any number N, it is possible to group the terms of the harmonic series as follows:

$$1 + \frac{1}{2} + \ldots + \frac{1}{a} > 1,$$
$$\frac{1}{a+1} + \ldots + \frac{1}{(a+1)^2} > 1,$$
$$\frac{1}{(a+1)^2 + 1} + \ldots + \frac{1}{((a+1)^2 + 1)^2} > 1,$$
$$\ldots$$

and so on until one obtains N groups. This means that the partial sums of series become greater than any number N[113] (see Bernoulli [P, 250–251]).

While these theorems from the *Positiones arithmeticae* show that Bernoulli paid much attention to convergence, others show that he did not hesitate to

---

[113]Bernoulli observed: "The sum of an infinite series whose final term vanishes perhaps may be both infinite and finite" [P, 252].

use formal manipulation. For instance, to sum $\sum_{n=1}^{\infty} \frac{2}{n(n+2)}$, Bernoulli [P, 252–253] used the same method used by Leibniz (see p. 30)[114] He set[115]

$$\sum_{n=1}^{\infty} \frac{1}{n} = A$$

and

$$\sum_{n=1}^{\infty} \frac{1}{n+2} = B$$

and derived

$$\sum_{n=1}^{\infty} \frac{2}{n(n+2)} = \sum_{n=1}^{\infty} \frac{1}{n} - \sum_{n=1}^{\infty} \frac{1}{n+2} = A - B = 1 - \frac{1}{2} = \frac{3}{2}.$$

Bernoulli was quite aware that the condition "a_n is an infinitesimal" for $n = \infty$ was necessary to sum a series $\sum a_n$. In 1689 he even showed what would happen if this was not satisfied. He set

$$S = \frac{2a}{c} + \frac{3a}{2c} + \frac{4a}{3c} + \frac{5a}{4c} + \dots$$

and

$$T = \frac{3a}{2c} + \frac{4a}{3c} + \frac{5a}{4c} + \dots$$

and observed that the application of the above method led to

$$Q = S - T = \frac{2a}{c},$$

whereas the sum is

$$Q = S - T = \frac{a}{2c} + \frac{a}{6c} + \frac{a}{12c} + \dots = \frac{a}{c}.$$

Bernoulli explained that the method failed because Q was equal to the first term of S (namely $\frac{2a}{c}$) minus the last term of T, which is not 0, but equals $\frac{a}{c}$.

Bernoulli limited himself to advising the use of such series with caution in these particular cases (an exhortation that is also found in the young Euler[116]). Bernoulli did not require a control of convergence of series as a

---

[114]Bernoulli explicitly referred to Leibniz and stated that, in his [1682], Leibniz had not explained how he derived his results; he therefore thought that it was appropriate to illustrate how certain sums could be determined (see Bernoulli [P, 252]).

[115]Equality $\sum_{n=1}^{\infty} \frac{1}{n} = A$ is found in the *Positiones arithmeticae* immediately after the proof that the sum of the harmonic series is infinite.

[116]See Chapter 15.

preliminary condition for the application of this procedure,[117] but simply verified the soundness of the results that unconditioned manipulation led to and, if necessary, provided an explanation of failures, which were considered as anomalies or exceptional cases. This approach prevented him from rejecting the use of divergent series in a clear and definitive way, even when it resulted in a contradiction concerning $\sum_{n=1}^{\infty} \frac{1}{\sqrt{n}}$. Jacob Bernoulli set

$$X = \sum_{n=1}^{\infty} \frac{1}{n^m} \quad \text{and} \quad Y = \sum_{n=1}^{\infty} \frac{1}{(2n-1)^m}.$$

By subtracting term by term, he obtained

$$X - Y = \sum_{n=1}^{\infty} \frac{1}{(2n)^m} = \frac{1}{2^m} \sum_{n=1}^{\infty} \frac{1}{n^m} = \frac{1}{2^m} X.$$

Therefore, $2^m X - X = 2^m Y$ and

$$Y : (Y - X) = \left(X - \frac{X}{2^m}\right) : \frac{X}{2^m} = \left(1 - \frac{1}{2^m}\right) : \frac{1}{2^m} = (2^m - 1) : 1.$$

Hence, he stated that the ratio between two series can be known even without knowing their sums. Bernoulli applied the last proportion to the series

$$\sum_{n=1}^{\infty} \frac{1}{\sqrt{n}}$$

(whose sum is infinite, he argued, since[118] its term $\frac{1}{\sqrt{n}}$ is larger than the term of $\sum_{n=1}^{\infty} \frac{1}{n}$) and noted that the ratio between the series of odd terms

$$Y = \sum_{n=1}^{\infty} \frac{1}{\sqrt{2n-1}}$$

and the series of even terms

$$X - Y = \sum_{n=1}^{\infty} \frac{1}{\sqrt{2n}}$$

---

[117]It was only Gauss who dealt with telescopic series in a way that can be considered satisfying (see Gauss [1812b, 143]).

[118]In his treatise Bernoulli applied what we term the "comparison test". However, he did not use it to establish whether a series was convergent, considering this as a preliminary condition for using a given series. Given two series $\sum_{n=1}^{\infty} a_n$ and $\sum_{n=1}^{\infty} b_n$ such that $a_n > b_n$, Bernoulli did not deduce the existence of $\sum_{n=1}^{\infty} a_n$ from the fact that $\sum_{n=1}^{\infty} b_n$, is finite. Rather he deduced that $A > B$ where $A = \sum_{n=1}^{\infty} a_n$ and $B = \sum_{n=1}^{\infty} b_n$ (independently of the fact whether A and B are finite or infinite).

is, unexpectedly,

$$Y : (Y - X) = (\sqrt{2} - 1) : 1$$

and, therefore,

$$\sum_{n=1}^{\infty} \frac{1}{\sqrt{2n-1}} < \sum_{n=1}^{\infty} \frac{1}{\sqrt{2n}},$$

while it is clear that

$$\sum_{n=1}^{\infty} \frac{1}{\sqrt{2n-1}} > \sum_{n=1}^{\infty} \frac{1}{\sqrt{2n}}$$

(see Bernoulli [P, 261–262]).

While in the first part of his treatise Bernoulli was mainly interested in summing numerical series, in the second he dealt with the expansion of analytical expressions (representing curves, such as hyperbola, etc.), into power series and used these expansions to square and rectify curves.

He first dealt with the method of long division (see p. 20). He then explained that there were three other "artifices" (*artificia*) for expanding analytically expressed quantities: Wallis's interpolation, binomial expansion, and the method of indeterminate coefficients. Moreover, in order to square curves, Bernoulli applied term-by-term integration of power series in an unproblematic way. In effect, Bernoulli's conception is very similar to Newton's and Leibniz's. Therefore, I shall not dwell upon it and I restrict myself to some observations concerning the expansion of the quantity $\frac{k}{m \mp n}$. Bernoulli stated that the method of long division applied to $\frac{k}{m \mp n}$ always yields a remainder: Only if $m > n$ does the remainder "decrease and is finally less than any given quantity" in which case one has:

$$\frac{k}{m \mp n} = \frac{k}{m} \pm \frac{kn}{m^2} + \frac{kn^2}{m^3} \pm \dots. \tag{53}$$

Bernoulli also proved that the sum of $\frac{k}{m} - \frac{kn}{m^2} + \frac{kn^2}{m^3} - \dots = \frac{k}{m+n}$ varies between $\frac{k}{m}$ and $\frac{k}{2m}$ for $0 < n < m$. He observed that if $m = n$, the series gives rise to the "*non inelegans*" paradox

$$\frac{k}{2m} = \frac{k}{m} - \frac{k}{m} + \frac{k}{m} - \frac{k}{m} + \dots. \tag{54}$$

He explained that in this case the remainder does not decrease but is always equal to $\pm\frac{k}{2m}$: The paradox actually originated in the fact that the result of division is not

$$\frac{k}{m} - \frac{k}{m} + \frac{k}{m} - \frac{k}{m} + \dots,$$

but
$$\frac{k}{m} - \frac{k}{m} + \frac{k}{m} - \frac{k}{m} + \dots (-1)^n \frac{k}{2m},$$
where n is the number of terms we added. Bernoulli, therefore, interpreted (53) as a relation that is legitimate under the condition $|m| < n$.[119]

To end this chapter I would like to point out an interesting result obtained by Bernoulli which is found in a letter written to Leibniz in 1703. It concerns the series solution to the equation

$$dy = y^2 dx + x^2 dx. \tag{55}$$

This equation is of interest because it is the first appearance of an equation of the Riccati type. Johann Bernoulli had already posed the problem of the solution to (55) in 1694. Jacob Bernoulli succeeded in solving it by transforming it into the simpler equation

$$\frac{d^2 z}{z} = -x^2 dx^2$$

by letting $y = -\frac{dz}{z dx}$ (see Leibniz [GMS, 3:74]). By using the method of indeterminate coefficients, it is easy to solve the equation $\frac{d^2 z}{z} = -x^2 dx^2$. Indeed one obtains

$$z = 1 - \frac{x^4}{3 \cdot 4} + \frac{x^8}{3 \cdot 4 \cdot 7 \cdot 8} - \frac{x^{12}}{3 \cdot 4 \cdot 7 \cdot 8 \cdot 11 \cdot 12} + \dots$$

By replacing this series into $y = -\frac{dz}{z dx}$, Bernoulli found the solution

$$y = \frac{\frac{x^3}{3} - \frac{x^7}{3 \cdot 4 \cdot 7} + \frac{x^{11}}{3 \cdot 4 \cdot 7 \cdot 8 \cdot 11} + \dots}{1 - \frac{x^4}{3 \cdot 4} + \frac{x^8}{3 \cdot 4 \cdot 7 \cdot 8} - \frac{x^{12}}{3 \cdot 4 \cdot 7 \cdot 8 \cdot 11 \cdot 12} + \dots}$$

to differential equation (55). This solution can be transformed into the series

$$y = \frac{x^3}{3} + \frac{x^7}{3 \cdot 3 \cdot 7} + \frac{2x^{11}}{3 \cdot 3 \cdot 3 \cdot 7 \cdot 11} + \frac{13x^{15}}{3 \cdot 3 \cdot 3 \cdot 3 \cdot 5 \cdot 7 \cdot 7 \cdot 11} + \dots.$$

(see Leibniz [GMS, 3:75]).

---

[119]Bernoulli's treatise ends this poem, which illustrated the difficulty of the notion of infinite series:

Even as the finite encloses an infinite series
And in the unlimited limits appear,
So the soul of immensity dwells in minutia
And in the narrowest limits no limit in here.
What joy to discern the minute in infinity!
The vast to perceive in the small, what divinity!
(Bernoulli [P, 306], translation by Walker in [Smith, 1:271])

6 The Taylor series

The Taylor series

$$f(x) = f(x_0) + \left.\frac{df}{dx}\right|_{x=x_0}(x-x_0) + \frac{1}{2!}\left.\frac{d^2 f}{dx^2}\right|_{x=x_0}(x-x_0)^2$$
$$+\frac{1}{3!}\left.\frac{d^3 f}{dx^3}\right|_{x=x_0}(x-x_0)^3 + \frac{1}{4!}\left.\frac{d^4 f}{dx^4}\right|_{x=x_0}(x-x_0)^4 + \dots \quad (56)$$

is one of the most important series in mathematics. It bears the name of Brook Taylor, who first published it in *Methodus incrementorum* [1715]. However, similar results were probably known to Gregory (see Chapter 1) and certainly to Newton, Johann Bernoulli (see Chapter 3), Leibniz (see Chapter 3), and de Moivre (see Wollenshläger [1933, 241–257]).

Newton's formulation of the Taylor series is found in the 1692 version of *De quadratura curvarum* (see Newton [MP, 7:96–98]). It is contained in the corollaries 3 and 4 to Proposition[120] 12. In modern terms, Corollary 3 stated that if the series that results from the application of the method contained in Proposition 12 is a power series

$$f(z) = \sum_{n=1}^{\infty} A_n z^n,$$

then

$$A_n = \frac{1}{n!}\left.\frac{d^n f}{dx^n}\right|_{z=0}.$$

In Corollary 4 Newton gave the more general series

$$f(x+w) = \sum_{n=1}^{\infty} \frac{1}{n!}\frac{d^n f}{dx^n}w^n.$$

(In both cases, Newton assumed that the first term of the series was equal to 0.) The brevity of Newton's treatment made it unclear as to the importance that he gave to this formula, which he removed from the later version of *De quadratura curvarum* and never published.

Taylor's derivation of the Taylor theorem was based on what he called the "method of increments."[121] This was substantially a theory of finite differences. Taylor thought that the method of increment provided "a perfect knowledge of the method of fluxions" (Taylor [1715d, 339]) and applied it to several mathematical and mechanical problems. Among the results he derived was the following:

---

[120] See Section 4.1, p. 62.

[121] On Taylor's contributions to the Taylor series, see Feigenbaum [1985] and Panza [1992, 364–375].

Proposition 7. Let z and x be two variable quantities, of which z increases uniformly with given increments $\dot{z}$. Let $nz = v$,

$v - z = \overset{\backslash}{v}$, $\overset{\backslash}{v} - z = \overset{\backslash\backslash}{v}$, and so on. Then I say that in the time that z increases to $z + v$, x will likewise increase to

$$x + x\frac{v}{\dot{\cdot}\,1 \cdot \dot{z}} + x\frac{\overset{\backslash}{v}\,v}{\dot{\cdot}\dot{\cdot}\,1 \cdot 2 \cdot \dot{z}^2} + x\frac{\overset{\backslash\backslash}{v}\,\overset{\backslash}{v}\,v}{\dot{\cdot}\dot{\cdot}\,1 \cdot 2 \cdot 3\dot{z}^3} + etc.$$

(Taylor [1715, 21], translation in Feigenbaum [1985, 40])

By using a more modern symbolism and the notion of a function, the previous theorem can formulated by stating that if $y(z)$ is a function of z, Δz and Δy are the increments of z and y, and $v_k = k\Delta z$, then

$$y(z + n\Delta z) = y(z) + \Delta y\frac{v_n}{1 \cdot \Delta z} + \Delta^2 y\frac{v_n v_{n-1}}{1 \cdot 2 \cdot (\Delta z)^2} + \Delta^3 y\frac{v_n v_{n-1} v_{n-2}}{1 \cdot 2 \cdot 3(\Delta z)^3} + \dots$$
$$(57)$$

To prove (57), Taylor observed that

$$\begin{aligned}
y(z) &= y(z), \\
y(z + \Delta z) &= y(z) + \Delta y, \\
y(z + 2\Delta z) &= y(z) + 2\Delta y + \Delta^2 y, \\
y(z + 3\Delta z) &= y(z) + 3\Delta y + 3\Delta^2 y + \Delta^3 y, \\
y(z + 4\Delta z) &= y(z) + 4\Delta y + 6\Delta^2 y + 4\Delta^3 y + \Delta^4 z, \\
&\dots
\end{aligned}$$
$$(58)$$

He stated that the coefficients of Δz in the table (58) "are formed in the same way as the coefficients of corresponding terms in the binomial expansion" (Taylor [1715, 22]). This made it possible to apply Newton's binomial expansion. In doing this Taylor passed from the finite series (58) to the infinite series[122]

$$y(z+n\Delta z) = y(z) + \frac{n}{1}\Delta y + \frac{n(n-1)}{1\cdot 2}\Delta^2 y + \frac{n(n-1)(n-2)}{1\cdot 2}\Delta^3 y + \dots \quad (60)$$

Through the substitution $k = \frac{v_k}{\Delta z}$, he obtained (57).

In the first corollary of this proposition, Taylor stated that, in the hypothesis of Proposition 7, if z decreases to $z - v$, x will decrease to

$$x - x\frac{v}{1\cdot z} + x\frac{v\dot{v}}{1\cdot 2\cdot z^2} - x\frac{v\dot{v}\,\ddot{v}}{1\cdot 2\cdot z^3} + \&c.$$

In Corollary 2, Taylor obtained the Taylor series merely by assuming $\Delta z = \dot{z}(t)o$, where o is an evanescent increment. To use his own terms:

> Corollary 2.   If we substitute for the evanescent increments the fluxions proportional to them, and if the $\dot{v}\,\ddot{v}$ are now made equal, then in the time that z, flowing uniformly becomes $z + v$, x will become
>
> $$x + \dot{x}\frac{v}{1\cdot \dot{z}} + \ddot{x}\frac{v^2}{1\cdot 2\cdot \dot{z}^2} + \dddot{x}\frac{v^3}{1\cdot 2\cdot \dot{z}^3} + \&c. \qquad (61)$$
>
> and likewise, with z decreasing to $z - v$, x will decrease to
>
> $$x - \dot{x}\frac{v}{1\cdot \dot{z}} + \ddot{x}\frac{v^2}{1\cdot 2\cdot \dot{z}^2} - \dddot{x}\frac{v^3}{1\cdot 2\cdot \dot{z}^3} + \&c.$$

(Taylor [1715, 23])

Taylor gave importance to (61) as a method for solving differential equations. According to him, one first had to seek a solution in closed form. If this was impossible, then one looked for a series solution. In this case, (61)

---

[122]Formula (57) is nothing but interpolation formula (17). A similar result was given by Newton in *Principia*. It can be formulated as follows. The nth degree interpolation polynomial $y(x)$, passing through any given $n + 1$ points with abscissas a, $a + h$,..., $a + nh$ and ordinates $y(a)$, $y(a + h)$, ..., $y(a + nh)$, is

$$\begin{aligned} y(x) \;=\;& y(a) + \frac{x}{h}\Delta y(a) + \frac{x(x-h)}{1\cdot 2\cdot h^2}\Delta^2 y(a) + \frac{x(x-h)(x-2h)}{1\cdot 2\cdot 3\cdot h^3}\Delta^3 y(a) + \dots \quad (59)\\ &+ \frac{x(x-h)(x-2h)\dots(x-(n-1)h)}{1\cdot 2\cdot 3\cdot \dots \cdot n\cdot h^n}\Delta^n y(a)\end{aligned}$$

In his [1985, 42–43], Feigenbaum notes that Newton's formula (59) explicitly provides a polynomial, whereas (57) was meant to be an infinite series. However, the early series theory was based on the systematic transition from the finite to the infinite, as Taylor's proof of the Taylor theorem confirms. For this reason it is difficult to consider (59) and (57) as being really different in the context of Taylor's proof.

offered a general method of solution. As an example Taylor [1715, 24–26] considered the equation that, in modern symbols, can be written as

$$xy'' + nyy'' - y' - (y')^2 = 0. \tag{62}$$

It is easy to derive

$$y'' = \frac{y' + (y')^2}{x + ny}$$

from (62). By repeated differentiation Taylor found

$$y''' = (2 - n)\frac{y'y''}{x + ny},$$

$$y^{(4)} = (3 - 2n)\frac{y'y^{(3)}}{x + ny},$$

$$y^{(5)} = (4 - 3n)\frac{y'y^{(4)}}{x + ny},$$

$$\cdots$$

If $c = y(a)$ and $c' = y'(a)$ are the initial conditions, the solution is given by the series

$$y(a + x) = y(a) + y'(a)x + y''(a)\frac{x^2}{2} + y'''(a)\frac{x^3}{3!} + \dots \tag{63a}$$

He also showed that, for particular values of a, the series (63a) was finite and then one obtained the preferred solution in closed form.

Among the other results contained in the *Methodus*, the following should be pointed out:

The fluent of $\dot{r}s$ can be expressed by either the series

$$\overline{\dot{r}s} = rs - \dot{r}\dot{s} + \overset{\backslash\backslash\cdots}{r}\overset{\cdot\cdot}{s} - \overset{\backslash\backslash\backslash\cdots}{r}\overset{\cdots}{s} + \&c. \tag{63}$$

or

$$\overline{\dot{r}s} = \dot{r}\overset{\backslash}{s} - \ddot{r}\overset{\backslash\backslash}{s} + \overset{\cdots}{r}\overset{\backslash\backslash\backslash}{s} + \&c.$$

(Taylor [1715, 38])

These formulas have often been considered as identical to the Bernoulli series.[123] For instance, (63) can be written as

$$\int s\,dr = rs - \int r\,dr \cdot \frac{ds}{dr} + \int dr \int r\,dr \cdot \frac{d^2s}{dr^2} - \int dr \int dr \int r\,dr \cdot \frac{d^3s}{dr^3} + \dots. \tag{65}$$

---

[123]Their proofs are also very similar, see [1715, 38–39].

Such a translation of (63) from Newtonian fluxional language into the Leibnizian one was the basis for Bernoulli's charge of plagiarism toward Taylor, which was first raised by Burchard, one of Bernoulli's students, in his *Epistola ad Virum Clarissimum Brook Taylor* [1721]. It is interesting noting that the charge of plagiarism only concerned series (63), not series (61). This is due to the fact that both Bernoulli and Taylor probably considered the Bernoulli and Taylor series as results differing from each other.[124]

In her [1985, 117–125] Feigenbaum notes that the interpretation that Taylor gave to (63) makes it more general than (65), so that the charge of plagiarism appears only partially justified. Based upon Taylor's explanation and the three examples he provided, Feigenbaum shows that (63) should be expressed in modern notation as follows:

$$\int s\,dr = \int s\frac{dr}{dw}dw$$
$$= rs - \frac{ds}{dw}\int r\,dw + \frac{d^2s}{dw^2}\int dw \int r\,dw$$
$$- \frac{d^3s}{dw^3}\int dw \int dw \int r\,dw + \ldots,$$

where s, r, and w are functions of the same variable z. The presence of the additional quantity w made it possible to expand $\overline{rs}$ into series in various ways. Moreover, an appropriate choice of w could make calculations easier, wich was an advantage of (63) with respect to the Bernoulli series.

In 1717, immediately after the publication of the Taylor's *Methodus*, Stirling gave a different proof of the Taylor series (56), or, to be precise, of the Maclaurin series, since he only considered the case $x_0 = 0$ (see Stirling [1717, 76–77]). In modern symbolism, Stirling's proof can be formulated as follows.[125] If one sets

$$f(x) = \sum_{i=0}^{\infty} A_i x^i$$

and repeatedly differentiates this equality term by term, then one obtains

$$\frac{df}{dx} = \sum_{i=0}^{\infty} A_i i x^{i-1},$$
$$\frac{d^2 f}{dx^2} = \sum_{i=0}^{\infty} A_i i \, (i-1) \, x^{i-2},$$
$$\frac{d^3 f}{dx^3} = \sum_{i=0}^{\infty} A_i i \, (i-1) \, (i-2) \, x^{i-3},$$
$$\ldots$$

---

[124]On this question, see Panza [1992, 406–421].
[125]This might have been Newton's proof. However, Newton was never explicit.

By setting $x = a$, one has

$$A_1 = \left.\frac{df}{dx}\right|_{x=0},$$
$$A_2 = \left.\frac{1}{2!}\frac{d^2 f}{dx^2}\right|_{x=0},$$
$$A_3 = \left.\frac{1}{3!}\frac{d^3 f}{dx^3}\right|_{x=0},$$
$$\cdots$$

Stirling illustrated this result with several examples, among which is the expansion for the cosine. According to Feigenbaum: "This would appear to be the first published instance of the explicit use of the Taylor theorem to generate a power series expansion of a well-known function" (see Feigenbaum [1985, 80]).

7 Quantities and their representations

An adequate understanding of 17th- and 18th-century series theory cannot avoid a discussion of the basic notion of quantity. In the first part of this chapter, I shall describe the evolution of the concept of quantity until the end of 18th century and will then examine the different ways - figural and symbolic- of representing quantities. In the second part, I shall illustrate the concepts of numbers and continuity and highlight the difficulties of translating the traditional notion of quantity into modern terms.

7.1 Quantity and abstract quantity

The 17th- and 18th-century notion of quantity had classic roots. In *Metaphysics* (V, 1020a7), Aristotle characterized quantity as "that which is divisible into two or more constituent parts". In *Physics* (VIII, 7, 260a7f) he stated that a thing changes in quantity if it is increased or decreased.

Greeks distinguished quantities as being continuous or discrete. Discrete quantity (or multitude) was made up of discontinuous parts, meaning there was no common boundary at which they joined. These parts formed a plurality, an aggregate of units. Number and speech were given as examples of discrete quantities (see Aristotle's *Categories*, 4b20–22). Aristotle stated:

> In the case of the parts of a number, there is no common boundary at which they join. For example: two fives make ten, but the two fives have no common boundary, but are separate; the parts three and seven also do not join at any boundary. Nor, to generalize, would it ever be possible in the case of number that there should be a common boundary among the parts; they are always separate. Number, therefore, is a discrete quantity. (*Categories*, 4b20)

A continuous quantity (or magnitude) consists of parts whose position is established in reference to each other, so that the limit of one is the limit of the next. Ancient Greeks considered several types of continuous quantities, such as time, movement, and various geometrical quantities. In his [1996, 363], Grattan-Guiness listed 10 different types of geometrical quantities in Euclid's *Elements*: straight lines, planar curved lines (arc of circle), planar rectilinear regions (rectangle), planar curvilinear regions (segment of circle), spatial rectilinear surfaces (pyramid), spatial curvilinear surfaces (sphere), rectilinear solids (cube), curvilinear solids (hemisphere), planar angles, solid planar angles.

All these differences between geometrical quantities were of importance since Greek mathematicians did not study quantity as an abstract entity with the capacity of decreasing and increasing; rather, they considered specific and determinate quantities in specific and determinate contexts. For the

Greeks, quantity always referred to an object: it was "quantity of". Quantity presupposed the material, or, at least, an idealization of the material (geometrical quantities, such as lines, were idealization of real objects)

Similar to geometrical quantities, numbers were also thought of as specific and definite objects.[126] According to Aristotle, "to be present in number" is to be some number of a given object (*Physics,* 4 221b14f). "[T]he assertion 'three trees' presupposes the assertion 'three', but what the assertion 'three' intends has no existence outside of the trees of which they are said to be three. For the number of the trees, i.e., 'three', has no proper, no independent 'nature'... " (Klein [1968, 101]). Number in Greek mathematics always meant "a definite number of definite objects".

These objects could be objects of sense or pure units or monads. It is worth noting that there was a difference between numbers, which were pluralities, and one, which was not a number but the unity by which the numbers were counted. According to Euclid's *Elements,* "a number is a multitude composed of units" (Book 7, Definition 2), while "a unit is that by virtue of which each of the things that exist is called one" (Book 7, Definition 1). Following Aristotle, Euclid considered unity as "unity of measurement".[127]

In the 17th and 18th centuries mathematicians continued to use quantity as the basic notion of mathematics —mathematics was regarded as the science of quantity— although the notion they used was markedly different from the Greek one. While Greek mathematicians had only dealt with specific and determinate quantities,[128] in the period around 1600, after a long evolution, mathematicians had finally become capable of investigating mathematical objects in a more general form. Quantities were now considered as the abstract objects of our thought. These abstract entities *were reified in appropriate symbolic representations (algebraic and geometric), which allowed their mathematical treatment.* The degree of abstraction and the forms of reification were subjected to further gradual changes from the 1590s to the 1740s and afterwards. The final result of this evolution was the mathematical treatment of quantity as such (general quantity).

To make this point clearer, I shall now investigate the nature of algebraic and geometric representations and describe the development of the relationship between algebraic and geometric representations during the 17th century.

$$*\quad*\quad*$$

The evolution of the concept of quantity underwent a turning point with

---

[126] In the quotation on p. 93, Aristotle speaks of "parts" of numbers that have no common boundary. This has a sense only if a number is deemed to refer to concrete or ideal objects.

[127] See Klein [1968, 108–112].

[128] It is worth noting that while Greek mathematicians had always dealt with specific and determinate quantities, Greek philosophers discussed the concept of quantity in an abstract sense.

the rise of algebraic or analytical symbolism. I specify that the decisive aspect of analytical or algebraic symbolism was not the use in itself of certain signs but the fact that those signs were the objects of manipulation in their own right. For instance, I can write

$$a \perp b$$

to indicate that the straight line a is perpendicular to b. However, if in the proof of a theorem of elementary geometry, for instance "Given a point A and a straight line a, there exists one and only one perpendicular b to a lead for A", I write $\perp$ in place of "perpendicular to", I do not really manipulate the symbol $\perp$ by itself, but work with the concept of "perpendicular to". The sign $\perp$ is employed as a mere shorthand symbol, unless one establishes a calculus upon and operates according to the rule of this calculus.

In symbolic expressions, such as

$$(a + b)^2 = (a + b)(a + b) = a^2 + 2ab + b^2,$$

the letters a and b, are used as the concrete objects of a *calculation* (see Panza [1992, 68-69]), namely they are manipulated according to certain rules, but the fact that they represented numbers or lines or other objects is of no importance. According to Leibniz, a calculation is *cogitatio caeca*, blind reasoning. Operating blindly was conceived of as moving pebbles mechanically in an abacus: What is of importance is that the concrete objects of manipulation (pebbles or graphical signs) are handled according to certain rules (syntactically, in modern terms), not their meaning.

Greeks did not manipulate algebraic symbols in their mathematical reasonings; rather, they reasoned upon figures. A figure is a symbolic representation as well; however, it has a different nature with respect to algebraic symbolism. It is iconic and imititative and reproduces the features of various real bodies by analogy. Figures can also be used in modern mathematics. However, there is a huge difference between the modern and the ancient use of figures. In modern mathematics, figures are dispensable tools for facilitating the comprehension: Their role is essentially pedagogical or illustrative. A modern theory is a conceptual system, composed of explicit axioms and rules of inference, definitions and theorems derived by means of a merely linguistic deduction. For instance, consider the proposition

> *Two equal circles of radius r intersect each other if the separation of their centers is less than 2r.*

In modern geometry one can state this proposition if an appropriate axiom (or a theorem based upon appropriate axioms) guarantees their intersection. Modern verbal formulation of geometry implies that terms such as circle, radius, and center, only have the properties that derive from their definitions and the axioms of the theory.

Instead, Greek geometry used figures as parts of reasoning (and not as a merely pedagogical or illustrative tool). Thus, in order to derive the existence of the intersection between two circles, say C and C', Greek geometers could instead refer to the evidence of Fig. 12 and simply say: "Look!" This is precisely what Euclid did in the proof of his very first proposition, where he constructed an equilateral triangle. There was no necessity to clarify

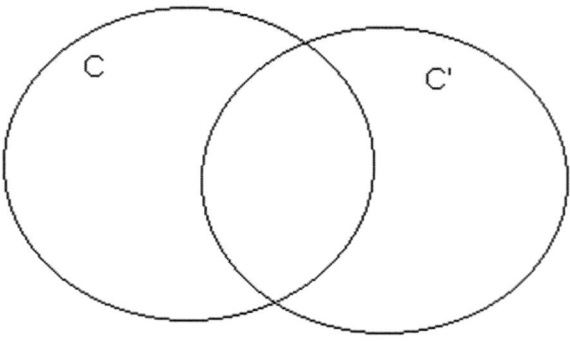

Fig. 12

precisely all the relationships between the objects of a theory, to make all axioms explicit and to define all terms. The mere inspection of figures provided information that we would now consider missing.[129]

Profound changes took place in the 17th and 18th centuries compared with the Greek world. The main change was that, whereas the diagram used by the Greeks represented a specific figure and only "this" figure could be the object of reasoning, 17th century figures were the concrete representations of abstract entities. A circle in a figure was the representation of the abstract concept of the circle and this abstract object that was the object of study.[130] This led a kind of dematerialization of the figures, which were conceived as the reification of the abstract notions under investigation (see Klein [1968, 206–207]).[131] Nevertheless, a figure still consisted of being the icon of a given geometrical object: this "imitative" aspect of figures continued to be used by 17th- and 18th-century mathematicians when they dealt with quantity from a geometric perspective.[132]

---

[129]In his *The Shaping of Deduction in Greek Mathematics* [1999a], Netz made a detailed analysis of the form of the proof in Greek geometry and concluded that figures did not just supplement the reasoning, but that the proof was reasoned using specific geometric quantities formalized by a specific figure. According to Netz [1999a, 240–270], Greek geometry was also based on repeatability, rather than generality.

[130]See p. 98.

[131]It even occurred that an area could be represented by means of a line.

[132]On the role of diagrams in 18th-century mathematics, see also Friedman [1992].

* * *

Viète's symbolism was a decisive step toward the mathematical treatment of abstract quantity. The subject of Viète's analysis was geometrical objects, but he conceived them in abstract entities that could be manipulated in a symbolic way. These abstract entities were not yet general quantities (which are characterized only due to the fact that they could be decreased and increased) but broad subclasses of general quantities. In his *In artem analyticem isagoge*, Viète distinguished between a *logistice numerosa*, which operated upon numbers, and a *logistice speciosa*, which operated "with species or forms of things, as, for example, with the letters of the alphabet" (Viète [1591, 328]). According to him, a species was a certain type of geometric quantity (magnitude) considered in the abstract and *logistice speciosa* was a general algebra that acted both on continuous "geometric" magnitudes and numbers divisible into "discrete units" (Klein [1968, 123]). This enabled Viète to write analytical expressions such as

$$\frac{(A\ plane) + (A\ in\ B)}{B}. \tag{66}$$

The expression *A plane* denoted a two-dimensional unknown, and *Z in B* was the product of two one-dimensional quantities[133] (see Viète [1591, 338]). Consequently, (66) stood for

$$\frac{a^2 + ab}{b}.$$

Although Viète's work marked a fundamental step in the process that led to a conceptualization of mathematics differing from the Greek one,[134] it still contained aspects that restricted its aspiration toward universality. Indeed, Viète attached dimension to the species in any given equation and thought that "only homogeneous magnitudes are to be compared with one another" (Viète [1591, 324–325]). Homogeneity is a form of determination that prevents the reduction of all quantities to only one type of quantity. For instance, the square $ABCD$ remained, for Viète, substantially different from the product of the measurements of the sides AB and CD and not be identified with the number $AB^2 = AB \times CD$.

Such a reduction was later used by Descartes by introducing a new definition of multiplication between segments. Taking an arbitrary line segment as the unit segment u, Descartes defined the product of two quantities a and b as the quantity c satisfying the proportion

$$u : a = b : c.$$

---

[133]In Viète's symbolic system, the unknown quantity was designated by the vowels A, E, I, O, U, or Y and the given quantities by the letters B, G, and D or other consonants (Viète [1591, 340]).

[134]For a hypothesis about the causes of this change in relation to Greek mathematics, see Klein [1968, 120–121]. A partially different viewpoint is offered by Netz [1999b, 43–45].

This allowed the powers to be interpreted appropriately: For example, x^2 was the quantity defined by

$$1 : x = x : x^2.$$

In this way dimensional homogeneity could be circumvented and any quantity could be reduced to a line.

It is clear that when any quantity is reduced to a line, the line itself assumed a symbolic character: in other words, it is the symbolic representation of the abstract notion of quantity. Descartes attempted to combine these symbolic figures and algebraic symbols. He thought that the symbols of algebra helped the understanding of geometrical figures and, vice versa, that geometrical figures helped the understanding of the symbols of algebra.[135] However, when a figure was used in reasoning, its iconic aspect continued to be employed.[136]

Newton and Leibniz continued to develop Descartes' abstract and symbolic conception, both on the mathematical level in the strict sense as well as on the epistemological level. However, they did not break the link between geometrical and algebraic representations. For instance, consider the following proposition:[137]

Given a curve of equation $y = ax^{m/n}$, its area is $\frac{na}{m+n} ax^{\frac{m+n}{n}}$.

In his [1711, 281–283], Newton[138] justified this proposition on the basis of various assumptions linked to the simple inspection of Fig. 13. Indeed, it is the figure that ensures the existence of the area ADB, the regular behavior of the curve (what is referred to today as the continuity of a curve), and the existence of the rectangle $BbKH$ whose area is equal to the trapezoid $BbdD$. Certainly, Newton's demonstration[139] gives considerable attention

---

[135]See Descartes [1637, 17–18 and 20].

[136]On Descartes, see Bos [1993] and [2001], and Klein [1968, 197–211].

[137]See Chapter 1, p. 68.

[138]Newton's proof runs as follows. He posed $AB = x$, area $ABD = z$, $DB = y$, $Ab = x + o$, area $Abd = z + ov$. He set

$$z = \frac{2}{3} x^{3/2}$$

(even if he considered a particular example, his reasoning is valid in general). He replaced x by $x + o$ and z by $z + ov$ into $z^2 = \frac{4}{9} x^3$ and derived

$$ov^2 + 2zv = \frac{4}{9} o^2 + \frac{4}{3} x^2 + \frac{4}{3} xo.$$

He supposed $Bb = o$ to be infinitely small and obtained $2zv = \frac{4}{3} x^2$. Since $z = \frac{2}{3} x^{3/2}$ and $v = y$, he had $y = x^{1/2}$ (Newton [1711, 281–283]).

[139]In this proof the role of the figure is, on the one hand, fundamental because it guarantees that the reasoning has a foundation but, on the other hand, only intervenes indirectly in the algorithmic game. The importance of the figure is clearly greater when dealing with the calculus with the synthetic method. For this method, my argument is valid *a fortiori*.

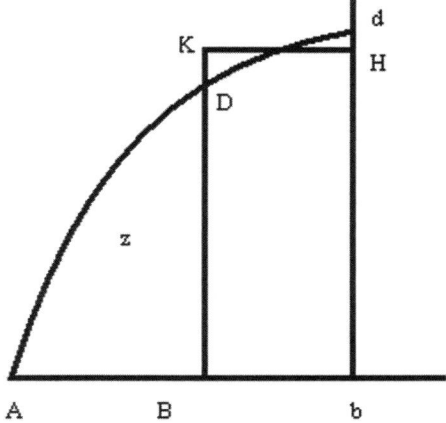

Fig. 13

to the relational aspect and shows a clear algorithmic structure. However, the figure cannot be eliminated, although it is only used to represent several properties of an entire class of curves, while the other characteristics are now entrusted to algebraic symbolism.

Newton's demonstration displays a mixture of figures and analytical symbolism. This mixture is typical of the early years of the calculus: It is also found in Leibniz and his school. Nevertheless, Leibniz's epistemological position[140] is worthy of mention because it heralds some of the important developments that definitely influenced mathematicians in the mid-18th century. Leibniz placed great insistence on the fact that the calculus could be understood as an algorithm that enabled operations that did not require lines or inspection of figures.[141] But it should be emphasised not only that Leibniz always embedded the algorithm into a geometrical interpretation[142] but also that his aim was considerably different from the 18th-century conception of a self-founding and self-meaning calculus. Leibniz aimed to exclude the inspection of the figure from certain algorithmic procedures concerning

---

[140]If one restricts oneself to mathematical practice, Newton seems, in certain aspects (I refer principally to Newton's *De Methodis Serierum et Fluxionum*, about which see Panza [1989, 168–170]), to anticipate 18th-century conceptions even more than Leibniz. If looked at from the epistemological point of view, the situation changes, since Newton gave priority to the synthetic method and this had an effect on the development of his school of thought.

[141]See, for example, Leibniz [GMS, 4:479].

[142]For instance, Guicciardini stated: "Leibniz always embedded the algorithm into a geometrical interpretation" (Guicciardini, [1999, 167–168]). See also Bos [1974, 8].

geometric objects, but he did not doubt that the algorithm had a meaning and could only be justified to the extent that it referred to such geometric objects. Analysis was an *ars inveniendi*, a method for treating entities that geometry offered (curves and geometric quantities related to them): It was not a theory that had its own objects, in the way that Eulerian analysis had functions. The nature of Leibniz's analysis clearly emerged when, for example, he dealt with the problem of the series $1 - 1 + 1 - \ldots$. In a letter to Wolff on July 13, 1712 (Leibniz [GBLW, 147]), he stated, among other things, that the relation $1-1+1-\ldots = \frac{1}{2}$ was well founded since a geometric demonstration (*demonstratio linearis*) of it existed.[143]

In conclusion, analysis, around 1700, developed as an algorithmic and symbolic method based on the notion of abstract quantity; nevertheless (and this is what distinguishes it from Eulerian and post-Eulerian analysis), it remained a method for studying geometry and did not constitute a self-founding theory: The idea that one could invent a mathematical theory whose aim was the study of quantity in an abstract sense, independent of any figural evidence, did not exist. The objects of Leibniz's and Newton's analysis remained the objects of geometry (analytically expressed), and the figure[144] continued to play one of the fundamental functions of the figure in Greek geometry: A part of the reasoning was unloaded on to it.

7.2   Continuous quantities, numbers and fictitious quantities

The preceding discussion of symbols and figures makes it possible to provide some clarifications concerning the terms "geometrical quantity", "analytical expression of geometrical quantity" (or "analytically expressed quantity"), and "general quantity", which I have so far used in a rather intuitive sense.

I refer to "geometrical quantities" or "figural quantities" when quantities were, or could be, referred to as a concrete and perceptible representation in a diagram so that it could be investigated, at least partially, by means of the diagram itself. Geometrical quantities were lines or other geometrical objects, e.g., the arclength, subtangent, and tangent of a curve, the area between given lines, and so on. I specify that 17th- and 18th-century mathematicians did not deal with geometrical quantities as specific lines in a given diagram, but as abstract entities of our thought, which were reified in the diagram.

I use the terms "analytical expressions of a geometrical quantity" or "analytically expressed quantities" to refer to analytical symbols associated with geometrical quantities (which therefore had or could have an explicit diagrammatic representation). Analytical expressions of a geometrical quan-

---

[143]See p. 124.

[144]This is true even if the figure sometimes had clearly a symbolic form, as in Leibniz's characteristic triangle.

tities were investigated symbolically or analytically (by manipulating symbols, according to given rules) and (if necessary) using figures.

I use the term "general quantity" to refer to an abstract entity[145] that can be increased and diminished (in a continuous way[146]). The investigation of general quantities consisted of the investigation of the relations between a certain quantity and other quantities (see Panza [1996, 241]). These relationships were only examined by symbolic notations (analytical expressions of general quantity or functions), without referring to figures (to emphasize the contrast with geometrical quantities, I also refer to them as "analytical quantities" or "nonfigural quantities"). The notion of general quantity lay at the basis of analysis after 1740.[147]

*   *   *

Indeterminacy was a crucial characteristic underlying the 17th- and 18th-century notion of quantity. A quantity (whether geometric or analytical) was an intrinsically indeterminate entity. Of course, it can be determined in infinitely different ways, but, as far as a quantity was considered in terms of its capacity to be increased and reduced, it was thought of as being indeterminate. For example, a segment is a quantity insofar as it has the capacity to become larger or smaller, whereas a fixed segment is only a determination of the quantity. The indeterminacy of a quantity gave rise to the notion of a variable quantity (or, more simply, a variable). A variable was merely a (continuous) quantity represented by the symbols x, y, z, ... In the first works on the calculus, a variable was a geometrical quantity.[148] After the 1740s, a variable was a general quantity.[149]

Another terminological specification is necessary at this point. Indeed, 17th- and 18th-century mathematicians used the term "quantity" not only to denote a variable entity capable of increasing or diminishing but also to indicate specific determinations of this variable entity (the values of the quantity). To avoid confusion, I shall hereafter use the term "quantity" (or, also, "indeterminate quantity" or "variable quantity") to refer to an entity in the sense of its capability to increase or diminish while I use the term "quantum" or "determinate quantity" to denote a specific determination of quantity.

---

[145] A general quantity was generated from particular quantities by means of a process of abstraction, which consisted of rendering as a general quantity what is common to all continuous quantities, just as the "greenness" consists of the specific shared attributes of all green individual objects, such as trees and grass. The shared attribute of all individual quantities was the capability of being increased or decreased.

[146] The general quantity was thought of as an abstraction drawn from continuous quantities and was assumed that it varied in a continuous way.

[147] See Chapter 18.

[148] For instance, in de l'Hôpital's *Analyse des infiniment petits, pour l'intelligence des lignes courbes*, variables were considered as lines denoted by the letters y, x, [1696, 1–2].

[149] See Chapter 18.

I stress the importance of the distinction between quantity and quantum in the calculus: The calculus referred to indeterminate quantities, subject to possible variations, whether increases or decreases, rather than to specific determinations of quantities or determinate quantities. A quantity could assume different values or determinations, although a quantity was not reduced to the enumeration of these values (see Ferraro [2000a, 108]).

The notion of quantity as an intrinsically indeterminate (or variable) entity did not prevent quantities from being divided into constants and variables.[150] However, this distinction did not depend on the nature of quantities but on specific questions, quantities being variable in themselves.[151]

Abstract general quantities, as well as concrete geometrical quantities, were regarded as continuous quantities. Mathematicians did not discuss the properties of continuous quantity explicitly: The continuous was substantially a primitive, undefined notion, based on physical reality (founded upon nature, as Newton says). It was tacitly assumed that continuous quantities behaved as a segment of a straight line or a piece of a curved line. In his *Les continu chez Leibniz*,[152] H. Breger has given a remarkable description of Leibniz's geometrical continuous quantity, which adequately clarifies 17th- and 18th-century mathematicians concept of the geometrical continuum. The main features of this commonly held concept are as follows.

First, a segment was divisible into parts, each of which was similar in kind to the original quantity, but it could not be reduced to an aggregate of points. Thus, the continuum was given as a whole and was not regarded as a set of points, even though it was possible to determine specific points in it.

Second, for the precise reason that a segment was not considered as a set of points it was impossible to distinguish between an open and a closed

---

[150]For instance, de l'Hôpital stated: "Variables quantities are called those which increase or decrease continually whereas constant quantities are those that remain the same while the others change" [1696, 1]. Similar definitions lasted during the whole 18th century, and Lacroix still wrote in 1797: "Quantities, considered as changing in value or capable of changing it, are called to be variables, and the name constants is given to those quantities that always maintain their value during the calculation" [1797–1800, 1:82].

[151]For instance, Euler explained: "[T]his calculus deals with variable quantities, even though every quantity, by its very nature, can be increased or diminished in infinitum; however, as long as the calculus is addressed towards a certain goal, some quantities are designed to maintain the same magnitude constantly while others are truly changed for each amount of increase and decrease: the former quantities are usually termed constants, the latter variables so that this difference is not expressed so much in terms of the nature of the thing as in the character of the question to which the calculus refers" [1755, 3]. As an example, Euler observed that the trajectory of a bullet was determined by four quantities: the amount of gunpowder, the angle of fire, the range, and the time. Each of them was a quantity in the sense that it could be increased or reduced. This property was never lost, though in certain calculations it was utilized and not in others: In this sense, a quantity could be imagined as a variable or constant according to the specific calculation.

[152]See Breger [1992a, 76–84].

segment. It is always thought of as including its endpoints: "One cannot, *e.g.*, consider the interval from 0 to 1 without the point zero. Imagine a metre long thread without the left extremity of the thread. It is clearly an absurdity. Precisely, in the same way, the point zero is not a part of continuum but its extremity on the left: the point cannot be suppressed, not even in thought" (Breger [1992b, 77]).

Third, a curve or a relation between quantities was not defined pointwise. An equation, such as $y = x^2$, was viewed as a relation that assigns an interval on the y-axis to an interval on the x-axis in an appropriate way (Breger [1992a, 77]). Curves were not plotted; they were generated by motion[153] (see also Mahoney's description of the notion of curve in Fermat in Mahoney [1994, 82]).

Unlike a geometrical quantity, a general quantity was not represented by a line in a diagram. It was, however, made up of what all the geometrical quantities have in common. This implies that the basic notions of continuous geometrical quantities were immediately transferred to general quantities and the analytical continuum. For instance,

1. the continuum was given as a whole and was not regarded as a set of elements (of any nature); however, it was possible to determine specific determinations of the continuum;

2. a variable quantity x always varied continuously and when it moved from a value x_1 to a value x_2, it was impossible to think of this variation without the initial and final values (or if one wishes, it was impossible to distinguish between an open and a closed interval);

3. a relation between general quantities was not defined pointwise.

Thus, throughout the 17th and 18th centuries the continuum used by mathematicians —even when they referred to general quantities— was substantially a geometrical continuum that reflected the properties of a segment. However, during this period, the substratum of the continuum changed. In other words, during the 17th and 18th centuries, the nature of the objects investigated by analysis underwent a significant change (from geometrical

---

[153]In *Tractatus de quadratura curvarum*, Newton expressed himself using the following terms: "I don't here consider Mathematical Quantities as composed of Parts extremely small, but as generated by a continual motion. Lines are described, and by describing are generated, not by any apposition of Parts, but by a continual motion of Points. Surfaces are generated by the motion of Lines, Solids by the motion of Surfaces, Angles by the Rotation of their Legs, Time by a continual flux, and so in the rest. These Geneses are founded upon Nature, and are every Day seen in the motion of Bodies" (see Newton [1704, 332] translation by Harris in Newton [QHW, 3]).

quantities into abstract quantities), whereas the concept of continuity remained unaltered.[154] I will return to the concept of continuity in Chapter 18.3.

*   *   *

Seventeenth- and 18th-century mathematicians went beyond the Greek concept of number as a multiplicity of unities (number of unities). This evolution, which developed gradually during the Middle Ages and Renaissance, underwent its turning point in the works of Stevin and Viète[155]. After 1600, the notion of number had two main characteristics.

1. It was abstract. Numbers were abstracted from counted things and became autonomous entities. A number, such as 7, was no longer considered as the attribute of a group of material or ideal objects; rather it was an abstract entity that expressed what all the things that are seven times the unity had in common.[156]

2. It was symbolic. The abstract concept of number was reified into symbols upon which one could operate according to given rules (the rules of arithmetic). Thus, the number 7 was also a symbol that reified an ideal entity into ciphers upon which one manipulates directly. In this sense, a number had a full significance only if it was considered an element of a set of signs governed by appropriate rules that allowed their manipulation (see Klein [1968, 193]). However, unlike modern mathematics, the rules governing the manipulation of this set of signs were not arbitrary (nor even potentially so).[157]

This abstract and symbolic concept of numbers made it possible to introduce new species of numbers in addition to natural numbers. Irrational numbers, negative numbers, and imaginary numbers entered the realm of numbers, although (and this is a very important point) they were *epistemologically different* from natural numbers.

To make this point clearer, I begin by observing that numbers were understood as the measurement of a given quantity with respect to another

---

[154]See also Chapter 32.

[155]On Renaissance notions of number and magnitudo, see Malet [1996]. On the notion of number from Stevin to Wallis, I refer to Klein [1968, 186–224]. On Euler, see Ferraro [2004, 39–43].

[156]Using Menninger's words, one can state that the relationship between numbers and counted objects is upset: It is not the operation of counting that originates numbers but numbers existed abstractly and the operation of counting assigned numbers to counted objects; "so long as there is no counting, it [the abstract number sequence] is merely there, detached from all concrete objects, unused but ready. But as soon as we count ... the number words become assigned to the objects, and ... the objects are placed in the empty boxes of the number sequences" (Menninger [1969, 7]).

[157]See Section 18.1.

taken with as a measure or unity. For instance, in *Universal Arithmetick*, Newton stated:

> By *number* we understand not so much a multitude of quantities, as the abstracted *ratio* of any quantity, to another quantity of the same kind, which we take for unity. (Newton [1720, 2])[158]

According to Newton:

> An integer is what is measured by unity, a fraction, that which a submultiple part of unity measures, and a surd, to which unity is incommensurable. (Newton [1707, 2])

Several decades later, Euler similarly stated that all the determinations or measures of any quantity are reduced to determining the relation that a given quantity has with a certain quantity of the same kind taken as a measure or unity:

> [This relation] is always indicated by numbers, so that a number is nothing but the relation of a quantity to another quantity, taken arbitrarily as a unity. (Euler [1770, 10])

Quantity was therefore considered as an entity that logically precedes number and number was viewed as a tool for treating quantity: According to Stevin, "Number is that by which the quantity of each thing is revealed" (Stevin [1585, 1]).

The concept of number as a measurement made it possible to embody the discrete within the continuous. Thus, even though the traditional distinction between discrete and continuous quantity was maintained, it lost its former importance. In effect, the discrete was thought of as originating from the continuous and was regarded as a particularization of the continuous. The sequence 1, 2, 3, . . . only identifies discrete determinations of a continuous quantity, depending on the choice of the unity.

A precise definition of measurement was not given. Mathematicians assumed implicitly that measuring meant repeating the operation of comparing with unity or one of its parts successively and finitely.

This concept of measurement, which assumes that the sequence of natural numbers[159] is intuitively known, made it possible to give an adequate treatment of fractions as numbers. For instance, one could have a clear idea of 7/3 by considering a segment 7 feet in length and by dividing it into 3 parts (see Euler [1770, 30]).

---

[158]See also Stevin [1585, 1] and Wallis [1657, 183].

[159]After Stevin, the idea that 1 was a number gained widespread acceptance. Klein views this development as one of the signs of the transition towards a more symbolic concept of numbers (see Klein [1968, 186–224]).

Difficulties arose with irrational numbers. During the 17th and 18th centuries, irrational numbers were accepted. Viète referred to them as *"numeri asymmetri"* (asymmetrical numbers); Stevin [1585, 30] stated "Every root is a number"[160] and, as we saw earlier, Newton considered numbers as having a threefold nature: integer, fractional, and surd. Nevertheless, irrational numbers were significantly different from natural and fractional numbers. Fractions had a meaning in terms of unity of measure and consequently were numbers in the strict sense of the term, or "true numbers". In contrast, irrational numbers were not true numbers since they did not represent a process of measurement in a precise sense. For instance, in his *Vollständige Anleitung zur Algebra,* Euler observed that the root of 12 is not a fraction; nevertheless, it is a determinate quantity, which is greater than

$$3, \; \frac{24}{7}, \; \frac{38}{11}, \; \frac{45}{13}, \; \cdots$$

and smaller than

$$4, \; \frac{7}{2}, \; \frac{52}{15}, \; \cdots :$$

Therefore, $\sqrt{12}$ is a new species of number. He then added that a correct idea of $\sqrt{12}$ can be gathered by observing that $\sqrt{12}$ is the number that, when multiplied by itself, makes 12 (this defines $\sqrt{12}$ operatively, as a symbolic entity) and that the value of $\sqrt{12}$ can be approximated as desired (Euler [1770, 50–51]).

The domain of true numbers (natural and rational numbers) was not sufficient to investigate quantity, nor was the addition of irrational numbers.[161] Other numbers apart from rational and irrational ones were necessary: negative numbers, zero, and imaginary numbers.[162]

Negative numbers[163] were usually introduced simply by stating that they were entities less than the nothing and that were represented by numbers with the sign − (in opposition, positive numbers were numbers greater than nothing and had the sign +).[164] This definition were accompanied with rules that made it possible to operate upon them as symbolic entities: It was these rules that transformed the symbols −1, −2, ... into numbers. As the other

---

[160]Stevin even refused to call them absurd, surd, or irrational numbers, because incommensurability was not a cause of absurdity (see Stevin [1595]).

[161]In 18th century, it is possible to grasp a distinction between irrational and transcendental numbers, where transcendental numbers are those numbers that derived from the application of transcendent operations (such as a logarithm) to a rational number. However, this distinction is not of importance for my purposes and I shall not take it into account.

[162]In this book I do not consider infinitesimal numbers. The following remarks also hold for them (see Ferraro [2004]).

[163]Their use came up against resistance. Viète did not yet acknowledge negative numbers.

[164]See, e.g., Newton [1720, 3].

above-mentioned species of numbers, they also had an intuitive meaning. For instance, Newton [1720, 3] gave the traditional interpretation of negatives as debits. He also stressed that a negative motion was a "regression", i.e., proceeding backwards. They could also be represented by directed segments. However, they did not correspond to a notion of measurement of a quantity in the strict sense of the term: They were not true numbers.

Zero was considered as the absence of quantity and was the name given to "nothing", from which numbers were subtracted to obtain negative numbers. In his *Arithmétique* [1585, 4] Stevin considered the zero to be the beginning of natural numbers (but not a number), in the same sense as the point is the beginning of the line (not an element of the line, see p. 102). Almost two centuries later, in *Vollständige Anleitung zur Algebra* Euler did not yet list zero as an integer (integers were the natural numbers $+1, +2, +3, \ldots$, which are greater than nothing, and negative numbers were $-1, -2, -3, \ldots$, which are less than nothing).[165]

Complex numbers had been introduced in the 16th century and gradually gained acceptance. Expressions such as $\sqrt{-1}$, $\sqrt{-2}$, $\sqrt{-3}$, ... were termed impossible or imaginary numbers: Nevertheless, mathematicians thought that they were useful for dealing with quantity and could have a role in our reasonings. Imaginary numbers were introduced as formal instruments for obtaining the root of certain numbers, in a similar way to irrational numbers. For instance, $\sqrt{-4}$ merely meant a number that multiplied by itself equals[166] -4. Imaginary numbers could not be reduced to the measurement of quantity, not even in an approximate sense. They differed from other numbers and were generated by the symbolic mechanism of analysis; they had no intuitive meaning on their own but assumed a meaning within the overall context of analysis.

At this point it is clear that even though all numbers were abstract and symbolic entities, only some adequately reflected the concept of number as the exact result of a process of measurement and were "true" numbers. Other types of numbers did not fit the notion of a number (although for different reasons). In the strict sense of the term they were not true numbers. I shall term them "fictitious numbers" or "fictions".

The idea of false or fictitious numbers is an old one. For instance, many mathematicians referred to negative numbers as false numbers.[167] In Sec-

---

[165]See Euler [1770, 14].

[166]See, e.g., Euler [1770, 56].

[167]Descartes wrote the following when commenting upon the roots of an equation: "But often it happens, that some of these roots are false, or less than anything, as if one supposes that x indicates also the defect of a quantity, which is 5, one has $x + 5 = 0$, which being multiplied by $x^3 - 9xx + 26x - 24 = 0$ is $x^4 - 4x^3 - 19xx + 106x - 120 = 0$, for an equation in which there are four roots, namely three true ones which are $2, 3, 4$, and a false one which is 5" (Descartes [1637, 56]). But, in 1545, Cardano had already stated: "If the square of a square is equal to a number and a square, there is always one true solution and another and fictitious solution equal to it." He gave the example $x^4 = 2x^2 + 8$, where

tion 2.1 we saw that Leibniz attempted to justify infinitesimal and infinite numbers as fictions similar to other fictions used in mathematics (imaginary numbers, the power whose exponents are not true numbers, etc.).[168] In effect, the idea of false numbers is at the basis of much of mathematical terminology regarding numbers, which we still partially retain today.

Unlike Leibniz, most mathematicians did not employ the term "fiction" explicitly. Nevertheless, I use this expression because it expresses the nature of the 17th- and 18th-century approach, in particular because it implies an *ontological* difference between that which is fictitious and that which is true. Seventeenth- and 18th-century mathematics effectively presents *an ontological difference* between natural and rational numbers (true numbers) and the other species of numbers (which did not correspond to the idea of numbers and therefore were fictitious numbers). To put it more clearly, nowadays $\sqrt{-1}$ is an element of the set of complex numbers C and exists in the same way as any other number in C, such as 1, 2, $\frac{1}{2}$, etc. During the 17th and 18th centuries, $\sqrt{-1}$ was a useful symbol for studying certain aspects of quantity; it did not have an existence in the same sense as true numbers. Similarly, 0 was the symbol that represented the absence of quantity, the nothing, the non-existence; it was not a number, because it did not measure quantity and did not denote anything; however, 0 could be treated as a number. *Mutatis mutandis*, the same holds for irrational (unspeakable, inexpressible) numbers and negative numbers.

I already pointed some aspects of fictitious quantities with reference to Leibniz at the end of Section 2.1. Now, in a more systematic and general way, I observe that fictions had the following five characteristics.

(a) Fictions were a useful tool for shortening the path of thought and arriving at new results.[169] It was of no importance whether fictions appeared in nature or not, namely if they represented physical or geometrical objects. Irrational numbers appeared in nature (they represented the length of a segment); imaginary numbers did not appear.

(b) Fictions, however, were always connected with reality, directly or indirectly. They were not arbitrary creations of the human mind but had to be well founded in reality and were needed for investigating reality (this is true even for imaginary numbers[170]).

By the phrase "well founded in reality or in nature",[171] I intend to highlight the fact that certain mathematical objects did not originate from

---

x "equals 2 or -2" (Cardano [1545, 11]).

[168] For instance, see the quotation on p. 35.

[169] See Leibniz's quotation on p. 35.

[170] See Leibniz [GMS, 4:92–93] and Euler [1770, 57].

[171] I derive this expression from Leibniz's statement that imaginary quantities have their foundation in nature (see Section 2.1, p. 35).

arbitrary definitions given in a theory based upon an arbitrary system of
axioms; instead, they originated

1. from the need to express certain properties of quantities and

2. from the need to manipulate objects that directly expressed quantities
   or properties of quantities (an example would be the *casus irreducibilus*
   of the equation of third degree).

In the first case, a well-founded object had an intuitively obvious interpretation (e.g., irrational numbers). For this reason I would say that it was
directly connected to reality.[172]

The second case was that of imaginary numbers, which did not have
an intuitively obvious meaning[173]. They were introduced in a merely formal way, but they made up for rational and irrational numbers when these
did not suffice: They were always connected to reality, even though only
indirectly.

In any case, well-foundedness, used in this sense, excludes the possibility
that mathematical objects could originate from a free act of will and required
them to be rooted in reality, directly or indirectly, as elements of a theory
that aimed to interpret the real.

It should also be emphasized that fictions were not of interest in themselves, but only insofar as they allowed one to solve problems concerning
quantities. They were auxiliary instruments for dealing with quantities.

(c) Fictions were manipulated as if they were true numbers. This means
    that a fiction was treated by analogical extensions of rules valid for true
    numbers or geometrical quantities.[174] Therefore, a fictitious number
    was more than a mere façon de parler or a shorthand way of denoting
    a certain operation upon true numbers: It was a symbolic entity that
    formed part of the symbolic nature of true numbers and quantities.

(d) An adequate theoretical construction for moving from fictions as a sign
    for shortening the path of thought to the analogical use of fictions as
    true numbers was completely lacking. Thus, well-foundedness in the
    above sense was the only justification for fictions.

---

[172]However, this does not mean that there exists a geometrical or physical object corresponding to it.

[173]See also footnote no. 263.

[174]This statement is to be understood as follows. The principle was assumed that if an
operation (not only an algebraic operation —sum, product, etc.— but also transcendental
operations —logarithm, etc.) had true numbers or geometrical quantities as operands
then it could have fictitious numbers as operands. Of course, some adjustments might be
necessary: They took the form of specific rules inherent to the peculiar nature of every
distinct species of fictitious numbers. For instance, the rule of signs was a specific rule
for negative numbers. These specific rules were what distinguished a calculation involving
a particular species of fictitious numbers from a calculation involving true numbers or a
different species of fictitious numbers.

(e) Even though quantity was an entity abstracted from geometrical quantity and had the same properties as lines, it could be determined by
fictitious values; in other words, one could assign fictitious numbers to
a variable x.

Seventeenth- and 18th-century mathematicians used the term "irrational
quantities" to refer to irrational numbers or irrational determinations of
quantity. They referred similarly to negative quantities, imaginary quantities, etc. I maintain this terminology and, more generally, I use the term
"fictitious quantities" by referring to fictitious determinations of quantity or
fictitious numbers.

A general quantity has some determinations that can be represented
by a nondirected segment, whereas others cannot. I use the term "real
quantity" to denote a quantity that only assumes these determinations and
that corresponds to the mental image of the geometrical or physical quantity.
I do not therefore intend this term in opposition to fictitious quantity, since
a real quantity can have both true numbers and certain fictitious numbers
as its determinations.[175]

Finally, I wish to make some simple consequences of the above-described
notions of quantity and numbers explicit.

First, since a single number was a specific determination of quantity, a
single number expressed a quantum rather than a quantity.

Second, even though each specific determination of real quantity can be
represented by means of numbers, the idea that quantity might be
reduced to a set of numbers was not taken into account.

Third, more generally, numbers were not conceived of as elements of a set, if
by "set" one means an extensional entity, which is arbitrarily defined,
entirely characterized by the list of its elements and having a certain
cardinality. Instead, 17th- and 18th-century mathematicians classified
numbers into different classes or species, where a "species" of numbers
was intended as an intensional entity, which could not be reduced to an
enumeration of objects: it was given by a nonarbitrary and nontrivial
property and was not necessarily associated with cardinality.

---

[175]Irrational numbers do not correspond to the idea of number and, therefore, are fictions; however, they have a very special nature with respect to other fictitious numbers
since they can be represented by means of a nondirected segment and answer the question: What is the measure of a given (real) geometrical quantity? Rational numbers
answer the same question (though in a more precise way), and thus these rational and
irrational numbers might be grouped together to form the class of (positive) real numbers
(and in effect rational and irrational numbers were often taken together; for instance, as
opposed to imaginary numbers, see (Euler [1748a, 18]). By so doing one obtains a different
classification, which considers the capacity of numbers to express the determinations of
geometrical quantities directly, but this is not relevant to my purposes in this book.

* * *

The preceding sections should make it clear that the transformation of 17th- and 18th-century mathematics into modern terms may cause a certain degree of stretching in the meaning of certain results and procedures. I would like to emphasize this effect of stretching by means of an example that highlights the difference between the 17th-century notion of limit and the modern one; it also aids in our understanding of Newton's concept of limit, which has been discussed in Chapter 4.

In *Analyse des Infiniment petits*, de l'Hôpital considered the following problem:

> *Let $x = AP$ and $y = PM$ be the abscissa and the ordinate of a given curve AMD. Suppose that $a = AB$ is a particular value of the abscissa and that the value of the ordinate $y = PM$ is expressed by a fraction, whose denominator and numerator both become equal to 0 when the abscissa x becomes equal to a. It is required to find the value of the ordinate y when $x = a$.* (de l'Hôpital [1696, 145])

By using the notion of a function, the problem can be formulated as follows:

> *Find the value of the function*
> $$y = f(x) = \frac{h(x)}{g(x)}$$
> *for $x = a$, when $h(a) = 0$ and $g(a) = 0$.*

The problem is solved by means the rule today named after de l'Hôpital

$$f(a) = \left(\frac{dh(x)}{dg(x)}\right)_{x=a}.$$

For instance, to find the value of

$$y = \frac{a^2 - x^2}{a - \sqrt{ax}}$$

for $x = a$, de l'Hôpital set

$$\left(\frac{a^2 - ax}{a - \sqrt{ax}}\right)_{x=a} = \left(\frac{a}{\frac{1}{2}\sqrt{\frac{a}{x}}}\right)_{x=a} = 2a$$

(see de l'Hôpital [1696, 146]). One may be struck by the similarity between this procedure for "finding" the value of and the modern problem of extending the function

$$F(x) = \frac{a^2 - ax}{a - \sqrt{ax}}$$

in a continuous way by setting

$$F(a) = \lim_{x \to a} \frac{a^2 - ax}{a - \sqrt{ax}}.$$

Indeed it is completely natural to translate

$$\left(\frac{a^2 - ax}{a - \sqrt{ax}} \right)_{x=a} = \left(\frac{a}{\frac{1}{2}\sqrt{\frac{a}{x}}} \right)_{x=a}$$

as

$$\lim_{x \to a} \frac{a^2 - ax}{a - \sqrt{ax}} = \lim_{x \to a} \frac{a}{\frac{1}{2}\sqrt{\frac{a}{x}}};$$

however, such a translation into the language of limits may produce misinterpretations. From a modern perspective, finding the value of $\frac{a^2-ax}{a-\sqrt{ax}}$ when $x = a$ (I assume $a > 0$) means that

a. one considers the function (function in the modern sense of the term, not as an analytical expression)

$$f(x) = \frac{a^2 - ax}{a - \sqrt{ax}}$$

   defined for $x > 0$ and $x \neq a$.

b. the domain of $f(x)$ has a point of accumulation at a so that we can attempt to calculate the limit as $x \to a$;

c. the application of l'Hôpital's rule, under whose hypotheses our case falls, makes it possible to state that such a limit exists and is equal to $2a$;

d. finally, we define a new function $f(x)$, which will be continuous at the point a, by setting

$$f(x) = \begin{cases} \frac{a^2-ax}{a-\sqrt{ax}} & \text{for } x \geq 0 \text{ and } x \neq a, \\ 2a & \text{for } x = a. \end{cases}$$

In this procedure we use notions such as the limit, value, and extension of a function, whose meaning are opportunely and explicitly defined. Indeed,

$$\lambda = \lim_{x \to x_0} F(x),$$

where $F(x)$ is a function with domain D in R, means

> *Given any $\varepsilon > 0$ there exists a $\delta > 0$ such that if x belongs to D and $|x - x_0| < \delta$ then $|F(x) - \lambda| < \varepsilon$.*

By $\lambda = F(x_0)$ we mean: λ is the number that the function F associates with the number x_0.

If $\lim_{x \to x_0} F(x)$ exists and is equal to λ, while the function $F(x)$ is not defined at the point x_0 or $F(x_0) \neq \lambda$, we can remove the discontinuity at x_0 by defining the new function

$$x \to \begin{cases} f(x) & x \in D - \{x_0\}, \\ \lambda & x = x_0. \end{cases}$$

These definitions presuppose knowledge of the notions of set, real numbers, function in the modern sense, continuity, etc. For this reason the above procedure is substantially meaningless for 17th and 18th century mathematicians. They did not consider a function as a pointwise correspondence between numerical sets but as a rule that linked two variables quantities and was embodied in one single analytical expression. They had no set of points or numbers, did not separate an interval of values (a segment) from its endpoints, etc., nor could they formulate the notion of extension of a function, but instead considered $\frac{a^2 - a^2}{a - \sqrt{a^2}} = 2a$ necessarily to be the value of $\frac{a^2 - ax}{a - \sqrt{ax}}$ when the variable x equals a.

8 The formal-quantitative theory of series

In this chapter, on the basis of the investigations carried out in the previous chapters, I would like to provide further clarification about the early theory of series.

First of all, I emphasize that all the mathematicians examined hitherto considered it obvious that the equality

$$Q = \sum q_k \qquad (67)$$

meant that $\sum q_k$ and Q denoted the same quantity. It also seemed obvious that the series $\sum q_k$ denoted the same quantity as Q if and only if the series $\sum q_k$ was *convergent* to the quantity denoted by Q, the sum of the series. For the sake of brevity, I shall refer to an equality of the kind $Q = \sum q_k$ as a *quantitative* equality if the series $\sum q_k$ converges to the quantity denoted by Q. (More generally, I shall refer to an equality of the kind $Q = P$ as a *quantitative* equality if Q and P represent the same quantity.)

The above investigations lead us to conclude that convergence was understood in the following sense:

(C) *A series $\sum_{k=0}^{\infty} a_k$ converges to a quantity S, the sum of the series, if and only if the sequence of nnth sums*

$$S_n = \sum_{k=0}^{n} a_k$$

approaches S indefinitely when n increases so that S_n is ultimately equal to S, when n is infinite.

The following property was a trivial consequence of the fact that the sequence S_n approaches S indefinitely:

(AP) *If the sequence of the nth sums $S_n = \sum_{k=0}^{n} a_k$ approaches S indefinitely when n increases, then the difference between S_n and S (in absolute value) becomes less than any given quantity.*

I would like to point out that the sum of a series was the ultimate value of the series and that (AP) was only an obvious property of the notion of approaching a quantity.

In Proposition (C) the symbols S and a_k can denote both determinate quantities as well as variable quantities; in modern terms, $\sum_{k=0}^{n} a_k$ can be both numerical and function series. Referring explicitly to function series, (C) can be reformulated as follows:

> (C_{FS}) *The series $\sum_{k=0}^{\infty} f_k(x)$ is said to be convergent to the function $f(x)$ on an interval[176] I_x, over which x varies, if and only if the sequence of the nth sum $\sum_{k=0}^{n} f_k(\alpha)$ approached $f(\alpha)$ indefinitely when n increases, for any value α of x belonging to I, and it is finally equal to $f(\alpha)$, when n is a infinite number.*

Some words of caution are necessary regarding (C_{FS}).

First, as should be clear from the previous chapters, function series around 1700 were constituted by power series $\sum_{k=0}^{\infty} a_k x^k$ or, at most, with series of the type

$$\sum_{k=0}^{\infty} a_k x^{-k} \quad \text{or} \quad \sum_{k=0}^{\infty} a_k x^{\alpha_k}$$

where a finite number of the exponents α_k were rational numbers. The early theory of function series was substantially a theory of power series. Only in the second part of the 18th century did other types of function series appear (but power series were always largely dominant). However, the use of the expression "function series" and the symbol $\sum_{k=0}^{\infty} f_k(x)$ in definition (C_{FS}) is useful in order to avoid repeating a similar definition of convergence with reference to other types of function series, when they come to be examined.

Second, the word "function" is an anachronism before the 1720s (or, perhaps, even before the 1740s). However, the use of this term is appropriate and simpler than other expressions, provided it is borne in mind that, when referring to the period prior to 1720, the word "function" should be understood either as an analytical expression — composed of algebraic, exponential and logarithmic operations— or directly as geometric quantities — this is the case with trigonometric quantities, which were not considered as analytical expressions, but whose expansions into series were known. (The meaning of the term "function" during the 18th century is illustrated in Section 18.2. This meaning largely justifies the anachronism.)

We saw that convergence was one of the two cornerstones of the early theory of series, the other one being formal manipulation. To stress the coexistence of these two factors, I shall later refer to the early theory of series as the "*formal-quantitative theory of series*". Some clarification is, however, appropriate with regard to the term "formal". A first good definition of "formal" is the following:

> *A procedure or rule is formal if it is applied to (finite or infinite) analytical expressions $A(x, y, \ldots)$, $B(x, y, \ldots)$, ... regardless of the actual meaning of such expressions. In other words, when formal procedures or rules are employed, the quantitative meaning of certain expressions is neglected.*

---

[176]The interval was not usually specified.

This meaning of the term "formal" can be made more precise when referred to 17th- and 18th-century mathematics. Indeed,

> *in the 17th and 18th centuries, procedures or rules were formal if they were based upon the* principle of the infinite extension from the finite to the infinite *or the* principle of the generality of algebra.

I refer readers to Section 18.2 for an analysis of the generality of algebra. I shall now dwell upon the principle of the infinite extension whose importance in the early series theory is immediately evident from the previous examination of the works of Leibniz, Newton, and other mathematicians. It consisted of the following assumption:

(IE) *If a rule R was valid for finite expressions or if a procedure P depended on a finite number n of steps S_1, S_2, S_3, ..., S_n, then it was legitimate to apply the rule R and the procedure P to infinite expressions and in an unending number of steps $S_1, S_2, S_3, ...$*

Principle (IE) stemmed from the lack of any distinction between infinite and finite series. For instance, since the following rules hold

$$\sum_{n=1}^{k} (a_n + b_n) = s_1 + s_2,$$

$$\sum_{n=1}^{k} \lambda a_n = \lambda s_1,$$

$$\left(\sum_{n=1}^{k} a_n \right) \left(\sum_{n=1}^{k} b_n \right) = s_1 s_2,$$

$$\sum_{n=2}^{k} a_n = s_1 - a_1,$$

$$\sum_{n=1}^{k} a_n = \sum_{n=1}^{k} a_{j(n)} = s_1,$$

when s_1 and s_2 are the sum of the finite series $\sum_{n=1}^{k} a_n$ and $\sum_{n=1}^{k} a_n$, λ is a number and $a_{j(n)}$ is any rearrangement of the finite sequence a_n, it seemed natural to extend these rules to infinite series as well.

Consequently, it was assumed that

$$\sum_{n=1}^{\infty} (a_n + b_n) = s_1 + s_2,$$
$$\sum_{n=1}^{\infty} \lambda a_n = \lambda s_1,$$
$$\left(\sum_{n=1}^{\infty} a_n\right)\left(\sum_{n=1}^{\infty} b_n\right) = s_1 s_2, \qquad (68)$$
$$\sum_{n=2}^{\infty} a_n = s_1 - a_1,$$
$$\sum_{n=1}^{\infty} a_n = \sum_{n=1}^{\infty} a_{j(n)} = s_1,$$

when s_1 and s_2 are the sum of the infinite series $\sum_{n=1}^{\infty} a_n$ and $\sum_{n=1}^{\infty} a_n$, λ is a number, and $a_{j(n)}$ is any rearrangement of the infinite sequence a_n.

It should be noted that even the symbolism masked the difference between finite and infinite series. In the 17th and 18th centuries, a series was generally denoted by a written expression like

$$a + b + c + \text{etc.} :$$

this could mean both $a + b + c + \ldots$ *ad infinitum* and $a + b + c + \ldots + p + q + r$ untill the term r.

I would also like to emphasize that formal procedures were not freely invented or created: In series theory, mathematicians only considered rules deriving from the infinite extension of the properties that were valid for finite expressions and a finite numbers of steps.

The analysis set out in the previous chapters makes it possible to group the standard procedure for expanding a quantity into series into the following types:[177]

(P_1) *The Mercator's expansions of fractions and square roots of polynomials.*

(P_2) *The binomial expansion for any (rational or irrational) exponent.*

(P_3) *Any expansion following the method of indeterminate coefficients.*

(P_4) *Any expansion deriving from contemporary differentiation or integration of both the sides of a given equality*

$$f(x) = \sum_{i=0}^{\infty} f_i(x)$$

—the operations on the series being performed term by term.[178]

---

[177]For more details on this classification, see Ferraro and Panza [2003].

[178]The procedures (P_4) depend on an infinite extension of the properties of linearity of differentiation and integration (today, we know that they do not follow from simple convergence). The Taylor series is an example.

(P₅) Any composition of the procedures (P₁), (P₂), (P₃), (P₄).[179]

The above investigations also make it possible to give an explicit explanation of how the formal and the quantitative were related to one other in the early theory of series. Indeed, the relationship between the formal and the quantitative was based upon the following three principles.

First, it was assumed that the usual procedures transformed a function $f(x)$ into a power series convergent to the function $f(x)$ at least over an interval I of values of variable x. This was an unproved assumption: It was merely derived from noting that it occurred in all known cases.

Second, the actual determination of the interval of convergence was an *a posteriori* question which intervened only when one wished to apply the expansion $\sum_{i=0}^{\infty} a_i x^i$ of a function $f(x)$, and, thus, one needed to compute the value of $f(x)$. In other terms, first, results were formally derived and then were subjected to reinterpretations that adapted them to concrete circumstances and fixed the bound of numerical validity of the formal equality $f(x) = \sum_{i=0}^{\infty} a_i x^i$. Convergence was not a preliminary condition to the manipulation of function series.

Third, even though the series $\sum_{i=0}^{\infty} a_i x^i$ was convergent to the function $f(x)$ over a certain interval I, mathematicians did not think that the manipulation of such an equality needed to be restricted to the values of x belonging to such an interval.[180] In other words, the relation $f(x) = \sum_{i=0}^{\infty} a_i x^i$ was considered as valid in manipulations independently of the value of x. Of course, it did not have a quantitative meaning outside the interval of convergence. For instance, though it was well known that the series $\sum_{i=0}^{\infty}(-1)^i x^i$ converges only for $|x| < 1$, the relation $\frac{1}{1+x} = \sum_{i=0}^{\infty}(-1)^i x^i$ was freely used in manipulations, without being restricted to the condition $|x| < 1$.

Finally, I would like to point out another important feature of the formal-quantitative notion of series. Nowadays, the sum of a function series $\sum_{k=0}^{\infty} f_k(x)$ is a function $f(x)$ that is defined in an appropriate subset A of $\Re$ by considering the sums of the convergent numerical series $f(\alpha) = \sum_{k=0}^{\infty} f_k(\alpha)$, where α belongs to the subset A. Function series are, in a sense, subordinated and reduced to numerical series whereas 17th- and 18th-century mathematicians expanded functions into series without referring to numerical series. For instance, Mercator's rule enabled the development of the function $\frac{1}{1+x}$ without being based upon numerical series.

---

[179]Mathematicians were open to the possibility of finding other procedures and, in effect, other, more particular procedures were applied in some particular cases.

[180]This is also an example of the generality of algebra; see Section 18.3.

This fact makes it possible to understand why the problem of the development of a function into a power series was viewed as a direct problem within the framework of the formal-quantitative theory of series, whereas the problem of the sum of a power series was thought to be an inverse problem.[181] In effect, summing a power series meant returning from the given series to the function whose expansion generated the power series. For instance, the function

$$\frac{1}{1+x}$$

was thought to be the sum of the series

$$\sum_{i=0}^{\infty}(-1)^i x^i$$

just because the application of Mercator's rule to $\frac{1}{1+x}$ made it possible to expand $\frac{1}{1+x}$ and obtain $\sum_{i=0}^{\infty}(-1)^i x^i$.

I would like to underline the difference between this way of conceiving the problem of the sum and the modern-day one. Nowadays, the problem of the sum is the direct problem, whereas the problem of development is the inverse problem. Indeed, we say that $\frac{1}{1+x}$ is the sum of $\sum_{i=0}^{\infty}(-1)^i x^i$ for $|x| < 1$, because

$$\lim_{n\to\infty}\sum_{i=0}^{\infty}(-1)^i x^i = \lim_{n\to\infty}\frac{1+(-1)^n x^{n+1}}{1+x} = \frac{1}{1+x}$$

for $|x| < 1$; instead, we say that $\sum_{i=0}^{\infty}(-1)^i x^i$ is the development of the function $\frac{1}{1+x}$ for $|x| < 1$, just because $\frac{1}{1+x}$ is the sum of $\sum_{i=0}^{\infty}(-1)^i x^i$.

In Chapter 33 we shall see that only when the 17th- and 18th-century methodology went into a state of crisis did the problem of the sum of a given series stop being viewed as the problem of returning to the generating function and instead became the direct problem of series theory.

---

[181]On this question I refer to Ferraro and Panza [2003].

9 The first appearance of divergent series

We have already seen that the early approach to series admitted the possibility of using divergent series as a means of deriving information about certain quantities. However, the problem of summing a divergent series, namely the problem of effectively associating a quantity with a divergent series, was not considered. It seemed obvious that a series such as $1 - 1 + 1 - 1 + \ldots$ had no sum, as Jacob Bernoulli had explicitly noted in 1696.[182]

The situation changed when Guido Grandi proved a theorem that can be formulated as follows (see Fig. 14).

> *Given a semicircle Γ with diameter IK, let us consider the tangent gI to circle at the point I, the intersection G of gI with the secant KZ. If GF_0 is equal and parallel to IK and the points F_n and GF_0 are such that*
> $$\frac{F_{n+1}G}{F_nG} = \frac{IG^2}{IK^2},$$
> *then we have*
> $$\sum_{k=0}^{\infty} F_{2k}F_{2k+1} = \sum_{k=0}^{\infty} (F_{2k}G - F_{2k+1}G) = KX = GD, \qquad (69)$$
> *where D is a point on the continuation of GF_0 such that $KX = GD$* (Grandi [1703, 27–28]).

If the point Z varies on the semicircle, the points F_n describe the parabolic curves IF_nA of degree $2n$ and the point D describes a curve KDd (today it is named the witch of Agnesi). By using the principle of continuity,[183] Grandi [1703, 29] stated that if GT coincides with the side gt of the square $IKgt$ and IG becomes equal to IK, then the point D coincides with d and the segment GD is equal to $gd = \frac{IK}{2}$. Therefore, the relation (69) gave rise to

$$\sum_{k=0}^{\infty} (IK - IK) = \frac{IK}{2} \qquad (70)$$

(see Grandi [1703, 29]). This series highlighted a contrast between geometric and algebraic principles. On the one side, the principle of continuity led to (70), which is the geometric version of

$$1 - 1 + 1 - 1 + \ldots = \frac{1}{2}; \qquad (71)$$

on the other hand, (70) could be read as the sum $0 + 0 + 0 + \ldots$, and it seemed trivial that $0 + 0 + 0 + \ldots$ was equal to 0.

---

[182]See Chapter 5.

[183]Leibniz formulated this principle as follows: "what is true up to the limit is true at the limit". The connection with the description of the continuous given in Section 7.2 is clear, and, in particular, the lack of distinction between closed and open segments.

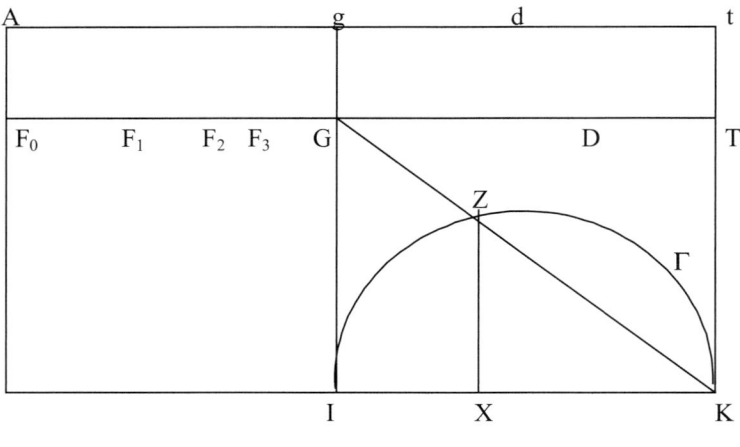

Fig. 14

Grandi's reasoning can be translated and summarized in more analytical terms (as 18th-century mathematicians did immediately) by stating that if the equation

$$y = \frac{a^3}{a^2 + x^2}$$

of the witch of Agnesi (where a is the diameter of the circle generating the witch) is expanded into the series

$$\sum_{n=1}^{\infty}(-1)^k \frac{x^{2k}}{a^{2k-1}},$$

then one obtains $1 - 1 + 1 - \ldots = 1/2$ by setting $a = x = 1$. From an analytical perspective, the problem derived from the passage from being a power series to a numerical series and posed the crucial question of the relationship between the formal and quantitative aspect of the notion of the sum.

The question of the series $1 - 1 + 1 - \ldots$ gave rise to a certain debate, of which there are many traces in Leibniz's correspondence. For example, on June 1, 1712, Hermann wrote to Leibniz, stating that Grandi's idea that infinite zeros could be gathered to form a finite quantity was clearly ridiculous. The mistake was due to treating

$$\sum_{k=0}^{\infty}(IK - IK)$$

as though it were a convergent series (see Leibniz [GMS, 4:369–370]).

Since (71) involved a basic notion of mathematics, namely the notion of continuity, Leibniz immediately realized that it could not be rejected for the reason that (71) was quantitatively false. Leibniz thought he had solved the question or had at least begun the solution in a paper[184] published in 1713. In this article, he first restated that (71) could be derived by using the same procedure commonly employed to derive valid results about convergent series.[185] Leibniz then reproposed Grandi's proof and regarded it as valid on the basis of the continuity principle [GMS, 5:385]. At the end of the article, he stated very clearly that if one takes G close to the segment IK, as desired, then the difference between gd and $\frac{IK}{2}$ can be made less than any given quantity and one therefore has $gd = \frac{IK}{2}$ by reasoning in the Archimedean manner.[186]

Nevertheless, Leibniz did not solve the crucial matter of the failure of algebraic rules at infinity. He limited himself to reaffirming that the rules of algebra are to be extended to infinity and that one must merely explain this exception to the general rule. The new science, he said, is not discredited by this paradox (Leibniz [GMS, 5:387]). However, he also sought to justify (71) in a different way so that an intuitively coherent meaning could be given from an algebraic viewpoint. The partial sums of the series $1 - 1 + 1 - \ldots$ are 0 or 1, attributing an even or odd number of terms to each sum, but if the series $1 - 1 + 1 - \ldots$ is continued up to the infinite, where the nature of the number vanishes, then the possibility of distinguishing between odd and even vanishes as well [GMS, 5:386]. According to Leibniz, the sum of the "whole" series cannot be either 0 or 1; since 0 and 1 have the same probability when the number of terms is finite, their arithmetic mean provides the sum of the series when the number of term is infinite [GMS, 5:386].

Leibniz refused to justify (71) in an exclusively algebraic manner and rejected certain merely formal methods that could be used to sum other divergent series such as $1 - 2 + 4 - 8 + 16 - 32 + \ldots$.

This is made clear by an exchange of letters with Wolff.[187] In a letter written on June 12, 1712, Wolff noted that if one considered

$$\frac{1}{1+x} = 1 - x + x^2 - x^3 + \ldots$$

---

[184]See *Epistola ad V. Cl. Christianum Wolfium, Professorem matheseos Halensem, circa scientiam infiniti* [1713]. In a letter to D. Bourget on May 3, 1715, by referring to (71), Leibniz stated: "it seems that there is a manifest absurdity. It is in the *Transactions* of Leipzig where I think I have given the solution to this enigma of the science of the infinite" [D, 6.1:219].

[185]He gave the following example. Since $y = \frac{1}{1+x} = 1 - x + x^2 - \ldots$ for $|x| < 1$, one obtained $\frac{1}{1+x^2} = 1 - x^2 + x^4 - \ldots$ and $\arctan x = \int_0^x \frac{1}{1+x^2} = x - \frac{x^3}{3} + \frac{x^5}{5} - \ldots$. For $x = 1$, the first series furnished $1 - 1 + 1 - 1 + \ldots = \frac{1}{2}$, whereas the third gave $\frac{\pi}{4} = 1 - \frac{1}{3} + \frac{1}{5} - \frac{1}{7} + \ldots$ (see [GMS, 5:382–383]).

[186]"[B]y assuming G close to V as desired, we can also show that GD will also be close to $\frac{1}{2}BV$ [$\frac{1}{2}gt$, with reference to Fig. 14], so that the difference can be made less than any given quantity. Whence, by inferring in the Archimedean manner, it also follows that VS [$= gd$] is equal to $\frac{1}{2}BV$" [GMS, 5:387].

[187]See Leibniz [GBLW, 143–149].

and set $x = 2$, then one obtained

$$\frac{1}{3} = 1 - 2 + 4 - 8 + 16 - 32 + \ldots,$$

namely the sum of the series $1 - 2 + 4 - 8 + 16 - 32 + \ldots$was equal to $\frac{1}{3}$. Wolff thought that it was possible to justify this as follows.

The sum of the "positive terms"

$$1 + 4 + 16 + \ldots + 2^{2n}$$

is equal to

$$\frac{8 \cdot 2^{2n-1} - 1}{3},$$

the sum of the "negative terms"

$$2 + 8 + 32 + \ldots + 2^{2n-1}$$

is equal to

$$\frac{4 \cdot 2^{2n-1} - 2}{3}.$$

Therefore,

$$s_{2n+1} = 1 - 2 + 4 - 8 + 16 - 32 + \ldots - 2^{2n-1} + 2^{2n} = \frac{4 \cdot 2^{2n-1} + 1}{3}.$$

In a similar way,

$$\begin{aligned} s_{2n+2} &= 1 - 2 + 4 - 8 + 16 - 32 + \ldots + 2^{2n} - 2^{2n+1} \\ &= \left(1 + 4 + 16 + \ldots + 2^{2n}\right) + \left(2 + 8 + 32 + \ldots + 2^{2n+1}\right) \\ &= \frac{4 \cdot 2^{2n} - 1}{3} - \frac{8 \cdot 2^{2n} - 2}{3} = \frac{-4 \cdot 2^{2n} + 1}{3}. \end{aligned}$$

The arithmetic mean of the partial sums $s_{2n+1}$and s_{2n+2} is equal to

$$\frac{2 \cdot 2^{2n-1} - 2 \cdot 2^{2n} + 1}{3}.$$

At this juncture, Wolff identified $2^{2n-1}$and 2^{2n} for $n = \infty$ and stated that the sum of the series was $\frac{1}{3}$.

Leibniz's answer[188] on July 13, 1712, seems to be a summary of the principles that had inspired his work. First of all, he stated that the summation of infinite series usually requires "decreasing terms" (see footnote no. 2). He then proposed three criteria to examine the correctness of a sum:

1. Is there a geometrical proof?

---

[188]See Leibniz [GBLW, 147].

2. Does the summation of a finite number of the terms provide anything that concerns the problem and that agrees with the infinite series or which at least approaches it continuously?

3. Is the demonstration sufficiently rigorous?[189]

   The first criterion underlines the strongly geometrical character of Leibnizian analysis. Effectively, an analytical expression such as $\frac{1}{1+x}$ was viewed as a geometrical quantity (represented symbolically) and, consequently, each analytical argumentation had to be reduced to a geometrical proof.
   The second criterion concerns the quantitative meaning of the sum: Partial sums had to give information about the sum of the series and ideally furnished the whole series at the infinite or, at least, approached it. In this sense partial sums of $1 - 2 + 4 - 8 - 16 + \ldots$ increase beyond any limit whereas the nth sums of $1 - 1 + 1 - \ldots$ stop at around $\frac{1}{2}$.
   As far as the third criterion is concerned, Leibniz observed that one cannot assume $A = 2A$ for $A = \infty$. The reference to a sufficiently rigorous proof (*satis accurate*) is especially interesting. Leibniz believed that analysis did not have a self-sufficient demonstrative power and it was necessary to resort to a geometric proof.
   In the years that followed, the debate about this kind of series continued; however, it should be specified that the question of the sum of (71), though epistemologically interesting, was not of great importance for the growth of mathematics at the beginning of the 18th century, prior to the evolution that I will describe in the second part.
   Divergent series were explicitly rejected by Varignon. In his *Précautions à prendre dans l'usage des suites ou series infinies*,[190] Varignon indeed considered the equality

$$\frac{a}{m \pm n} = \sum_{i=0}^{\infty} (\mp 1)^i \frac{a n^i}{m^{i+1}} \tag{72}$$

for $m > 0$ and $n > 0$ and stated that

1. if $m > n$, then equality (72) was true;

2. if $m < n$, then equality (72) was false;

3. if $m = n$ and the denominator is the difference $m - n$, then equality (72) is simply $\infty = \infty$: It is true but does not provide information;

4. if $m = n$ and the denominator is the sum $m + n$, then equality (72) was false.

---

[189]This is Knobloch's formulation of Leibniz's criteria (see Knobloch [1991, 290]).
[190]See Varignon [1715, 203–225].

Varignon's proof consists in verifying that (72) is quantitatively valid in case 1 (and in case 3) and that the first and the second sides of (72) are not equal in the other cases. In particular, in case 4, Varignon states that the sum of $1 - 1 + 1 - \ldots$ can never be $1/2$.[191]

In the 1720s, Goldbach and D.Bernoulli discussed divergent series in a series of letters dating from July 23 to October 12, 1724.[192] Goldbach opened the discussion by asking Bernoulli's opinion about the series

$$1 - 2 + 4 - 8 + 16 - 32 + \ldots.$$

He recalled that Varignon disapproved of the use of such a series as it was derived from an invalid division, whereas he thought that it was to be tolerated (Fuss [1843, 2:210]).

In response, Daniel Bernoulli stated that the sum of $1 + 2 + 4 + 8 + \ldots$ was -1. He calculated the sum according to the following scheme:

$$
\begin{array}{lll}
A & 1 + 2 + 4 + 8 + \ldots & = \quad s, \\
a & \phantom{1 + {}}2 + 4 + 8 + \ldots & = \quad s - 1, \\
B & \phantom{1 + {}}1 + 2 + 4 + \ldots & = \quad \dfrac{s-1}{2} = s.
\end{array}
$$

The equation

$$\frac{s-1}{2} = s$$

provides the sum $s = -1$.[193]

Daniel Bernoulli used the same reasoning as Leibniz and Jacob Bernoulli did in summing[194]

$$\sum_{n=1}^{\infty} \frac{2}{n(n+1)} = 2.$$

However, he neglected the constraint $a_\infty = 0$.

It is very interesting to highlight similarities and differences between Daniel Bernoulli's and Jacob Bernoulli's behavior on this point. We saw

---

[191]It is worth noting that Varignon's view is the same as Jacob Bernoulli's. As we saw in chapter 5, p. 84, in the third part of his *Positiones arithmeticae de seriebus infinitis* (see [1689–1704, 749–752]), published in 1696 (before Grandi posed the question of the series $1 - 1 + 1 - \ldots$, Bernoulli had examined the *non inelegans* paradox (54) and had provided an explanation.

[192]See Fuss [1843, 2:210–226]. Their main findings were later published (see Goldbach [1720], [1727], [1728] and Bernoulli [1728] and [1771]).

[193]See Fuss [1843, 2:214–216] and Bernoulli [1771, 72]. Goldbach criticized this procedure, observing that the series A and B, though similar, did actually differ. He considered two bodies that moved according to the laws expressed from series A and B (that is to say, in the first hour A covered the space 1 and B covered no space, in the second hour A covered 2 and B covered 1, etc.) and noted that the distance between the two bodies became increasingly larger (see Fuss [1843, 2:220–221]).

[194]See Chapters 2 and 5.

that mathematicians summed series by applying the rules (68) formally; only later did they verify whether the result was a convergent series. Thus Jacob Bernoulli did not consider $a_\infty = 0$ as a preliminary condition; it rather explained why the procedure succeeded in summing a convergent series. In other words, Jacob Bernoulli applied the rules (68) independently of convergence; however, he was interested in convergent series, and this made the constraint $a_\infty = 0$ crucial.

The context of Daniel Bernoulli's derivation is different. He intended to give a meaning to the sum of a divergent series, a problem that derived from the observation that the relation $a - a + a - a + \ldots = \frac{a}{2}$ had a geometric meaning. In order to do this, the condition $a_\infty = 0$ had to be neglected. Daniel Bernoulli showed that the mere transformation of a divergent series by the usual rules (68) allowed it to be summed and that the possibility of giving a sum to a divergent series was a natural consequence of the usual procedures in series theory.

In a letter to Goldbach, Daniel Bernoulli also illustrated another method for summing divergent series (see Fuss [1843, 2:216]). He considered the partial sums

$$D(x) = \sum_{n=1}^{x} a_n$$

of a given series $\sum_{n=1}^{\infty} a_n$, where $x = 2m + 1$ is an odd number, and the partial sums

$$P(x) = \sum_{n=1}^{x} a_n,$$

where $x = 2m$ is an even number. He stated that the arithmetic mean of $P(x)$ and $D(x)$ is equal to the sum of series *ad infinitum*. For instance, Bernoulli considered the series $1 - 2 + 4 - 8 + \ldots$ and found

$$D(x) = \sum_{n=0}^{x-1} (-2)^n = \frac{1 - (-2)^x}{1 - (-2)} = \frac{1 - (-2)^x}{3},$$

which is equal to

$$\frac{1 + 2^x}{3}$$

since x is an odd number. Similarly, he derived

$$P(x) = \sum_{n=0}^{x-1} (-2)^n = \frac{1 - (-2)^x}{1 - (-2)} = \frac{1 - 2^x}{3}.$$

Consequently, the sum of the series is

$$S = \frac{D(x) + P(x)}{2} = \frac{1}{3}.$$

In the same way, Bernoulli calculated the sum of the following series

$a - a + a - a + \ldots$	$D(x) = a$	$P(x) = 0$	$S = \frac{D(x)+P(x)}{2} = \frac{a}{2}$,
$1 - 2 + 3 - 4 + 5 - \ldots$	$D(x) = \frac{x+1}{2}$	$P(x) = -\frac{x}{2}$	$S = \frac{D(x)+P(x)}{2} = \frac{1}{4}$,
$1 - 3 + 5 - 7 + 9 - \ldots$	$D(x) = x$	$P(x) = -x$	$S = \frac{D(x)+P(x)}{2} = 0$,
$1 - 4 + 9 - 16 + 25 - \ldots$	$D(x) = \frac{x^2+x}{2}$	$P(x) = -\frac{x^2+x}{2}$	$S = \frac{D(x)+P(x)}{2} = 0$.

This procedure cannot be translated into modern terms by stating that Bernoulli, *de facto*, calculated

$$\lim_{x \to \infty} \frac{s_x + s_{x+1}}{2},$$

where s_x is the partial sum of the series $\sum_{n=1}^{\infty} a_n$, because if x is odd and $D(x) = s_x$, then $P(x)$ is different from $s_{x+1} = P(x+1)$.[195] In a sense, Bernoulli took the mean of the analytical expressions $P(x)$ and $D(x)$ and the result represented the expectation that the sum assumed the form $P(x)$ or the form $D(x)$.

Daniel Bernoulli's procedure was essentially based upon the fact that the sum of the above series can be expressed as

$$H + (-1)^x f(x),$$

for finite x, where H is a constant and $f(x)$ is an appropriate function of the index x. He takes

$$(-1)^x f(x)$$

equal to zero when x is infinite. Daniel Bernoulli presented his ideas on the summation of series in a paper published only in 1771, where he observed that the sum of $1 - 1 + 1 - 1 + \ldots$ had the form

$$s_x = \frac{1}{2} - \frac{(-1)^x}{2}.$$

He stated that if one assumed that the sum of series to be $\frac{1}{2} + k$, then one could equally assume it to be $\frac{1}{2} - k$; therefore, one must assume $k = 0$ to avoid contradictions (D. Bernoulli [1771, 72–73]).

---

[195]This procedure is therefore different from Hutton's method, which defines the sum of a series $\sum_{n=1}^{\infty} a_n$ as follows. If we take

$$S_n^{(k)} = \frac{S_{n-1}^{(k-1)} + S_n^{(k-1)}}{2} \qquad (n \geq 0),$$

with $S_n^{(0)} = \sum_{i=1}^{n} a_i$ and $S_{-1}^{(k)} = 0$ $(k \geq 0)$, and if

$$\lim_{n \to \infty} S_n^{(k)}$$

exists, then the series is said to be (Hu, k)-summable and its sum is

$$S = \lim_{n \to \infty} S_n^{(k)}.$$

On Hutton's method, see Hardy [1949, 21–22].

In this paper, Daniel Bernoulli also observed that the partial sums of periodic series of the type

$$a_{n+k} = a_n,$$

with $a_0 + a_1 + \ldots + a_{k-1} = 0$, assumes solely k values, S_0, S_1, $\ldots$, S_{k-1}. In this simple case he took

$$\sum_{n=1}^{\infty} a_n = \frac{S_0 + S_1 + \ldots + S_{k-1}}{k}$$

(see D. Bernoulli [1771, 77–80]).

In his correspondence with Bernoulli, Goldbach, in his turn, suggested a method for summing divergent series which was based on the formal transformations of a divergent series using (68). He published his findings in *De transformatione serierum* [1727]. In this paper he considered a series $\sum_{n=1}^{\infty} b_n$ such that

a. the "last" term b_n is infinitesimal,

b. $\sum_{n=1}^{\infty} b_n = 1$.

Given a series $\sum_{n=1}^{\infty} a_n$, he multiplied $\sum_{n=1}^{\infty} a_n$ by $\sum_{n=1}^{\infty} b_n$ so to obtain the series

$$\sum_{n=1}^{\infty} c_n = \left(\sum_{n=1}^{\infty} a_n \right) \left(\sum_{n=1}^{\infty} b_n \right).$$

The sum of $\sum_{n=1}^{\infty} c_n$ is precisely equal to the sum of $\sum_{n=1}^{\infty} a_n$. Therefore, if one was able to calculate $\sum_{n=1}^{\infty} c_n$, then one also determined

$$\sum_{n=1}^{\infty} a_n = \sum_{n=1}^{\infty} c_n.$$

In particular, $\sum_{n=1}^{\infty} b_n$ could be considered as being of the form

$$1 + d_1 - d_1 + d_2 - d_2 + d_3 - d_3 + d_4 - d_4 + \ldots.$$

Goldbach applied his method to the geometric series $\sum_{n=0}^{\infty} (-1)^n m^n$, $m > 1$. He chose[196]

$$d_1 = \frac{m^2 + m - 1}{m + 2}, \quad d_n = \frac{d_{n-1}}{m + 2},$$

rearranged the product appropriately, and obtained

$$\sum_{n=0}^{\infty} (-1)^n m^n = \sum_{n=0}^{\infty} \left(\frac{1}{m + 2} \right)^n.$$

---

[196]The method actually consists of adding fictitious terms to a given series. This idea was probably the cue for Euler's *Delucidationes in capita postrema calculi mei differentialis de functionibus inexplicabilibus* (see [EDel]).

Since

$$\sum_{n=0}^{\infty} \left(\frac{1}{m+2} \right)^n = \frac{1}{m+1},$$

Goldbach derived

$$\sum_{n=0}^{\infty} (-1)^n m^n = \frac{1}{m+1}.$$

It is worthwhile noting that Goldbach did not use

$$\sum_{n=1}^{\infty} a_n = \left(\sum_{n=1}^{\infty} a_n \right) \left(\sum_{n=1}^{\infty} b_n \right) = \sum_{n=1}^{\infty} c_n$$

as a definition of the sum: He tackled the question of the sum of $\sum_{n=1}^{\infty} a_n$ as if the symbol $\sum_{n=1}^{\infty} a_n$ had a natural meaning that he merely had to discover. He never considered the idea that it was necessary to define the sum of $\sum_{n=1}^{\infty} a_n$ before manipulating it.

Part II

From the 1720s to the 1760s:
The development
of a more formal conception

In Part I we saw that the early theory of series was characterized by the simultaneous presence of the quantitative and the formal: The quantitative goals were obtained using formal rules. Formal manipulation meant that there was no *a priori* control of the single steps. However, a series had intrinsic meaning in quantitative terms, which derived not from formal manipulation but from its convergence. Divergent series were not taken into consideration *per se*[197] but could be used as tools for deriving results about convergent series.

By the 1720s, the early theory of series was enriched by many new results. In particular, de Moivre began his investigation of recurrent series, Stirling derived the series named after him, Euler and Maclaurin obtained the Euler–Maclaurin sum formula, continued fractions and infinite products underwent significant development, series were used in number theory, and new applications were discovered in numerical analysis.

The new results were derived by using the same procedures and principles that had been used until 1720; in a sense, they were a natural development of the early theory. However, many of them could not be reduced to the original concept, since they used series to count objects or connect different entities or provide approximate values of quantities even though they diverged.

Furthermore, the geometric context of the early theory of series disappeared. Previously, series had been a part of a procedure for solving geometric problems that consisted of two phases: In the first phase, the series expansion of a quantity was sought, while in the subsequent phase the expansion was applied to specific problems and the interval of convergence was adapted to a specific situation. The two phases still remained, although they were no longer two stages of a geometric procedure. The first belonged to analysis, which was now separated from geometry, whereas the second belonged to other fields of mathematics: geometry, mechanics, numerical analysis.

From the 1730s onward certain difficulties arose and mathematicians thought they could be solved by applying a merely formal notion of series.

---

[197]The fact that a series could represent a quantity not only in the sense of convergence appears implicitly in Wallis's work (see p. 11). However, this kind of series seemed to be a hesitant product of the problem of quadrature, resolved by the development of integral calculus. At the beginning of the 18th century, the problem of the sum of divergent series was completely marginal.

But the 18th-century formal notion differs considerably from the modern one: It included the quantitative insofar as the formal consisted of the generalization of procedures, rules, and results that were valid for finite sums or for certain intervals of the values of variables. Thus, the quantitative remained related to the formal, perhaps even beyond the intention of 18th-century mathematicians.

Part II is divided into 10 chapters. In the first chapter (Chapter 10), I illustrate the rise of the theory of recurrent series and its application to the solution of algebraic equations. Chapter 11 is devoted to the problem of the acceleration of series and the emergence of asymptotic series. Maclaurin's contributions are the subject matter of Chapter 12. Chapters 13 to 17 mainly concern Euler. Euler's influence on 18th-century mathematics and, in particular, on the theory of series was enormous, in terms of both the depth and the range of his writings. I describe Euler's use of the series up to the 1760s: In particular, I deal with the problem of the search for the general term, the relation between sequences and functions, Euler's use of earlier procedures, infinite products, continued fractions, the application of series to number theory. In Chapter 18, I briefly discuss the new concept of analysis, which reached its maturation in the 1740s and within whose framework the evolution of series theory should be viewed. Finally, in Chapter 19, I illustrate what was the final result of this evolution: the formal theory of series.

10 De Moivre's recurrent series and Bernoulli's method

In the first edition of his *The Doctrine of Chance* [1718, 127–134] A. de Moivre dealt with a peculiar type of power series, which he later named as recurrent series in his *De fractionibus algebricis* [1722, 175–176]. According to de Moivre, the terms of a recurrent series are "so related to one another that each of them may have to the same number of preceding terms a certain given relation, always expressible by same index" [1718, 133].[198]

Even though this definition is very general, de Moivre in reality considered only power series and linear laws of relation; in other words, he considered a power series $\sum a_n x^n$ as recurrent or recurring series if the coefficient a_n of its general term $a_n x^n$ was a linear combination of a fixed number of antecedent terms:

$$a_n = b_1 a_{n-1} + b_2 a_{n-2} + \ldots + b_s a_{n-s}. \tag{73}$$

The constants

$$(b_1, \quad b_2, \quad \ldots, \quad b_s)$$

were called the scale of series, and the polynomial

$$1 - b_1 x - b_2 x^2 - \ldots - b_s x^s$$

was termed the differential scale of series.[199] For instance,

$$1 + 3x + 7x^2 + 17x^3 + 41x^4 + 99x^5 + \ldots.$$

is a recurrent series, its scale is $(2, 1)$, and its differential scale is

$$1 - 2x - x^2.$$

Recurrent series were a topic of remarkable interest during the 18th century. Even though they were power series, they were not of interest for their capability of representing analytical expressions in the quantitative sense. Mathematicians began to use them because they offered the possibility of investigating the combinations of objects. In the theory of recurrent series, questions of convergence were non-existent. The letter x was treated in a merely combinatorial way, namely as a mere sign, a placeholder, and one operated upon series merely by combining and rearranging letters and numbers. Here are some examples. In his *Miscellanea analytica*, de Moivre formulated the following theorem:

---

[198]In the second and third editions of *The Doctrine of Chance*, De Moivre gave the following definition: "a recurring series ... is so constituted, that having taken at pleasure any number of its terms, each following term shall be related to the same number of preceding terms, according to a constant law of relation." [1738, 183 and 1756, 220].

[199]Although 18-century mathematicians generalized this notion (see, for instance, Goldbach [1728, 164]), they maintained de Moivre's use of the term "recurrent series" for series given by the law of recurrence (73).

If the series

$$\sum_{n=0}^{\infty} a_n x^n$$

has the scale $(b_1, -b_2)$ and the equation

$$x^2 - b_1 x + b_2 = 0$$

has the roots $\xi_1 \neq \xi_2$, then the given series can be divided into two geometric series with ratios $\xi_1 x$ and $\xi_2 x$. (de Moivre [1730, 27–28])

The proof was based upon the fact that the relation between the terms of the geometric series $\sum_{n=0}^{\infty} C_n x^n$, whose ratio is μ, can be written not only in the form $C_i = \mu C_{i-1}$ but also in the form

$$C_i = (\mu + \lambda)C_{i-1} - \mu\lambda C_{i-2},$$

for any number λ. Therefore, if ξ_1 and ξ_2 are the roots of the equation $x^2 - b_1 x + b_2 = 0$, the relation between the terms of each of the progressions

$$A_i \sum_{n=0}^{\infty} \xi_i^n x^n \quad (i = 1, 2)$$

can be expressed by

$$A_i \xi_i^n = (\xi_1 + \xi_2)A_i (\xi_i)^{n-1} - \xi_1\xi_2 A_i (\xi_i)^{n-2},$$

and the sum of these series

$$A_1 \sum_{n=0}^{\infty} \xi_1^n x^n + A_2 \sum_{n=0}^{\infty} \xi_2^n x^n = \sum_{n=0}^{\infty} (A_1\xi_1^n + A_2\xi_2^n)x^n$$

has a scale $(\xi_1 + \xi_2, -\xi_1\xi_2)$. Of course,

$$b_1 = \xi_1 + \xi_2, b_2 = -\xi_1\xi_2,$$

and therefore the sum

$$A_1 \sum_{n=0}^{\infty} \xi_1^n x^n + A_2 \sum_{n=0}^{\infty} \xi_2^n x^n$$

is exactly

$$\sum_{n=0}^{\infty} a_n x^n.$$

De Moivre considered the analogous result for series whose scale has length 3 and 4, this allowed for the realization that a recurrence series, whose differential scale has no multiple roots, can generally be reduced to the sum of geometric series (see de Moivre [1730, 29–31]).

De Moivre also showed that certain operations on series corresponded to operations upon differential scales. For instance, in his [1738, 193],[200] he proved that if the series

$$\sum_{n=0}^{\infty} a_n x^n \quad \text{and} \quad \sum_{n=0}^{\infty} b_n x^n$$

have the differential scales

$$1 - \alpha_1 x + \alpha_2 x^2 \quad \text{and} \quad 1 - \beta_1 x + \beta_2 x^2,$$

then the series

$$\sum_{n=0}^{\infty} c_n x^n = \sum_{n=0}^{\infty} a_n x^n + \sum_{n=0}^{\infty} b_n x^n$$

has the differential scale

$$(1 - \alpha_1 x + \alpha_2 x^2)(1 - \beta_1 x + \beta_2 x^2).$$

The heart of the theory of recurrent series is the observation that *any rational function with a numerator whose degree is less than the denominator can be expanded into a recurrent series and, vice versa, that any recurrent series can be summed and that sum is a rational function with a numerator whose degree is less than the denominator* (see de Moivre [1730, 27–35]).

In order to prove that the sum of a recurrent series $z = \sum_{n=0}^{\infty} a_n x^n$, with scale $(b_1, \ldots, b_s)$, is a rational function de Moivre operated according to the following scheme:[201]

---

[200]See also de Moivre [1756, 228].

[201]The general scheme is a reconstruction. As often occurred, de Moivre considered some particular cases as arbitrary examplifications. An "arbitrary exemplification" is a demonstrative technique that consists of using an appropriate example rather than a general demonstration. It is based on the repeatability of the procedure employed, which is invariable with respect to the specific exemplification.

$$z = \sum_{n=0}^{\infty} a_n x^n = \sum_{n=0}^{s-1} a_n x^n + \sum_{n=s}^{\infty} a_n x^n$$

$$= \sum_{n=0}^{s-1} a_n x^n + \sum_{n=s}^{\infty} (b_1 a_{n-1} + b_2 a_{n-2} + \ldots + b_s a_{n-s}) x^n$$

$$= \sum_{n=0}^{s-1} a_n x^n + b_1 \sum_{n=s}^{\infty} (a_{n-1} x^{n-1}) x$$

$$+ b_2 \sum_{n=s}^{\infty} (a_{n-2} x^{n-2}) x^2 + \ldots + b_s \sum_{n=s}^{\infty} (a_{n-s} x^{n-s}) x^s$$

$$= \sum_{n=0}^{s-1} a_n x^n + b_1 \sum_{n=0}^{\infty} (a_{n+s-1} x^{n+s-1}) x$$

$$+ b_2 \sum_{n=s}^{\infty} (a_{n+s-2} x^{n+s-2}) x^2 + \ldots + b_s \sum_{n=0}^{\infty} (a_n x^n) x^s$$

$$= \sum_{n=0}^{s-1} a_n x^n + b_1 x (z - \sum_{n=0}^{s-2} a_n x^n) + b_2 x^2 (z - \sum_{n=0}^{s-3} a_n x^n) + \ldots + b_s x^s z.$$

By solving the equation

$$z = \sum_{n=0}^{s-1} a_n x^n + b_1 x (z - \sum_{n=0}^{s-2} a_n x^n) + b_2 x^2 (z - \sum_{n=0}^{s-3} a_n x^n) + \ldots + b_s x^s z$$

with respect to the unknown z, we have that z is equal to fraction whose denominator is $1 - b_1 x - b_2 x^2 - \ldots - b_s x^s$ and the numerator has a degree less than s (see de Moivre [1730, 27–35])

The inverse statement (the expansion of any rational function is a recurring series) could be easily derived by applying the method of indeterminate coefficients. For instance, in [1730, 35] de Moivre posed

$$\frac{1}{1 - fx + gx^2 - hx^3 + kx^4} = P + Qx + Rx^2 + Sx^3 + \ldots.$$

Hence,

$$\left(1 - fx + gx^2 - hx^3 + kx^4\right)\left(P + Qx + Rx^2 + Sx^3 + \ldots\right) = 1$$

and

$$Q = fP,$$
$$R = fQ - gP,$$
$$S = fR - gQ + hP,$$
$$T = fS - gR + hQ - kP,$$
$$\ldots$$

Consequently, $P + Qx + Rx^2 + Sx^3 + \ldots$ is a recurrence series with the scale of relation $(f, -g, h, -k)$.

These results pave the way for the identification of any recurrent series with a function (later referred to as a generating function). In his [1748a, 1:175], Euler made this explicit and defined the sum of a recurrent series to be the function that generated it.

De Moivre and other 18th-century mathematicians attached importance to the determination of the general term of a recurring series. They observed that the general term a_n of a recurrent series could be written in the form

$$a_n = \sum_{i=1}^{r} P_i(n) q_i^n,$$

where q_i are the roots of equation

$$1 - b_1 x - b_2 x^2 - \ldots - b_s x^s = 0$$

and $P_i(n)$ is a polynomial of degree equal to 1 less the multiplicity of q_i (see de Moivre [1730], Bernoulli [1728], and Euler [1748a]). For instance, if we consider the Fibonacci sequence

$$0, \ 1, \ 1, \ 2, \ 3, \ 5, \ 8, \ 13, \ 21, \ 34, \ \ldots$$

defined by

$$a_n = a_{n-1} + a_{n-2},$$

since the scale of recurrence is $(1, 1)$ and the roots of equations $x^2 = x + 1$ are $\frac{1 \pm \sqrt{5}}{2}$, the general term of the Fibonacci sequence is

$$\alpha \left(\frac{1 + \sqrt{5}}{2} \right)^n + \beta \left(\frac{1 - \sqrt{5}}{2} \right)^n.$$

The numbers α and β are to be determined by taking into account $a_0 = 1$ and $a_1 = 1$: So one has

$$\alpha = \beta = \frac{1}{\sqrt{5}}$$

(see Bernoulli [1728, 89–90]).

In *Observationes de seriebus*, Daniel Bernoulli reversed the last result and, using several examples, showed how recurrent series could arise in numerical computations, involving only combinatorial properties of series. Indeed, if

$$|r_1| > |r_2| > \ldots > |r_n|$$

are the roots of the equation

$$1 = a_1 x + a_2 x + \ldots + a_n x^n,$$

D. Bernoulli explained that in order to determine r_1, one had to write a recurrent sequence b_n with the scale $a_1, a_2, \ldots, a_n$ and whose first n terms are arbitrary. Then the ratio

$$\frac{b_n}{b_{n+1}}$$

"is nearly equal to a sought root". For instance, consider $1 = -2x + 5x^2 - 4x^3 + x^4$ and form the sequence with the scale $(-2, 5, -4, 1)$ and the first four terms equal to 1, namely the series

$$1, \ 1, \ 1, \ 1, \ 0, \ 2, \ -7, \ 25, \ -93, \ 341, \ -1254, \ \ldots.$$

Then one of the roots of the equation is approximately equal to

$$\frac{-341}{1254}$$

(see D. Bernoulli [1728, 92]).

   This method differed from other already known techniques for determining the roots (which consisted of the approximation of a numerical value by means of the nth partial sum of a convergent series). It offered a sequence whose terms had a numerical value only indirectly connected to the result: These were the steps of a transformation that enabled one to arrive at the final result. It was evident that the divergence of the sequence was of no importance for determining the numerical value of the root.

   Bernoulli's method was later improved by Euler and Lagrange. In his [1748a, 1:339–361], Euler reformulated Bernoulli's method as follows. Let us consider the equation

$$F(x) = 1 - b_1 x - b_2 x^2 - \ldots - b_s x^s = 0,$$

If we suppose that the roots q_i are simple, then $F(x)$ can be thought of as the denominator of a rational function

$$\frac{G(x)}{F(x)} = \sum_{i=1}^{s} \frac{A_i}{1 - q_i x},$$

where $G(x)$ is an arbitrary polynomial with degree less than s. (This corresponds to the arbitrary choice of the first n terms of the recurrent series in Bernoulli's formulation.) Since each addend can be expanded into a power series

$$\sum_{h=0}^{\infty} A_i q_i^h x^h,$$

by rearranging we have

$$\frac{G(x)}{F(x)} = \sum_{h=0}^{\infty} \left(\sum_{i=1}^{s} A_i q_i^h\right) x^h = \sum_{h=0}^{\infty} Q_h x^h,$$

where

$$Q_h = \sum_{i=1}^{s} A_i q_i^h.$$

From the theory of recurrent series, we know that this is a recurrent series with the scale $(b_1,\ldots,b_s)$. At this point Euler supposed that the root $|q_1|$ was greater than $|q_i|$, $i = 2,\ldots,s$ and stated that if h is an infinite number, then we have $Q_h = A_1 q_1^h$; "but if h is only a very large number, we have very nearly" $Q_h = A_1 q_1^h$.

Similarly, Q_{h+1} approximates $A_1 q_1^{h+1}$ for large h, and so

$$\frac{Q_{h+1}}{Q_h}$$

is an approximation of q_1.

The polynomial $G(x)$ has to be chosen so that the coefficient A_1 is different from 0 [then $1 - q_1 x$ is not a factor of $G(x)$]. Once q_1 has been determined, one can choose $G(x)$ so that it contains the factor $1 - q_1 x$ and in this way it becomes possible to determine the root $|q_2|$ where $|q_2|$ is the greatest in the set $\{|q_i|, i = 3,\ldots,s\}$.

If there are two roots q_1 and q_2 such that $q_1 = -q_2$ and

$$|q_1| = |q_2| > |q_i|, \quad i = 3,\ldots,s,$$

Euler showed that $\frac{Q_{h+1}}{Q_h}$ did not approximate q_1; however,

$$\frac{Q_{h+2}}{Q_h}$$

approximated $|q_1|^2$. He also dealt with the case where there are two complex conjugate roots and proposed a method for accelerating the convergence when there are multiple roots (the convergence might be very slow in this case).

Euler even applied Bernoulli's method to infinite equations. Indeed, he sought the largest root of

$$\frac{1}{2} = z - \frac{1}{6}z^3 + \frac{1}{120}z^5 - \frac{1}{5040}z^7 + \ldots$$

(it is the equation $\frac{1}{2} = \sin z$). He wrote

$$\frac{1}{1 - 2z + \frac{1}{3}z^3 - \frac{1}{60}z^5 + \dots} = 1 + \sum_{h=1}^{\infty} \left(2z - \frac{1}{3}z^3 + \frac{1}{60}z^5 - \dots\right)^h$$

$$= 1 + 2z + 4z^2 + \frac{23}{3}z^4 + \frac{44}{3}z^4$$

$$+ \frac{1681}{60}z^5 + \frac{2408}{45}z^6 + \dots$$

Hence, he obtained

$$z = \frac{1681}{60} \cdot \frac{2408}{45} \approx 0.52356$$

(see Euler [1748a, 360]).

At the end of the 18th century, Lagrange also made a contribution to Bernoulli's method. In *Traité de la resolution des èquations numeriques* [1798], he assumed that the numerator $G(x)$ of the fraction $\frac{G(x)}{F(x)}$ was chosen to be the derivative of the denominator $F(x)$. In this way the denominator of the fraction $\frac{F'(x)}{F(x)}$ had no multiple roots.

11   Acceleration of series and Stirling's series

In the first part of this book, we saw that power series were used with a quantitative purpose, but mathematicians separated the investigation of convergence from the search of the law of formation of coefficients. Once this law was found, it seemed easy to establish the interval of convergence (given the very simple nature of series employed), which consisted of fitting analytical results to the specific problem by making an appropriate choice of the parameters. From this point of view, the crucial problem was not convergence *per se* but achieving *fast convergence*. Researchers indeed endeavored to speed up the convergence of series. However, this search was also undertaken through purely algebraic manipulations, and the usefulness of the results was only tested *a posteriori*. In this way, it led to results that could not be reduced to the formal-quantitative concept of series. These results are the subject matter of the present chapter.

The acceleration of series underwent extensive investigation by James Stirling. In the preface to his *Methodus differentialis* [1730], Stirling explained that series often converged so slowly that they were *de facto* no more useful than if they had been divergent. For this reason, he provided various methods in the first part of his treatise by means of which the sum of these series could be quickly obtained.

I shall focus on three of Stirling's theorems. The first proposition concerns the sum of finite series. The second concerns the transformation of infinite series into a more rapid convergent series. The third proposition is Stirling's derivation of Stirling's series,[202] the first example of an asymptotic series (as they were termed later[203]). To prove this last theorem, he used the same methodology as in the other two propositions.   He obtained a series

---

[202]Today, the expression

$$\sum_{n=1}^{\infty} \frac{B_{2n}}{2n(2n-1)z^{2n-1}} = \frac{1}{2}\log 2\pi + \left(z - \frac{1}{2}\right)\log z - z + \frac{1}{12z} - \frac{1}{360z^3} + \frac{1}{1260z^3} - \dots \quad (74)$$

is called Stirling's series.

[203]An asymptotic series is a series expansion of a function in a variable z that may converge or diverge but whose partial sum can be made an arbitrarily good approximation to a given function for large enough z. For instance, $\sum_{i=0}^{\infty} a_i x^{-i}$ is the asymptotic expansion of $f(x)$, if

$$\lim_{x \to \infty} x^n \left[f(x) - \sum_{i=0}^{\infty} a_i x^{-i} \right] = 0$$

for fixed n. The asymptotic series for the gamma function $\Gamma(z)$ is given by

$$e^{-z}z^{z-1/2}\sqrt{2\pi}\left(1 + \frac{1}{12z} + \frac{1}{288z^2} - \frac{139}{51840z^3} + \dots\right).$$

Stirling's series is the asymptotic expansion of $\log\Gamma(z)$.

that was very effective for approximate calculations, but, unlike the series in Proposition 2, was not convergent. This result was, in a sense, unexpected and difficult to be set against in the formal-quantitative series theory.

Proposition 1. The sum of the finite series $\sum_{n=1}^{z} a_n$, where

$$a_n = b_0 + b_1 n + b_2 n(n-1) + b_3 n(n-1)(n-2) + \ldots,$$

is

$$\sum_{n=1}^{z} a_i = b_0 z + \frac{1}{2} b_1 (z+1)z + \frac{1}{3} b_2 (z+1)z(z-1)$$
$$+ \frac{1}{4} b_3 (z+1)z(z-1)(z-2) + \ldots$$

(Stirling [1730, 20]).

In the proof, Stirling assumes that the thesis is true and lets $S(z) = \sum_{n=1}^{z} a_i$ and $S(0) = a_0 = 0$. Then

$$S(z-1) = \sum_{n=1}^{z-1} a_i = b_0(z-1) - \frac{1}{2} b_1 z(z-1) - \frac{1}{3} b_2 z(z-1)(z-2)$$
$$- \frac{1}{4} b_3 z(z-1)(z-2)(z-3) + \ldots$$

Since

$$[(z+1)z(z-1)\ldots(z-n+2)] - [z(z-1)\ldots(z-n+1)]$$
$$= nz(z-1)\ldots(z-n+2),$$

he has

$$S(z) - S(z-1) = a_z$$
$$= b_0 z + \frac{1}{2} b_1 (z+1)z + \frac{1}{3} b_2 (z+1)z(z-1)$$
$$+ \frac{1}{4} b_3 (z+1)z(z-1)(z-2) + \ldots$$
$$- b_0(z-1) - \frac{1}{2} b_1 z(z-1) - \frac{1}{3} b_2 z(z-1)(z-2)$$
$$- \frac{1}{4} b_3 z(z-1)(z-2)(z-3) - \ldots$$
$$= b_0 + b_1 z + b_2 z(z-1) + b_3 z(z-1)(z-2) + \ldots$$

for all $z > 0$. Then he stated that since

$$S(z) = S(z-1) + a_z,$$

if a_z is given, one inversely obtains the sum $S(z)$.[204]

Using this theorem Stirling obtained

$$1 + 2 + \ldots + z = \frac{z(z+1)}{2}$$

and

$$1 + 8 + 27 + \ldots + z^3 = \frac{z^2}{4}(z+1)^2$$

[since $z^3 = z + 3z(z-1) + z(z-1)(z-2)$].

Stirling employed a similar procedure to accelerate the convergence of an infinite series.

Proposition 2. The sum $S(z)$ of the infinite series $\sum\limits_{n=z}^{\infty} a_n$, where

$$a_n = \frac{b_0}{n(n+1)} + \frac{b_1}{n(n+1)(n+2)} + \frac{b_2}{n(n+1)(n+2)(n+3)}$$
$$+ \frac{b_3}{n(n+1)(n+2)(n+3)(n+4)} + \ldots,$$

is

$$S(z) = \frac{b_0}{z} + \frac{b_1}{2z(z+1)} + \frac{b_2}{3z(z+1)(z+2)} \qquad (75)$$
$$+ \frac{b_3}{4z(z+1)(z+2)(z+3)} + \ldots$$

(Stirling [1730, 25]).

If the thesis were true, we would have

$$S(z+1) = \frac{b_0}{z+1} + \frac{b_1}{2(z+1)(z+2)} + \frac{b_2}{3(z+1)(z+2)(z+3)}$$
$$+ \frac{b_3}{4(z+1)(z+2)(z+3)(z+4)} + \ldots.$$

Since

$$\frac{1}{z(z+1)\ldots(z+k-1)} - \frac{1}{(z+1)\ldots(z+k)} = \frac{k}{z(z+1)\ldots(z+k)}$$

we obtain

$$a_z = S(z) - S(z+1) = \frac{b_0}{z(z+1)} + \frac{b_1}{z(z+1)(z+2)}$$
$$+ \frac{b_2}{z(z+1)(z+2)(z+3)} + \frac{b_3}{z(z+1)(z+2)(z+3)(z+4)} + \ldots.$$

---

[204] Stirling supposes that if the thesis were true for $S(z-1)$, it would be true for $S(z) = S(z-1) + a_z$.

Conversely if this is the general term, the sum is given by (75) (Stirling [1730, 25]).[205]

Stirling applied this theorem to determine the value of various sums. In particular, he found

$$\sum_{i=1}^{\infty} \frac{1}{n^2} \approx 1.644934065.$$

The third proposition that I would like to consider is the following.

Proposition 3. *If the numbers* $x + n$, $x + 3n$, $x + 5n$, ..., $z - n$ *are in an arithmetic progression, then*

$$\lg(x+n) + \lg(x+3n) + \lg(x+5n) + \ldots + \lg(z-n) \qquad (76)$$
$$= \left(\frac{z \lg z}{2n} - \frac{az}{2n} - \frac{an}{12z} + \frac{7an^3}{360z^3} - \frac{31an^5}{1260z^5} + \frac{127an^7}{1680z^7} - \frac{511an^9}{1188z^9} + \ldots \right)$$
$$- \left(\frac{x \lg x}{2n} - \frac{ax}{2n} - \frac{an}{12x} + \frac{7an^3}{360x^3} - \frac{31an^5}{1260x^5} + \frac{127an^7}{1680x^7} - \frac{511an^9}{1188x^9} + \ldots \right)$$

where $a = \frac{1}{\log 10}$ *and* $\lg X$ *and* $\log Y$ *denote the logarithm to the base 10 and natural logarithm of the numbers X and Y, respectively* (Stirling [1730, 135]).

To prove this proposition, Stirling considered

$$A = \frac{z \lg z}{2n} - \frac{az}{2n} - \frac{an}{12z} + \frac{7an^3}{360z^3} - \frac{31an^5}{1260z^5} + \ldots$$

and

$$B = \frac{(z-2n)\lg(z-2n)}{2n} - \frac{a(z-2n)}{2n}$$
$$- \frac{an}{12(z-2n)} + \frac{7an^3}{360(z-2n)^3} - \frac{31an^5}{1260(z-2n)^5} + \ldots$$

The terms of the series B were reduced to the same form as the terms of the series A "by performing a suitable division". This means that he considered

---

[205] An alternative procedure can be found in Stirling [1719].

the expansion of $\frac{(z-2n)\lg(z-2n)}{2n}$:

$$
\begin{aligned}
\frac{(z-2n)\lg(z-2n)}{2n} &= \frac{z\lg(z-2n)}{2n} - \lg(z-2n) \\
&= \frac{z}{2n}\left(\lg z - \frac{2an}{z} - \frac{4an^2}{2z^2} - \frac{8an^3}{3z^3} - \cdots\right) \\
&\quad - \left(\lg z - \frac{2an}{z} - \frac{4an^2}{2z^2} - \frac{8an^3}{3z^3} - \cdots\right) \\
&= \frac{z}{2n}\lg z - a - \frac{an}{z} - \frac{4an^2}{3z^2} - \cdots \\
&\quad - \lg z - \frac{2an}{z} - \frac{4an^2}{2z^2} - \frac{8an^3}{3z^3} - \cdots
\end{aligned}
$$

and the expansions of $-\frac{a(z-2n)}{2n}$, $-\frac{an}{12(z-2n)}$, $\frac{7an^3}{360(z-2n)^3}$, $-\frac{31an^5}{1260(z-2n)^5}$, $\cdots$:

$$
\begin{aligned}
-\frac{a(z-2n)}{2n} &= -\frac{az}{2n} + a, \\
-\frac{an}{12(z-2n)} &= -\frac{an}{12z}\left(1 + \frac{2n}{z} + \frac{4n^2}{z^2} + \frac{8n^3}{z^3} + \cdots\right), \\
\frac{7an^3}{360(z-2n)^3} &= \frac{7an^3}{360z^3}\left(1 + 3\frac{2n}{z} + 6\frac{4n^2}{z^2} + 10\frac{8n^3}{z^3} + \cdots\right), \\
-\frac{31an^5}{1260(z-2n)^5} &= -\frac{31an^5}{1260z^5}\left(1 + 5\frac{2n}{z} + 20\frac{4n^2}{z^2} + \cdots\right), \\
&\cdots
\end{aligned}
$$

Then he subtracted these expansions from A and obtained

$$
\begin{aligned}
D_1 &= A - B = \lg z - \frac{an}{z} - \frac{an^2}{2z^2} - \frac{an^3}{3z^3} - \frac{an^4}{4z^4} - \cdots \\
&= \lg z - \lg(1 - \frac{n}{z}) = \lg(n - z).
\end{aligned}
$$

In a similar way, he found that $\log(z - 3n)$ is equal to $D_2 = B - C$, where

$$
\begin{aligned}
C &= \frac{(z-4n)\lg(z-4n)}{2n} - \frac{a(z-4n)}{2n} - \frac{an}{12(z-4n)} \\
&\quad + \frac{7an^3}{360(z-4n)^3} - \frac{31an^5}{1260(z-4n)^5} + \cdots.
\end{aligned}
$$

So, generally, one can express any of the logarithms

$$
\lg(z - 5n), \quad \cdots, \quad \lg(x + 3n), \quad \lg(x + n)
$$

by the difference of the type $D_1, D_2, \ldots$. The sum of these differences furnishes (76).

In a subsequent example, Stirling obtained the relation

$$\lg x! \;=\; \frac{1}{2}\lg 2\pi + \left(x + \frac{1}{2}\right)\lg\left(x + \frac{1}{2}\right) - a\left(x + \frac{1}{2}\right) \qquad (77)$$
$$-\frac{a}{2\cdot 12\left(x + \frac{1}{2}\right)} + \frac{7a}{8\cdot 360\left(x + \frac{1}{2}\right)^3} - \cdots$$

by setting $x = \frac{1}{2}$, $n = \frac{1}{2}$, and $z = x + \frac{1}{2}$.

Series (77) is equivalent to series (74), and its first few terms provide extremely accurate approximations to $\log x!$ for large x; however, it is divergent. I emphasize that Stirling's proof employed the usual modality of manipulation of series:[206] Asymptotic series emerged in a rather unconscious way.[207] This means that, although series theory was designed to determine convergent series, the fact that the conditions of convergence were not determined *a priori* and manipulations were unrestricted made it possible to go from convergent to divergent series. This also means that, *while today the notions of asymptotic series and convergent series are different and grounded on different definitions, in the 18th century they were not effectively distinct.* Both notions were reducible to the idea that series were infinite expressions derived from the finite analytical expressions of given quantities by formal manipulations; only under certain conditions could the values of the quantities be approximated by series that were generated by the quantities themselves. In the years that followed, Euler made this concept explicit.

---

[206] "Stirling does not perceive any difficulty" (Guicciardini [1989, 35]). In effect, none of this enables us to understand whether Stirling understood the real nature of what today is known as Stirling's series.

[207] The same thing occurred for de Moivre, who gave a similar result in his [1730, 127].

12   Maclaurin's contribution

Nowadays, the name Maclaurin in series theory is linked to the Maclaurin series and the Euler–Maclaurin summation formula. The first is merely a special case of the Taylor series, which in any case was explicitly dealt with by various mathematicians. The latter was a new result, which he derived independently and almost at the same time as Euler. Maclaurin's proofs of these results are interesting because they were part of an extreme attempt at laying the foundations of calculus on a geometrical framework.

In 1742, Maclaurin published the *Treatise of Fluxions*, a work partly written in response to an attack by Berkeley on the principles of the infinitesimal calculus.[208]   Maclaurin's aim was to obtain a secure proof of Newton's results using the ancient method of exhaustion. The treatise is divided into two books. In the first, he presented a *geometry of fluxions*: He investigated fluents and fluxions in a purely geometric way, by referring to diagrammatic representations and without using any specific signs or characters. Maclaurin thought that this would provide a complete and clear description of the theorems and would avoid the suspicion that the symbols employed in the calculus served to cover defects in the principles and the demonstrations (see Maclaurin [1742, 575]).

Maclaurin[209] provided a kinematic model of Newtonian theory. He considered motion, space, and velocity as primitive notions and gave the following definition of fluxions: "The velocity with which a quantity flows, at any term of the time while it is supposed to be generated, is called its Fluxion, which is therefore measured by the increment or decrement that would be generated in a given time by the motion, if it was continued uniformly from that term without acceleration or retardation" [1742, 57].

Maclaurin represented the measures of the fluxions geometrically, by means of segments or finite surfaces. For instance, he proved that (see Fig. 15), given the curve AKP, if KL is the tangent at the point K, then BC measures the fluxion of AB, IL measures the fluxion of the ordinate KB,[210] and KL measures the fluxion of the curve[211] (see Maclaurin [1742, 181]). Moreover, the rectangles $BCDE$ and $KILM$ measure the fluxions of the trapezoid $ABEJ$ and the rectangle $BCIK$, respectively.

This way of representing (the measures of) fluxions made it possible to consider the theorems on fluxion as geometrical theorems, which Maclaurin proved by *reductio ad absurdum* inspired by the method of exhaustion (to

---

[208]In the *Preface* of the *Treatise of Fluxions*, Maclaurin stated: "A letter published in the year 1734, under the title of *The Analyst,* first gave occasion to the ensuing treatise, and several reasons concurred to induce me to write on this subject at so great length."

[209]On Maclaurin's *Treatise of Fluxions*, see Panza [1989, 213–240] and Guicciardini [1989, 47–51].

[210]In modern terms, IL represents the derivative of the function $y(x)$ represented by the curve AKP, provided AD flows uniformly.

[211]The triangle KIL is Maclaurin's version of Leibniz's characteristic triangle.

prove that $A = P$, where A and P are geometric quantities, he proved *geometrically* that neither $A > P$ nor $A < P$).

In the second book, Maclaurin dealt with the *calculus of fluxions*. Here, he investigated quantities abstractly using the analytical symbolism[212] (obviously, he employed Newtonian symbolism, and specifically the symbolism used in *De quadratura*). His aim was to provide the analytical version of the geometric results given in the first part. He attempted to give an analytical version of the double *reductio ad absurdum* (to prove that $A = P$, where A and P are analytical quantities, he proved *analytically* that neither $A > P$ nor $A < P$). In this context, he often made remarkable use of the technique of algebraic inequalities whose influence on later mathematics should not be underestimated. For example, in order to demonstrate that if the fluxion of A is $\dot{A} = a$, then the fluxion of A^n is $\dot{A}^n = naA^{n-1}$, Maclaurin stated that if one supposed that the fluxion of A^n to be greater than naA^{n-1} or less than naA^{n-1}, then one arrived at a contradiction. Indeed, he first showed that, if A assumes the values[213] $A - u$, A, $A + u$, then the fluxion $\dot{A}^n$ of A satisfies the condition

$$A^n - (A - u)^n < \dot{A}^n < (A + u)^n - A^n. \qquad (78)$$

If the fluxion of A^n is greater than naA^{n-1}, then it is equal to $naA^{n-1}+\sigma$, where $\sigma > 0$. If

$$o = \sqrt[n-1]{A^{n-1} + \frac{\sigma}{nA}} - A,$$

then one obtains

$$naA^{n-1} + \sigma = na(A + o)^{n-1}.$$

Hence, the ratio between $\dot{A}^n$ and $\dot{A}$ is $n(A + o)^{n-1}$. Suppose now that u is any increment of A less than o; according to Maclaurin, since a is to u as $n(A + o)a$ to $n(A - o)u$, it follows[214] that if the fluxion of A should be represented by u, the fluxion of A^n would be represented by $n(A + o)^{n-1}$, which is greater than $nu(A + u)^{n-1}$. Since

$$\frac{(A + u)^n - A^n}{u} < n(A + u)^{n-1},$$

---

[212]He explicitly referred to "quantities considered abstractly, or as represented by general characters in algebra" [1742, 575]. In the second book, fluxions are defined in the following way: "By the fluxions of quantities we shall therefore now understand any measure of their respective rates of increase or decrease, while they vary (or flow) together" [1742, 579].

[213]A and u are positive quantities.

[214]In [1742, Section 706] Maclaurin stated: "As the fluxion of quantities are any measures of the respective rates according to which they increase or decrease, [...] so it is of no importance how great or small soever those measures are [...] Therefore if the fluxions of A and B may be supposed equal to a and b, respectively, they may be likewise supposed equal to $\frac{1}{2}a$ and $\frac{1}{2}b$, or to $\frac{m}{n}a$ and $\frac{m}{n}b$."

for $A > 0$ and $u > 0$; one easily obtains $\overset{..n}{A} > (A + u)^n - A^n$, contrary to (78). Similarly, he shows that the fluxion of A^n is not less than naA^{n-1} (see Maclaurin [1742, Sections 704 and 713])

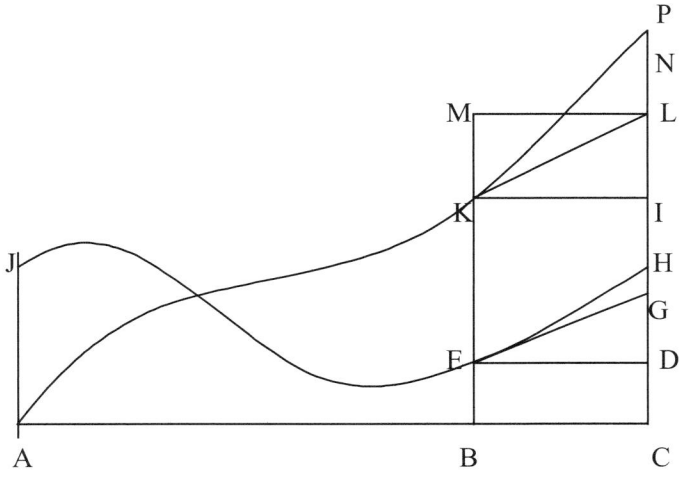

Fig. 15

I shall now move on to consider the Taylor series. In the first part of the *Treatise*, Maclaurin gave the geometrical version of the Taylor theorem as a corollary of the following proposition.

> *Proposition 20. The trapezoid $ABEJ$ and the rectangle $BCIK$ have the same area if, and only if, $BE = IL$; namely: the curve JE, whose ordinate BE is the fluxion of the curve AK, whose ordinate is KB if, and only if, the area under the curve JE is equal to the ordinate KB.*

In modern terms, if $BE = f(x)$ and $KB = F(x)$, Proposition 20 can be formulated by stating that

$$F'(x) = f(x)$$

if, and only if,

$$F(x) = \int_0^x f(x)dx.$$

One of the corollaries of Proposition 20 asserts the following:

    a. *If the arc EH is a segment, then the increment IP of the ordinate of the curve AK can be divided into two parts, $IL = BE$ and $LP = \frac{1}{2}DG$, where DG is (or, to use a more precise term, measures) the fluxion of the fluxion.*

    That is to say, let ΔF be the increment of the ordinate $F(x) = \int_0^x f(x)dx$ of the curve; if one assumes that $F(x)$ is of the type $ax^2 + bx + c$, then $\Delta F = f'(x) + \frac{1}{2}f''(x)$.

    The next corollary concerns the curves of the type $F(x) = ax^3 + bx^2 + cx + d$.

    b. *If the arc EH is a parabola, then the increment IP of the ordinate of the curve AK can be divided into three parts, $IL = BE$, $LN = \frac{1}{2}DG$, $NP = \frac{1}{3}GH$, where DG is the fluxion of BE (namely, DG is the second fluxion of BK), and GH is $\frac{1}{2}$ of the second fluxion of BE (namely, GH is the third fluxion of BK).*

In other terms, if

$$F(x) = ax^3 + bx^2 + cx + d,$$

then

$$\Delta F = f'(x) + \frac{1}{2}f''(x) + \frac{1}{3}\cdot\frac{1}{2}f'''(x).$$

    Maclaurin then generalized these corollaries. By assuming that the curve AKP could be expressed by a polynomial of degree n or by an infinite polynomial, he stated:

    [T]he increment of a fluent can be approximated continually by adding continually together the right line that measure the first fluxion [of BK,] ... $\frac{1}{2}$ of that which measures the second fluxion of the ordinate, $\frac{1}{6}$ of that which measures its third fluxion, $\frac{1}{24}$ of that which measures its fourth fluxion, and so on (Maclaurin. [1742, 223–224])

    In the second volume of the *Treatise*, Maclaurin, following Newton, used series to solve differential equations or, to use his terminology, to find the fluent. He stated:

    When a fluent cannot be represented exactly by in algebraic terms, it should then be expressed by a convergent series. (Maclaurin [1742, 604])

In this context, he also provided an analytical proof of the Taylor theorem that reproduced Newton's and Stirling's proofs; it corresponds exactly to the above geometrical proof. Maclaurin assumed that a function could be expanded into a power series $A + Bx + Cx^2 + \ldots$ and then determined the coefficients A, B, C by means of step-by-step differentiation (see Maclaurin [1742, 610–612]).

Maclaurin's interest in the use of convergent series to express a fluent led him to examine the problem of the velocity of convergence. Indeed, he noted that "when an area or a fluent is reduced to a series ... these series, in some cases, converge at so slow a rate as to be of little use for finding the area" [1742, 670]. As an example, Maclaurin considered the expansion of $y = \frac{1}{1+x}$ and stated:

> [T]his series converges so that the sum of the first 1000 terms of it was found deficient from the true value of the area in the fifth decimal; and other examples similar to this might be brought, wherein the area may be more easily computed by the inscribed polygons than from series. Some further artifice is therefore necessary in order to compute the areas in such cases. [1742, 672]

Thus, he determined a formula that might "be of use for this purpose": the Euler–Maclaurin summation formula, which he formulated so:

> *Suppose that the base $AP = z$ [see Fig. 16], the ordinate $PM = y$, and the base being supposed to flow uniformly, let $\dot{z} = 1$. Let the first ordinate AF represented by a, $AB = 1$, and the area $ABEF = A$. Since A is the area generated by the ordinate y, so let B, C, D, E, etc. represent the areas upon the same base AB, generated by the respective ordinates*

$$\dot{y}, \ \ddot{y}, \ \dddot{y}, \ \ddddot{y}, \ etc.$$

> *Then*

$$AF = a = A - \frac{B}{2} + \frac{C}{12} - \frac{E}{740} + \frac{G}{30240} - etc. \qquad (79)$$

(Maclaurin [1742, 672]).

In order to obtain this formula, Maclaurin [1742, 672–673] applied the

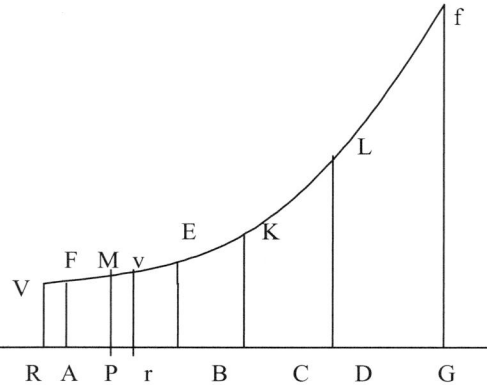

Fig. 16

Maclaurin series repeatedly. In modern symbols:

$$A = \int_0^1 y(z)dz = y(0) + \frac{1}{2}y'(0) + \frac{1}{6}y''(0) + \frac{1}{24}y'''(0) + \frac{1}{120}y^{(4)}(0) + \dots,$$

$$B = \int_0^1 y'(z)dz = y(1) - y(0) + y'(0) + \frac{1}{2}y''(0) + \frac{1}{6}y'''(0) + \frac{1}{24}y^{(4)}(0) + \dots,$$

$$C = \int_0^1 y''(z)dz = y'(1) - y'(0) = y''(0) + \frac{1}{2}y'''(0) + \frac{1}{6}y^{(4)}(0) + \dots,$$

$$D = \int_0^1 y'''(z)dz = y''(1) - y''(0) = y'''(0) + \frac{1}{2}y^{(4)}(0) + \dots,$$

$$E = \int_0^1 y^{(4)}(z)dz = y'''(1) - y'''(0) = y^{(4)}(0) + \dots,$$

Maclaurin rearranged these equations, "exterminated" (to use his term) $y'(0)$, $y''(0)$, $y'''(0)$, $y^{(4)}(0)$, ... from the first equation (namely, he made successive substitutions starting from the last written equation). In this way, he obtained

$$y(1) \;=\; a = \int_0^1 y(z)dz - \frac{y(1) - y(0)}{2} + \frac{y'(1) - y'(0)}{12}$$
$$- \frac{y''(1) - y''(0)}{740} + \frac{y'''(1) - y'''(0)}{30240} - \dots$$

By repeating the same reasoning for $y(2)$, $y(3)$, ..., he derived

$$\sum_{i=0}^{n-1} y(i) = \int_0^n y(z)dz - \frac{y(n)-y(0)}{2} + \frac{y'(n)-y'(0)}{12}$$

$$-\frac{y''(n)-y''(0)}{740} + \frac{y'''(n)-y'''(0)}{30240} - \cdots$$

(80)

This is the Euler–Maclaurin summation formula in one of its asymmetrical forms.[215]

Series (80) is not necessarily convergent, but Maclaurin did not consider this fact. Like Stirling, Maclaurin came up against the problem of divergent series while he manipulated series to improve their convergence. Even though his aim was to estimate an area, the formal methodology he used led him unintentionally to stumble on divergent series.

---

[215] Maclaurin also gave some variants of (80); see [1742, 292–293].

13 The young Euler between innovation and tradition

Series were one of Euler's favourite topics. In his *Opera omnia* there are three volumes, the last of which was published in two parts, which are devoted to series (totaling more than 1800 pages). In reality, many other articles (which were not explicitly classified as belonging to series theory by the editors of *Opera omnia*) as well as his textbooks contain much material concerning series. The extent of his results is impressive and his influence on 18th-century mathematics was enormous.

The treatment of Euler's contributions is subdivided into various chapters in Part II and also Part III. In this chapter, I examine the main findings of Euler's early work which appeared in issues 5 to 7 of the *Transactions of the Saint-Petersburg Academy*. (I explicitly note that these issues were dated 1730–1731, 1732–1733 and 1734–1735, respectively, but their actual publication occurred several years later, from 1738 to 1740.)

13.1 The search for the general term

Euler's first published article about series concerns the problem of finding the general term of a series. This problem had been tackled by various mathematicians in the 1720s. In Chapter 10 we saw that Daniel Bernoulli determined the general term of a recurrent series and applied this result to an approximate determination of the roots of equations.[216] In *Specimen methodi ad summas serierum* [1720] and *De terminis generalibus serierum* [1728], Christian Goldbach expressed the general terms and the partial sums of certain series by means of finite differences. He also tried to generalize the results by examining the series derived by means of a variable "law of progression".[217] In [1728] Goldbach considered the series $\sum_{n=1}^{\infty} a_i$ given by a recurrence relation and, using the formula (59), expressed its nth term as

$$a_n = a_1 + \frac{n-1}{1}\Delta a_1 + \frac{(n-1)(n-2)}{1 \cdot 2}\Delta^2 a_1$$
$$+\frac{(n-1)(n-2)(n-3)}{1 \cdot 2 \cdot 3}\Delta^3 a_1 + \dots, \qquad (81)$$

where $\Delta a_n = a_{n+1} - a_n$ and $\Delta^n a_n = \Delta^{n-1}a_{n+1} - \Delta^{n-1}a_n$. His procedure was usual for this period:[218] It consisted of giving non-integral values to n in (81) and so producing an infinite expression. Goldbach said that it

---

[216]Cf. Bernoulli [1728].

[217]"A law of progression is a formula by means of which, given one or more terms of the series we can find another term antecedent or successive"; for instance, $u_{n+1} = ku_n$ and $u_{n+2} = \frac{ku_{n+1}+ru_n}{u_{n+1}+u_n}$ are two constant laws of progression for k and r constants, whereas $u_{n+1} = nu_n$ is variable (Goldbach [1728, 164]).

[218]See, e.g., Chapters 4 and 8.

approximated the general term a_n as desired and was highly suitable for interpolation.[219] For instance, in order to interpolate $a_n = \frac{1}{n!}$ for $n = \frac{3}{2}$, he expressed $\frac{1}{n!}$ by means of (81) as

$$\frac{1}{n!} = 1 - \frac{1}{2}(n-1) + \frac{1}{6}\frac{(n-1)(n-2)}{2} + \frac{1}{24}\frac{(n-1)(n-2)(n-3)}{6} + \cdots,$$

where

$$a_1 = 1, \quad \Delta a_1 = -\frac{1}{2}, \quad \Delta^2 a_1 = \frac{1}{6}, \quad \Delta^3 a_1 = \frac{1}{24}, \quad \cdots$$

For $n = \frac{3}{2}$ he had

$$\frac{1}{\frac{3}{2}!} = 1 - \frac{1}{2 \cdot 2} - \frac{1}{6} \cdot \frac{1}{2 \cdot 4} + \frac{1}{24} \cdot \frac{1}{2 \cdot 4 \cdot 6} + \cdots$$

Goldbach also observed that if one interpolated the sequence $n!$ for $n = \frac{3}{2}$, (81) provided the series

$$1 + \frac{1}{2} - \frac{3}{8} + \frac{11}{6} - \frac{265}{128} + \cdots,$$

which is divergent. He thought, however, that one could assume

$$\frac{3}{2}! = \left(\frac{1}{\frac{3}{2}!}\right)^{-1} = \frac{1}{1 - \frac{1}{2 \cdot 2} - \frac{1}{6} \cdot \frac{1}{2 \cdot 4} + \frac{1}{24} \cdot \frac{1}{2 \cdot 4 \cdot 6} + \cdots}.$$

By using (81), Goldbach simplified the procedure of interpolation that had enabled Wallis to obtain $4/\pi$ as infinite products.

This sort of problem attracted the attention of the young Euler, who at once obtained several important results. In a letter dated October 13, 1729 (see Fuss [1743, 1:3–7]), the first in a long-lasting correspondence, Euler communicated to Goldbach the formula

$$n! = \prod_{k=1}^{\infty} \frac{k^{1-n}(k+1)^n}{k+n}, \qquad (82)$$

which he was about to publish in his *De progressionibus transcendentibus* [1730–31a] and asked him to give his opinion on it.

As Euler made clear, however, he did not consider satisfactory the expression of the general term of a series by means of another series. He wrote to Goldbach (Fuss [1743, 1:4]) stating that, if one expresses the general term of a series by means of other series, then the intermediate terms, i.e., those

---

[219]On the problem of interpolation, see Chapter 1, p. 14 and Chapter 4.1, p. 57.

with a nonintegral index, are determined only approximately. For this reason, he stopped treating the matter using series and became interested in a method that would enable him to determine the real (and not approximate) intermediate terms. Furthermore, in *De progressionibus transcendentibus* [1730–31a, 3], he stated that he chose not to dwell upon (82) because he already had more suitable ways to express the nth term of the hypergeometric sequences $n!$. So doing, he posed the so-called problem of integration: to express the general term of a series by an integral, namely, by a formula of the kind[220]

$$\int_0^b p(x, n)dx.$$

Euler explained that the function p depends on x and certain constants (of which one is n) and that the integration of $p(x, n)$ from 0 to the real number b yields a "function of the index n and constant quantities," namely, the general term (Fuss [1843, 1:12]). At this point, two observations are necessary.

First, Euler regarded the notions of the general term and the function as formulas. In [1730–31a] he offered this definition of the general term:

> *A general term is a formula that consists of constant quantities or any other quantities like n, which gives the order of terms; thus, if one wishes the third term, 3 can be set in the place of n.* (Euler [1730–31a, 4])

Second, Euler considered infinite expressions as unsuitable for providing general terms. In *De summatione* and in his letter to Goldbach, Euler conceived (82) only as a tool for computing the terms of series approximately. Euler suggested that integration provided the "true" result, since integration was interpreted geometrically as a quadrature, i.e., the result was exact insofar as it was geometrically conceived.

Euler's method for solving the problem of integration involved considering a certain integral and finding the series whose general term corresponded to it. He later [1732–33] termed such a procedure a *synthetic* method, since *a priori* knowledge of the result was supposed and one simply verified that a certain integral (which was already known) expressed the general term of

---

[220]It should be emphasised that when, in the Dedication of *Arithmetica infinitorum*, Wallis tackled the problem of interpolating series of numbers, such as 1, 6, 30, 140, 630, ... [namely, $a_0 = 1$, $a_n = a_{n-1}(4 + \frac{2}{n})$, $n = 1, 2, 3, \ldots$], the absence of integral formula at that time led him to use the complicated procedures I mentioned in Chapter 1. Following Euler, one may state that the sequence 1, 6, 30, 140, 630, ... can be interpolated by means of the integral $\frac{1}{\int_0^1 (x-x^2)^n dx}$, $n = 0, 1, 2, \ldots$ However, I would underline that the translation of Wallis's procedure by means of integral formulas strains it.

a series.[221] For example, Euler [1730–31a, 7–12] considered

$$\int_0^1 x^q (1-x)^n dx.$$

By expanding $(1-x)^n$, he derived

$$\int_0^1 x^q (1-x)^n dx = \int_0^1 x^q \sum_{h=0}^{\infty} \binom{n}{h}(-x)^h dx = \int_0^1 \sum_{h=0}^{\infty} \binom{n}{h}(-1)^h x^{h+q} dx$$

and, by a term-by-term integration, he obtained

$$\int_0^1 x^q (1-x)^n dx = \sum_{h=0}^{\infty} \binom{n}{h} \int_0^1 (-1)^h x^{h+q} dx = \sum_{h=0}^{\infty} \binom{n}{h} \frac{(-1)^h}{q+h+1}. \quad (83)$$

Euler verified that the right-hand side of (83) equals $\frac{n!}{(q+1)(q+2)\ldots(q+n+1)}$ for $n = 0,\ 1,\ 2,\ 3$ and concluded that $\int_0^1 x^q (1-x)^n dx$ is the nth term of the series

$$a_n = \frac{n!}{(q+1)(q+2)\ldots(q+n+1)}.$$

In order to determine an integral expression of $n!$, Euler multiplied

$$\int_0^1 x^q (1-x)^n dx = \frac{n!}{(q+1)(q+2)\ldots(q+n+1)}$$

by $(q+n+1)$ to get

$$(q+n+1)\int_0^1 x^q (1-x)^n dx = \frac{n!}{(q+1)(q+2)\ldots(q+n)}.$$

If $q = f/g$, then

$$a_n = \frac{f+(n+1)g}{g^{n+1}} \int_0^1 x^{f/g}(1-x)^n dx = \frac{n!}{(f+g)(f+2g)\ldots(f+ng)}, \quad (84)$$

which becomes

$$\frac{1}{0^{n+1}} \int_0^1 x^{1/0}(1-x)^n dx = n!, \quad (85)$$

for $f = 1$ and $g = 0$.

---

[221] See Section 13.2.

At this point Euler sought the value of $\frac{1}{0^{n+1}} \int_0^1 x^{1/0}(1-x)^n dx$.[222] The problem is similar to problems such as "determining the value" of fractions $\frac{f(x)}{g(x)}$, when they assume the form $0/0$ for a certain x, which was examined at the end of Chapter 7, p. 111. Euler's conception did not differ from the one illustrated in that chapter. Indeed, he observed that the previous result also holds if we replace x by any function $f(x)$ [provided $f(0) = 0$ and $f(1) = 1$] in the above integrals (and obviously with df in place of dx). For $f(x) = x^{\frac{g}{f+g}}$, the left-hand side of (84) becomes

$$\frac{f + (n+1)g}{g^{n+1}} \int_0^1 x^{\frac{g}{f+g}} \left(1 - x^{\frac{g}{f+g}}\right)^n dx,$$

from which

$$\int_0^1 \frac{\left(1 - x^0\right)^n}{0^n} dx = n!,$$

for $f = 1$, $g = 0$. Euler interpreted $\frac{1-x^0}{0}$ as

$$\left(\frac{1 - x^z}{z}\right)_{z=0}$$

and found that it is equal to $-\log x$, by applying the so-called l'Hôpital rule. Consequently, we have the formula

$$n! = \int_0^1 (-\log x)^n dx, \tag{86}$$

which enables us to attribute a value to sequence $a_n = n!$ for n nonintegral. Historically, (86) was the first integral expression of the function[223]

$$\Pi(x) = x! = \int_0^\infty t^x e^{-t} dt.$$

From the modern viewpoint, however, its derivation is problematic as is the formulation of the question (the interpolation) that leads to (86) or (82). Furthermore, the final result (86) was not viewed as a function in its own right; it was seen merely as a tool for evaluating and representing $n!$.

In *De summatione innumerabilium progressionum*, Euler also verified (86) in certain particular cases and noted that (86) did not allow for an easy calculation of the value of $n!$. It is possible, however, to reduce the

---

[222] The problem is similar to problems such as "determining the value" of fractions $\frac{f(x)}{g(x)}$, when they assume the form $0/0$ for a certain x, which was examined at the end of chapter 7, p.7.2. Euler's conception did not differ from the one illustrated in that chapter.

[223] The factorial function $\Pi(x) = x! = \int_0^\infty t^x e^{-t} dt$ is related to the gamma function $\Gamma(x) = \int_0^\infty t^{x-1} e^{-t} dt$ by the equation $x\Gamma(x) = \Pi(x)$.

calculation of $\int_0^1 (-\log x)^n dx$ to the quadrature of certain algebraic curves by formulas of the type

$$\int_0^1 (-\log x)^n dx = \frac{(f+g)(f+2g)\dots(f+(n+1)g)}{g^{n+1}} \int_0^1 x^{f/g}(1-x)^n dx,$$
(87)

i.e., in more modern notation,

$$\Gamma(n+1) = \frac{(f+g)(f+2g)\dots(f+(n+1)g)}{g^{n+1}} B\left(\frac{f}{g}+1, n+1\right),$$

where $B(x,y)$ is the beta function.[224] The latter formulas enable us to compute some values of $n!$ for noninteger n, if we assume the quadrature of the given algebraic curve to be known (Euler [1730–31a, 13–14]).

In *De progressionibus transcendentibus*, Euler found other relations involving these integrals. For instance, he wrote (87) in the form

$$\prod_{i=1}^n (f+ig) = \frac{g^{n+1}\int_0^1 (-lgx)^n dx}{(f+(n+1)g)\int_0^1 x^{\frac{f}{g}}(1-x)^n dx}$$

and obtained

$$\frac{\prod_{i=1}^n (f+ig)}{\prod_{i=1}^n (h+ik)} = \frac{g^{n+1}(h+(n+1)k)\int_0^1 x^{\frac{h}{k}}(1-x)^n dx}{k^{n+1}(f+(n+1)g)\int_0^1 x^{\frac{f}{g}}(1-x)^n dx},$$
(88)

where n is a whole number (see Euler [1730–31a, 13–15]).

13.2   Analytical and synthetical methods in series theory

In his *Methodus generalis summandi progressiones* [1732–33] Euler made a remarkable observation about analytical and synthetic methods. He stated:

> Last year I proposed a method for summing innumerable progressions,[225] which not only covers the series having an algebraic sum but also provides the sum of the series dependent on the quadrature of curves, which cannot be summed algebraically. I then used a synthetic method; indeed taking any general formula I asked myself what the series could be whose sums are expressed by that formula. In this way I obtained several series, whose sums I had been able to assign. In order to make it easier and clearer to find the sum of any proposed series, provided this can be achieved, I communicate this analytical method, which allows the discovery of the summation term by the nature of the series. (Euler [1732–33, 42])

[224]It is defined by $B(m+1,n+1) = \int_0^1 x^m(1-x)^n dx$.
[225]He refers to *De summatione innumerabilium progressionum* [1730–31b].

The synthetic method used in *De summatione innumerabilium progressionum* [1730–31b] started with a *determinate* formula [specifically, Euler considers an integral formula of the kind $\int_c^b p(n,x)dx$] and derived a (finite or infinite) series whose sum term $S(n,x)$ was equal to $\int_c^b p(n,x)dx$ [$\int_c^b p(\infty,x)dx = S(\infty,x)$ was a special case]. Euler started from the result, supposed to be entirely determined, and arrived at the summation term. In the analytic method of *Methodus generalis* [1732–33], to which the previous quotation refers, Euler supposed that an *unknown* formula $f(x)$ was the sum of the series and, operating on $f(x)$, tried to derive $\sum_{n=0}^{\infty} f_n(x) = f(x)$. In both [1730–31b] and [1732–33], Euler operated on the sum $f(x)$, deriving certain results concerning the series $\sum_{n=0}^{\infty} f_n(x)$, but in the synthetic method the function sum was already known, guessed at in some way, while for the analytic method the sum $f(x)$ was unknown.

In [1732–33] Euler adopted Pappus's classical terminology[226] to the analysis of the infinite. The synthetic method of *De summatione* yields a synthetic solution (in Pappus's sense) of the converse problem [i.e., given a function $f(x)$ find the series whose summation term is $f(x)$], which, read backwards, becomes the solution to the direct problem. With regard to the theory of series, the use of Pappus's terminology implies the following elaborate scheme:

		To sum			To develop	
Analytical procedure	AS			AD		
Synthetic procedure		S_1S	S_2S		S_1D	S_2D

In the above scheme:

- AS is the analytical method of the sum. One operates upon the indeterminate object $f(x)$,[227] satisfying the condition *of being the sum of the given series* $\sum_{n=0}^{\infty} f_n(x)$ (this is the method used in [1732–33]).

- AD is the analytical method of development. Given $f(x)$, one operates upon the indeterminate series[228] $\sum_{n=0}^{\infty} f_n(x)$, which satisfies the condition of being the development of $f(x)$, and obtains a known series which is the development of $f(x)$.

---

[226]Such terminology already seems residual and relative to a phase in which Euler's formalism was not yet entirely developed. It no longer appears in his later papers about series, where the term "analytical" has a different meaning (see Chapter 18).

[227]Usually, the indeterminate object on which one operates in any application of the analytical method is denoted by a symbol. For instance, we can denote the unknown sum of $\sum_{n=0}^{\infty} f_n(x)$ simply by the symbol $f(x)$. This fact is irrelevant to my argument. Of course, we can directly operate by means of $\sum_{n=0}^{\infty} f_n(x)$, which is always an unknown object as long as the sum of the series is unknown.

[228]Even if the series is indeterminate, its form is assumed to be of a particular kind, usually a power series.

- $S_1 S$ is the synthetic method of the sum. The series $\sum_{n=0}^{\infty} f_n(x)$ is given, the sum $f(x)$ is guessed at in some way, and one shows $f(x) = \sum_{n=0}^{\infty} f_n(x)$. As Euler put it in [1732–33, 42], to sum a series, one needs to compare it with known formulas and to investigate whether the series is derived from one of them (this is the method used in [1730–31b]).

- $S_1 D$ is the synthetic method of development. One guesses at the expansion of the given function $f(x)$ in some way and proceeds to derive $f(x) = \sum_{n=0}^{\infty} f_n(x)$.

- $S_2 S$ is the synthetic method of the sum. The series is given, and one derives $f(x)$.

- $S_2 D$ is the synthetic method of development. The function $f(x)$ is given and one derives $\sum_{n=0}^{\infty} f_n(x)$.

In *De summatione innumerabilium progressionum*,[229] Euler applied the synthetic method of the sum as follows. Since

$$\int_0^1 \frac{1 - x^n}{1 - x}\, dx = 1 + \frac{1}{2} + \frac{1}{3} + \ldots + \frac{1}{n}$$

for n an integer, this integral can be used to interpolate the series

$$1 + \frac{1}{2} + \frac{1}{3} + \ldots + \frac{1}{n}$$

[since it makes up its nth term] and to sum a finite number of the terms of the harmonic series (since it expresses the summation term of $\sum_n \frac{1}{n}$). More generally, Euler denoted a function of x by $P(x)$ and considered

$$\int_0^k \frac{1 - P^n(x)}{1 - P(x)}\, dx = k + \int_0^k P(x)dx + \int_0^k P^2(x)dx + \ldots + \int_0^k P^{n-1}(x)dx$$

to be the summation term of the series

$$a_n = \int_0^k P^n(x)dx.$$

In particular, he studied $P(x) = \frac{x^\alpha}{a^a}$ and reduced

$$\sum_{n=1}^{\infty} \frac{b^{(n-1)i+1}}{c + (n-1)a}$$

---

[229] See Euler [1730–31b, 26–30].

to this case. If $b = 1$, the latter becomes

$$\sum_{n=1}^{\infty} \frac{1}{c + (n-1)a},$$

whose summation term is expressed by

$$\int_0^1 \frac{1 - x^{na/c}}{(1 - x^{a/c})c}dx.$$

Euler then[230] generalized this result, stating, for example, that the sum term of the series

$$\sum_{n=1}^{\infty} \frac{c}{(c + (n-1)a)\,(c(\alpha + 2) + (n-1)a)}$$

is

$$\int_0^1 x^\alpha dx \int_0^1 \frac{1 - x^{na/c}}{(1 - x^{a/c})c}dx.$$

He noted that

$$\frac{1 - x^{na/c}}{1 - x^{a/c}} = \frac{1}{1 - x^{a/c}}$$

if $n = \infty$ and $(0 <)x < 1$. This equality was also considered valid when $x = 1$ because of the continuity of the quantity x, and therefore Euler derived the sums of the series whose summation terms had been determined earlier. For instance,[231]

$$\sum_n \frac{1}{c + (n-1)a} = \int_0^1 \frac{1}{(1 - x^{a/c})c}dx$$

and

$$\sum_{n=1}^{\infty} \frac{c}{(c + (n-1)a)\,(c\alpha + (n-1)a)} = \int_0^1 x^{\alpha-2}dx \int_0^1 \frac{1}{(1 - x^{a/c})c}dx.$$

Euler used such results to transform slow convergent series into fast convergent series in his [1730–31b, 38–41]. For instance,

$$1 + \frac{1}{4} + \frac{1}{9} + \ldots = 1 + \frac{1}{8} + \frac{1}{36} + \frac{1}{128} + \frac{1}{400} + \ldots + (\log 2)^2$$
$$= 1.164481 + 0.480453 = 1.644934.$$

[230] See Euler [1730–31b, 34–36].
[231] See Euler [1730–31b, 36–37].

In contrast to the synthetic method usually used in *De progression-ibus transcendentibus* [1730–31] and *De summatione innumerabilium pro-gressionum* [1730–31b], in *Methodus generalis* [1732–33] Euler proposed two analytical techniques for summing (finite or infinite) series. One of these is the summation formula which will be examined in Chapter 14. The other involves seeking the summation term (expressed by an integral or algebraic expression) by manipulating the series through appropriate (algebraic or differential) operations on its terms which reduce the given series to either another series (which can be more easily summed) or again to itself (which yields an equation providing the sum) [1732–33, 44]. Two examples will clarify this technique. The first does not involve the calculus. Euler set[232]

$$S(m, x) = \sum_{n=1}^{m} x^{(n-1)b+a}$$

and, by the elementary technique still used today, derived

$$S(m, x) = \frac{1 - x^{mb}}{1 - x^b} x^a.$$

From this, under the condition $0 < x < 1$, he obtained

$$S(\infty, x) = \frac{1}{1 - x^b} x^a.$$

In the second example, he set

$$S(m, x) = \sum_{n=1}^{m} \frac{x^{(n-1)\alpha+\beta}}{an + b}$$

and manipulated this equation as follows. He multiplied it by px^r (p and r appropriate constants), differentiated it, and rearranged and summed the finite geometric progression on the right-hand side. He then integrated both sides of the final equation to obtain

$$S(m, x) = \frac{\beta}{\alpha} x^{(\alpha a - a\beta - b\beta)/a} \int x^{(\alpha\beta + b\beta - \alpha)/a} \frac{1 - x^{m\beta}}{1 - x^\beta} dx. \tag{89}$$

He imposed the condition $S(m, 0) = 0$. When $m = \infty$, (89) becomes

$$S(\infty, x) = \frac{\beta}{\alpha} x^{(\alpha a - a\beta - b\beta)/a} \int x^{(\alpha\beta + b\beta - \alpha)/a} \frac{1}{1 - x^\beta} dx$$

since Euler was concerned with calculating $S(\infty, 1)$ and could limit himself to $0 < x < 1$.

---

[232] See footnote no. 227.

13.3 The manipulation of the harmonic series and infinite equations

The last examples in the previous section showed that, in *De progressionibus transcendentibus* and *Methodus generalis*, Euler obtained results about infinite series by imposing limitations on the range of variables. He calculated the sum of (finite or infinite) numerical series by using power series and setting $x = 1$. In modern terms, he calculated

$$\lim_{x \to 1^-} S(m, x)$$

if the series is finite, and

$$\lim_{x \to 1^-} S(\infty, x)$$

[having previously determined $S(\infty, x)$ for $0 < x < 1$] if the series is infinite. Euler considered the series only for $0 < x < 1$. For this reason, the integrals are calculated in $(0, 1)$ where x^n is infinitesimal for $n = \infty$ and can be neglected.

Many other examples prove that Euler had mastered the quantitative concept of the sum and was aware of the problem of convergence, following the approach of earlier mathematicians. This is particularly evident in *De progressionibus harmonicis observationes* [1734–1735b], where Euler gave a new proof of the divergence of the harmonic series. The proof is extremely interesting because it was based on a principle that can be considered a forerunner of Cauchy 's criterion for convergence. Euler stated:

> A series which, after its continuation to the infinite, has a finite sum, does no experience any increase, even if it is continued twice as far, but that which is added in thought after infinity is reached will actually be infinitely small. For, otherwise, the sum of the series continued *in infinitum* would be unbounded and, consequently, not finite. From this follows that, if what originates from the continuation beyond an infinitesimal term is of finite magnitude, then the sum of the series must necessarily be infinite. Thus, from that principle we shall be able to judge whether the sum of any given series is infinite or finite. [1734–35b, 88]

The context makes it clear that Euler referred to series of positive terms and that the statement "even if it is continued twice as far" must not be taken literally: The number 2 is an arbitrary exemplification and can be changed into any natural or fractional number k, as Euler himself did in applying the principle to

$$\sum_{n=0}^{\infty} \frac{c}{a + bn^p}.$$

In a modern notation, the principle therefore states:

If $\sum_{n=1}^{\infty} a_n$ ($a_n > 0$) has a finite sum, then

$$\sum_{n=i+1}^{h} a_n$$

is infinitesimal for infinitely large h and i; vice versa, if $\sum_{n=i+1}^{h} a_n$ is a finite quantity for appropriate h and i, then

$$\sum_{n=1}^{\infty} a_n$$

has an infinite sum.

It is evident that Euler had a clear idea of convergence; however, following previous mathematicians, he thought the expression $\sum_{n=1}^{\infty} a_n$ could be manipulated even if the series was not convergent. Here are two remarkable examples. The first is taken from the same paper, *De progressionibus harmonicis*, where Euler expounded the above principle. It shows how Euler obtained the approximate value of the constant named today after Euler and Mascheroni by using divergent series. In *De progressionibus harmonicis observationes* [1734–35b, 90], Euler had found the relation

$$\sum_{j=1}^{i} \frac{1}{j} = C + \log(1+i), \text{ for } i = \infty, \tag{90}$$

where C is a constant.[233] Hence, he obtained

$$\log n = \log \frac{ni+1}{i+1} = \log(ni+1) - \log(i+1) = \sum_{j=1}^{ni} \frac{1}{j} - \sum_{j=1}^{i} \frac{1}{j}, \text{ for } i = \infty.$$

By giving particular values to n and rearranging the harmonic series, he derived

$$
\begin{aligned}
\log 2 &= 1 - \frac{1}{2} + \frac{1}{3} - \frac{1}{4} + \frac{1}{5} - \frac{1}{6} + \frac{1}{7} - \frac{1}{8} + \cdots, \\
\log 3 &= 1 + \frac{1}{2} - \frac{2}{3} + \frac{1}{4} + \frac{1}{5} - \frac{2}{6} + \frac{1}{7} + \frac{1}{8} - \cdots, \\
\log 4 &= 1 + \frac{1}{2} + \frac{1}{3} - \frac{3}{4} + \frac{1}{5} + \frac{1}{6} + \frac{1}{7} - \frac{3}{8} + \cdots, \\
&\text{etc.}
\end{aligned}
$$

---

[233]On the procedures Euler used to determine Equation (90), see Chapter 14, p. 167.

Moreover, by expanding $\log(1 + x)$, he obtained

$$\frac{1}{j} = \log\left(1 + \frac{1}{j}\right) + \sum_{h=2}^{\infty} \frac{(-1)^h}{hj^h}$$

Hence,

$$\sum_{j=1}^{i} \frac{1}{j} = \sum_{j=1}^{i} \left(\log\left(\frac{1+j}{j}\right) + \sum_{h=2}^{\infty} \frac{(-1)^h}{hj^h}\right).$$

Euler freely rearranged and obtained

$$\sum_{j=1}^{i} \frac{1}{j} = \log(1 + i) + \sum_{h=2}^{\infty} \left[\frac{(-1)^h}{h} \sum_{j=1}^{i} \frac{1}{j^h}\right].$$

He observed that the series $\sum_{j=1}^{i} \frac{1}{j^h}$ "are convergent and, if approximately summed, they generate

$$1 + \frac{1}{2} + \frac{1}{3} + \ldots + \frac{1}{i} = \log(1 + i) + 0.577218"$$

(see Euler [1734–35b, 94]).

The second example concerns Euler's solution of the Basel problem in *De summis serierum reciprocarum* [1734–35a]. In this paper, Euler succeeded in summing the series $\sum_{n=1}^{\infty} \frac{1}{n^{2k}}$ by dealing with infinite algebraic equations as if they were finite. This was simply another application of the principle of infinite extension, although it came in for some criticism, as we shall see in Section 19.1. Euler observed that the sine $PM = y$ of the arcs $AM = s$ (see the circle in Fig. 17, where $AC = 1$) could be expressed as

$$y = \sin s = s - \frac{1}{3!}s^3 + \frac{1}{5!}s^5 - \frac{1}{7!}s^7 + \ldots$$

Hence, he obtained the equation

$$0 = 1 - \frac{s}{y} + \frac{1}{3!y}s^3 - \frac{1}{5!y}s^5 + \frac{1}{7!y}s^7 - \ldots \tag{91}$$

He factored

$$1 - \frac{s}{y} + \frac{1}{3!y}s^3 - \frac{1}{5!y}s^5 + \frac{1}{7!y}s^7 - \ldots$$

as

$$\left(1 - \frac{s}{A_1}\right)\left(1 - \frac{s}{A_2}\right)\left(1 - \frac{s}{A_3}\right)\cdots,$$

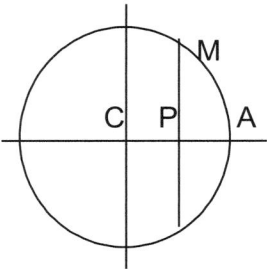

Fig. 17

where A_1, A_2, A_3 are the roots of Equation (92). At this point Euler considered some well-known properties of the roots of finite equations. Indeed, it is possible to prove that if one sets

$$S_{-k} = \sum_{i=o}^{n} \alpha_i^{-k},$$

where α_i are the roots of the equation

$$a_0 + a_1 x^1 + a_2 x^2 + \ldots + a_n x^n = 0,$$

then

$$a_0 S_{-1} + a_1 = 0,$$
$$a_0 S_{-2} + a_1 S_{-1} + 2a_2 = 0,$$
$$\ldots \qquad\qquad (92)$$
$$a_0 S_{-n} + a_1 S_{-n+1} + \ldots + a_{n-1} S_{-1} + na_0 = 0,$$
$$\ldots .$$

Euler considered the series $1 - \frac{1}{3!y} s^2 + \frac{1}{5!y} s^4 - \ldots$ as an infinite polynomial

$$a_0 + a_1 x^1 + a_2 x^2 + \ldots + a_n x^n + \ldots$$

and applied formulas (92). Hence,

$$S_{-1} = -\frac{a_1}{a_0} = \frac{1}{y},$$
$$S_{-2} = \frac{a_1 S_{-1} + 2a_2}{a_0} = \frac{1}{y^2},$$
$$\ldots .$$

Then he set $y = 1$ so that Equation (91) became

$$0 = 1 - \frac{s}{1} + \frac{1}{3!}s^3 - \frac{1}{5!}s^5 + \frac{1}{7!}s^7 - \ldots = 1 - \sin s,$$

The roots A_n of the equation $1 = \sin s$ are

$$\frac{\pi}{2}, \quad -\frac{\pi}{2}, \quad \frac{3\pi}{2}, \quad -\frac{3\pi}{2}, \quad \frac{5\pi}{2}, \quad -\frac{5\pi}{2}, \quad \frac{7\pi}{2}, \quad -\frac{7\pi}{2}, \quad \ldots;$$

therefore, the sum

$$S_{-1} = \sum_{n=1}^{\infty} \frac{1}{A_n} = 2\sum_{n=0}^{\infty} (-1)^n \frac{2}{(2n+1)\pi}$$

is equal to $-\frac{a_1}{a_0} = \frac{1}{y} = 1$; namely,

$$2\sum_{n=0}^{\infty} (-1)^n \frac{2}{(2n+1)\pi} = 1.$$

Hence,

$$\sum_{n=0}^{\infty} (-1)^n \frac{1}{2n+1} = \frac{\pi}{4}.$$

Similarly,

$$\sum_{n=1}^{\infty} \frac{1}{A_n^2} = 2\sum_{n=0}^{\infty} (-1)^n \frac{4}{(2n+1)^2\pi^2} = 1,$$

namely,

$$\sum_{n=0}^{\infty} \frac{1}{(2n+1)^2} = \frac{\pi^2}{8}.$$

Since

$$\sum_{n=1}^{\infty} \frac{1}{n^2} = \sum_{n=0}^{\infty} \frac{1}{(2n+1)^2} + \sum_{n=1}^{\infty} \frac{1}{(2n)^2},$$

he had

$$\sum_{n=1}^{\infty} \frac{1}{n^2} = \frac{4}{3}\sum_{n=1}^{\infty} \frac{1}{(2n+1)^2} = \frac{\pi^2}{6}.$$

This procedure allowed Euler to find the sum of

$$\sum_{n=0}^{\infty} (-1)^n \frac{1}{(2n+1)^{2k+1}}, \quad \sum_{n=0}^{\infty} \frac{1}{(2n+1)^{2k}}, \text{ and } \sum_{n=1}^{\infty} \frac{1}{n^{2k}}, \text{ for } k = 1, 2, 3, \ldots$$

14 Euler's derivation of the Euler–Maclaurin summation formula

Euler first mentioned the summation formula in *Methodus generalis summandi progressiones* [1732–33] and proved it, for the first time, in *Inventio summae cuiusque seriei ex dato termino generali* [1736b]. It was again called analytical, but this term had a meaning different from that given in *Methodus generalis*. In *Inventio summae*, Euler spoke of an analytical method in contrast to the geometric method used in *Methodus universalis serierum convergentium summas quam proxime inveniendi* [1736a]:

> When I gave more precise consideration to the mode of summing which I had dealt with by using by the geometrical method in the above dissertation[234] and investigated it analytically, I discovered that what I had derived geometrically could be deduced from a peculiar method for summing that I mentioned three years before in a paper[235] on the sum of series. (Euler [1736b, 108])

In *Methodus universalis*, Euler based the determination of an approximating evaluation of the sum of a series $\sum a_n$ on a geometric representation. He wrote the series as $a+b+c+d+e+$etc. and denoted the nth and $(n+1)$st terms by x and y, respectively. He considered the diagram shown in Fig. 18, where

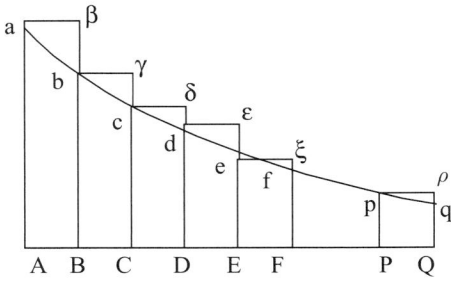

Fig. 18

$$aA = a_1, \ bB = a_2, \ cC = a_3, \ dD = a_4, \ \ldots, \ pP = a_n, \ qQ = a_{n+1},$$

and

$$AB = BC = CD = DE = \ldots = PQ = 1.$$

---

[234] *Methodus universalis serierum convergentium summas quam proxime inveniendi* [1736a].
[235] *Methodus generalis summandi progressiones* [1732–33].

Since "x is a quantity composed of n and constants," $pP = x$ provides, according to Euler, the equation between AQ and qQ that expresses the nature of the curved line ap, i.e., the equation of the curve is

$$x = a(n) = a_n.$$

Of course,

$$\sum_{i=1}^{n} a_i > \int_{1}^{n+1} a(n)dn$$

or, as Euler said,

$$s_n = \sum_{i=1}^{n} a_i > \int a(n+1)dn,$$

with the condition that the value of $\int a(t)dt$ is equal 0 at s_0.[236]

In order to improve this approximation, he observed that the curvilinear triangles $ab\beta$, $bc\gamma$,..., $pq\rho$, which have been neglected, are greater than the rectilinear triangles $ab\beta$, $bc\gamma$,..., $pq\rho$ (the curved line aq is convex, at least for large enough n). Since the sum of the areas of the rectilinear triangles $ab\beta$, $bc\gamma$,..., $pq\rho$ is $(Aa - Qq)AB : 2$, we have[237]

$$\sum_{i=1}^{n} a_i > \int a(t+1)dt + \frac{a_1}{2} - \frac{a_{n+1}}{2}.$$

Finally, Euler considered the secant bc (Fig. 19) and approximated the arc ac by an appropriate arc of a parabola. Thus, the median bm is close to the tangent to the curve, and

$$S_{ab} = \frac{1}{3}T_1 = \frac{1}{6}T_2,$$

where the area between the curved line and the segment ab is denoted by S_{ab}, and the areas of triangles abm and abn are denoted by T_1 and T_2. Since

---

[236] Euler called this inequality an upper limit of series. Similarly, he derived

$$s_n = \sum_{i=1}^{n} a_i < \int_{1}^{n+1} a(t-1)dn \ (= \int a(t)dt,$$

with the condition that the value of $\int a(t)dt$ is equal 0 at the origin), i.e., a lower limit.
[237] Analogously, he obtained

$$\sum_{i=1}^{n} a_i < \int_{1}^{n+1} a(t-1)dt - \frac{a_1}{2} + \frac{a_{n+1}}{2}.$$

$na = Aa - 2Bb + Cc,$[238] we have

$$T_2 = \frac{Aa - 2Bb + Cc}{2} = \frac{a_1 - 2a_2 + a_3}{2}$$

and

$$S_{ab} = \frac{a_1 - 2a_2 + a_3}{12}.$$

Of course, the sum of all the areas

$$S_{ab} + S_{bc} + S_{cd} + \ldots + S_{pq},$$

which is neglected in the last inequality, is approximately equal to

$$\frac{a_1 - a_2}{12} - \frac{a_{n+1} - a_{n+2}}{12}.$$

Therefore, Euler determined the approximating formula:

$$\sum_{i=1}^{n} a_i = \int a(t+1)dt + \frac{a_1}{2} - \frac{a_{n+1}}{2} + \frac{a_1 - a_2}{12} - \frac{a_{n+1} - a_{n+2}}{12}. \qquad (93)$$

In proving (93), Euler implicitly assumed the convexity of the arc ab and hence of the curve. Furthermore, the errors are small when n is large and a_n is small. He actually applied (93) to sum the first 1,000,000 terms of the

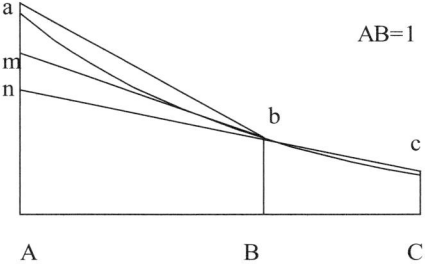

Fig. 19

harmonic series. For $n = 990,000$, (93) yields

$$\frac{1}{11} + \frac{1}{12} + \ldots + \frac{1}{1{,}000{,}000}$$

$$= \log\frac{1{,}000{,}001}{11} + \frac{1}{22} + \frac{1}{132}$$

$$- \frac{1}{144} - \frac{1}{2{,}000{,}002} - \frac{1}{12{,}000{,}012} + \frac{1}{12{,}000{,}024},$$

---

[238]Indeed, in the trapezoid $ACcn$, if a segment Bb is drawn parallel to the sides An and Cc, and bisecting AC, then $bB = \frac{An + Cc}{2}$.

and therefore

$$1 + \frac{1}{2} + \frac{1}{3} + \ldots + \frac{1}{1,000,000} = 14.392669.$$

When $n = \infty$, (93) becomes

$$\sum_{i=1}^{\infty} a_i = \int a(t+1)dt + \frac{7a_1}{12} - \frac{a_2}{12},$$

and Euler illustrated this in the context of the example:

$$\begin{aligned}
\sum_{i=1}^{\infty} \frac{1}{i^2} &= \sum_{i=1}^{10} \frac{1}{i^2} + \sum_{i=11}^{\infty} \frac{1}{i^2} \\
&= 1.549768 + \frac{1}{11} + \frac{7}{12 \cdot 121} - \frac{1}{12 \cdot 144} = 1.644920.
\end{aligned}$$

The geometrical method of *Methodus universalis* hinged on using appropriate geometrical figures; some steps of the deduction were inferred by scrutinizing these figures. The analytical method of *Inventio summae*, i.e., the summation formula, dispensed with the geometrical representation.[239] If, however, it is true that the analytical method actually dispensed with figures, this does not mean that it also lacked geometrical references. Indeed, if we look carefully at *Methodus universalis* and *Inventio summae*, we note that both papers are based upon the same principles and, in effect, it would be easy to translate *Methodus universalis* into analytical symbols. Thus, in *Methodus universalis*, the sequence a_n was viewed as a curved line whose equation is $y = a(n)$ and was assumed to be a continuous curve (both in the sense that it makes no jumps and in Euler's sense[240]) possessing a tangent at each point. In *Inventio summae*, the sequence a_n was viewed as a continuous and infinitely differentiable function $a(n)$ (in modern sense). Both articles substantially turn the study of the series $\sum a_n$ into the study of the function $a(n)$ and its integral $\int a_n dn$. As Euler put it, "I reduced summation to integration" [1736a, 101].[241]

With respect to *Methodus universalis* and *Inventio summae*, Euler stated that the analytical method was superior to the geometric one not only because it included the geometrically determined formula (93) but also because it led to an improvement of the sum so that the true result could be obtained through the addition of other terms (which were unlikely to be resolved by the geometric method). The power of the analytical method of *Inventio summae* derives from the concept of the summation term $S_x = \sum_{n=1}^{x} a_n$ as a continuous and differentiable function of the index x.

---

[239] See Chapter 7.
[240] See Chapter 18.
[241] This integral expression of the summation differs from those of the type $s_n = \int f(x,n)dx$ examined above (see Section 13.1).

   This concept appeared for the first time in *De summatione innumerabil-ium progressionum* [1730–31b, 29–30]. Euler took the summation term of the series

$$a_n = \frac{b^{(n-1)i+1}}{c + (n-1)q}$$

to be A and said that we can consider n and A as flowing quantities (*quantitates fluentes*) when n is almost infinitely greater than 1, and therefore the differential dn is to dA as 1 is to $\frac{b^{ni+1}}{c+nq}$. He thus obtained the differential equation

$$dA = \frac{b^{ni+1}}{c + nq} dn$$

whose integral gives A as a function of n. For $i = 0$, Euler actually solved this equation and found

$$A = \frac{b}{q} \log(C(c + nq)),$$

C being an indeterminate constant.

   This conception was later developed in *De progressionibus harmonicis*, where Euler considered the series

$$\sum \frac{c}{a + (i-1)b}$$

and stated that, when i increases by 1, the sum term s of the series $\frac{c}{a+ib}$ increases by $\frac{c}{a+ib}$ and therefore

$$\frac{di}{ds} = \frac{1}{\frac{c}{a+ib}}.$$

This equation furnishes the summation term

$$s = C + \frac{c}{b} \log(a + ib)$$

(Euler [1734–35b, 90]).[242]

   In *Methodus universalis* [1736a] Euler explicitly enunciated the general principle upon which the above results are based:

$$dn : ds_n = 1 : a_n, \tag{94}$$

---

[242]For $a = b = c = 1$, we have

$$\sum_{j=1}^{i} \frac{1}{j} = C + \log(1 + i) = C + \log i,$$

for $i = \infty$. In this way Euler proved that $\sum_{j=1}^{i} \frac{1}{j}$ and $\log(1 + i)$ differed for a constant, whose values he determined as described on p. 167.

i.e., $ds_n = a_n dn$ (s_n is the summation term of the series $\sum a_n$), and he derived $s_n = \int a_n dn$ from this. He assumed the validity of (94) under the condition that n is large and the increment of s_n is very small.[243]

In *Inventio summae* [1736b], Euler proved the summation formula by using the Taylor series in a decisive way. He held that a function $y(x)$ could be expanded in Taylor series "if y is given in whatever way by means of x and constants" [1736b, 109]. The general term $X = X_i$ of a series and its summation term $S(x) = \sum_{i=1}^{x} X_i$ could also be expanded in Taylor series, because "both S and X, in the case that the series is determined, are composed of x and constants" [1736b, 112]. As a consequence, he wrote

$$S(x-1) = \sum_{i=1}^{x-1} X_i = S(x) - \frac{dS}{1!dx} + \frac{d^2 S}{2!dx^2} - \frac{d^3 S}{3!dx^3} + \frac{d^4 S}{4!dx^4} - \cdots$$

and

$$S(x) - S(x-1) = X = \frac{dS}{1!dx} - \frac{d^2 S}{2!dx^2} + \frac{d^3 S}{3!dx^3} - \frac{d^4 S}{4!dx^4} - \cdots, \qquad (95)$$

where (95) expressed the general term as a function of the summation term. Now, Euler wanted to derive S as a function of X. Setting[244]

$$\frac{dS}{dx} = \sum_{n=0}^{\infty} a_n \frac{d^n X}{dx^n}$$

yielded

$$S = a_0 \int X dx + \sum_{n=1}^{\infty} a_n \frac{d^{n-1} X}{dx^{n-1}}$$

and

$$\frac{d^h S}{dx^h} = \sum_{n=0}^{\infty} a_n \frac{d^{n+h-1} X}{dx^{n+h-1}}.$$

Replacing the last formulas in (95), Euler inferred that

$$\begin{aligned} X &= \sum_{h=1}^{\infty} (-1)^{h+1} \frac{d^h S}{h! dx^h} = \sum_{h=1}^{\infty} \frac{(-1)^{h+1}}{h!} \sum_{n=0}^{\infty} a_n \frac{d^{n+h-1} X}{dx^{n+h-1}} \\ &= \sum_{n=0}^{\infty} \frac{d^n X}{dx^n} \sum_{h=1}^{n+1} \frac{(-1)^{h+1} a_{n+1-h}}{h!}. \end{aligned} \qquad (96)$$

---

[243] In later papers Euler used (94) independently of this condition.
[244] Of course, $\frac{d^0 X}{dx^0} = X$.

By comparing the left-hand and the far right-hand sides of (96), he derived

$$a_0 = 1, \ a_n = \sum_{i=1}^{n} \frac{(-1)^{i+1} a_{n-i}}{(i+1)!}$$

and the summation formula

$$S(x) = \int X\,dx + \frac{X}{2!} + \frac{dX}{3!2dx} - \frac{d^3 X}{5!6dx^3} + \frac{d^5 X}{7!6dx^5} - \frac{d^7 X}{9!10dx^7}$$
$$+ \frac{5d^9 X}{11!6dx^9} - \frac{691 d^{11} X}{13!210 dx^{11}} + \frac{35 d^{13} X}{15!2 dx^{13}} - \frac{3617 d^{15} X}{17!30 dx^{15}} + \dots \ (97)$$

The indefinite integral yields a constant that is determined by the condition $S(0) = 0$.

Immediately following the completion of *Inventio summae*, Euler returned to (97) in *Methodus universalis series summandi ulterius promota* [1736c], where he modified the summation formula (now called the universal method) in order to make the calculation easier. He [1736c, 125] asserted that difficulties arose from the fact that the index of the general term increases by one unit at a time. He therefore considered a series

$$\sum_i X_{a+ib}$$

and, setting $S(x) = \sum_i^x X_{a+ib}$, obtained

$$S(x - b) = \sum_{h=0}^{\infty} (-1)^h \frac{b^h d^h S}{h! dx^h}$$

and

$$S(x) = \int \frac{X\,dx}{b} + \frac{X}{2!} + \frac{bdX}{3!2dx} - \frac{b^3 d^3 X}{5!6dx^3} + \frac{b^5 d^5 X}{7!6dx^5} - \frac{3b^7 d^7 X}{9!10dx^7}$$
$$+ \frac{5b^9 d^9 X}{11!6dx^9} - \frac{691 b^{11} d^{11} X}{13!210 dx^{11}} + \frac{35 b^{13} d^{13} X}{15!2 dx^{13}} - \frac{3617 b^{17} d^{15} X}{17!30 dx^{15}} + \dots$$

with $S(a) = X_a$. Euler also added another term to this formula, but it is incorrect.

Euler's research into the sum formula continued throughout his life, and he applied it to numerous series. In *Institutiones calculi differentialis*, he explicitly linked the coefficients a_n with the Bernoulli numbers B_r [245] (already studied in his *De seriebus quibusdam considerationes* [1740] and related to the sum of some remarkable series) and found

$$S(x) = \int z(x)dx + \frac{1}{2}z + \sum_{n=1}^{\infty} (-1)^{n-1} \frac{B_{2n}}{(2n)!} \frac{d^{2n-1}z}{dx^{2n-1}}. \tag{98}$$

---

[245] Bernoulli numbers are defined by the relation $\frac{t}{e^t-1} = 1 + \sum_{r=1}^{\infty} (-1)^{[r/2]+1} \frac{B_r}{r!} t^r$, where $[x]$ denotes the integral part of x.

He specified that in $\int z(x)dx$ "a constant must be added so that, if setting $x = 0$, we shall have $S = 0$".

The demonstration of the *Institutiones calculi differentialis* is remarkable because it indicates an evolution in Euler's thought toward a more formal conception and procedures that seem to be a prelude to the calculus of operations. In [1755, 2: Sections 167–168], Euler denoted $\sum y_x$ by Sy [with $S(0) = 0$] and interpreted S as a symbolic operation that enjoys certain formal properties:

1. finite and infinite additivity, which he explicitly formulated in these terms: If

$$y_x = p_x + q_x + r_x + \dots,$$

then

$$Sy = Sp + Sq + Sr + \dots,$$

that is,

$$\sum_{n=1}^{x} y_n = \sum_{n=1}^{x} p_n + \sum_{n=1}^{x} q_n + \sum_{n=1}^{x} r_n + \dots;$$

2. the commutativity of the operations S and $\frac{d^n}{dx^n}$, namely[246]

$$\frac{d^n}{dx^n}(Sy) = S\frac{d^n y}{dx^n}.$$

Euler set $v = y(x - 1) = y_{x-1} = v_x$ $(x = 2, 3, \dots)$ and $v(1) = A$ to get

$$Sv = \sum_{n=1}^{x} v_n = A + \sum_{n=2}^{x} y_n = A + Sy - y.$$

By applying the additivity of the operation S to

$$v_x = y(x - 1) = \sum_{n=1}^{\infty}(-1)^n \frac{1}{n!}\frac{d^n y}{dx^n}$$

and rearranging it, he derived

$$S\frac{dy}{dx} = y - A + \sum_{n=2}^{\infty}(-1)^n S\frac{d^n y}{n!dx^n} \quad \text{and}$$

---

[246]From a modern viewpoint, this formula only has meaning if we interpret S as an integral. In this case, it corresponds to $\frac{d^n}{dx^n}(\int y dx) = \int \left(\frac{d^n y}{dx^n}\right) dx$.

$$Sz = \int z\,dx - A + \sum_{n=1}^{\infty}(-1)^{n+1}S\frac{d^n y}{(n+1)!dx^n}$$

[where $z = \frac{dy}{dx}$ and the condition $S(0) = 0$ holds]. By differentiating and applying property (2), Euler found

$$S\frac{d^h z}{dx^h} = \frac{d^{h-1}z}{dx^{h-1}} + \sum_{n=2}^{\infty}\frac{(-1)^n}{n!}S\frac{d^{h+n-1}z}{dx^{h+n-1}}, \quad h = 0,1,2,\ldots$$

(here $\frac{d^{-1}z}{dx^{-1}} = \int z\,dx$). He then expressed Sz as

$$Sz = \int z\,dx + \sum_{n=1}^{\infty}a_n\left(\frac{d^{n-1}z}{dx^{n-1}}\right)$$

and, proceeding as above, derived (98).

Later, Euler provided other proofs of the sum formula. The basic principles did not change; however, there are some significant differences.[247]

---

[247] See, for instance, Euler [1761].

15　On the sum of an asymptotic series

We have previously seen the problematic nature of asymptotic series within the framework of 18th century mathematics. It is probable that, in deriving (77), neither Stirling nor de Moivre noted the divergence of series; however, the existence of series that had an anomalous behavior with respect to formal-quantitative notion of the sum was already clear to Euler in his *Consideratio progressionis cuiusdam ad circuli quadraturam inveniendam idoneae* [1739c], where he pointed out some difficulties of the theory of series. In this paper, Euler considered $\arctan t = \int_0^t \frac{dt}{1+t^2}$ as a sum of the type $\sum \frac{dt}{1+t^2}$ and, using relations concerning the sums of powers of integer numbers, derived

$$\pi = \frac{1}{n} + \sum_{k=1}^{n} \frac{4n}{n^2 + k^2} + \sum_{h=0}^{\infty} (-1)^h B_{4h+2} \frac{1}{2^{2h}(2h+1)n^{4h+2}}. \tag{99}$$

Euler understood the nature of asymptotic series. He observed that, although (99) seems to become increasingly convergent when n increases, it actually "converges" (i.e., its terms decrease in absolute value) only up to a certain term, after which it begins to diverge[248] This statement is based upon the observation that (for $h > 0$)

$$\frac{B_{2h+2}}{B_{2h}} > \frac{(h-1)(2h-3)}{2\pi^2},$$

so that, if $h = \infty$, this ratio is equal to $\frac{h^2}{\pi^2}$ (i.e., $\frac{B_{2h+2}}{h^2 B_{2h}}$ approaches asymptotically $\frac{1}{\pi^2}$). Euler suggested using (99) by calculating its first terms (according to Euler's terminology: the terms that "converged") and neglecting them as soon as they began to "diverge". More specifically, the terms of (99) had to be considered up to the index $s \approx \frac{\pi n}{\sqrt{2}}$ (approximately $s \approx 2n$) (Euler [1739c, 357]). Indeed, the ratio of the successive terms of the series

$$\sum_{h=0}^{\infty} (-1)^h B_{4h+2} \frac{1}{2^{2h}(2h+1)n^{4h+2}},$$

for large enough h, is approximately equal to $-\frac{4h^4}{n^4\pi^4}$ (if we take b_h as the coefficient of the series, $-\frac{b_{h+1}}{h^4 b_h}$ is asymptotic to $\frac{4}{n^4\pi^4}$). Euler also suggested approximating the terms a_n of (99) when the index n is greater than s by the geometric series

$$\sum_{r=0}^{\infty} P \left(-\frac{4s^4}{n^4\pi^4} \right)^r = \frac{\pi^4 n^4 P}{n^4\pi^4 + 4s^4} \approx \frac{P}{1 + \left(\frac{16s^4}{363n^4} \right)},$$

---

[248]For this reason, these series were termed "semiconvergent series" by Legendre in his [1811–17].

181

where P is the first omitted term. By giving the values 1, 3, and 5 to n, Euler determined some approximations of π. He compared these approximations with a known approximation of π and observed that one moved away from the truth (*a veritate*). In his opinion, this fact was worthy of note because there was no error in the proof and formula (99) allowed π to be approximated easily. He thought that this departure from the truth was due to the divergence of series and advised using divergent series with caution [1739c, 359-360], as Jacob Bernoulli had also done in [1689–1704].

He even gave an example that showed a contradiction deriving from the use of divergent series. Given a series $\sum_{n=-\infty}^{\infty} a_n$, Euler assumed the validity of formula (81) for any number n and wrote

$$
\begin{aligned}
a_0 &= a_1 + (a_1 - a_2) + (a_1 - 2a_2 + a_3) + \dots, \\
a_{-1} &= a_1 + 2(a_1 - a_2) + 3(a_1 - 2a_2 + a_3) + \dots, \\
a_{-2} &= a_1 + 3(a_1 - a_2) + 6(a_1 - 2a_2 + a_3) + \dots, \\
&\dots\dots
\end{aligned}
\tag{100}
$$

By summing column by column, and using the formal relations

$$
\begin{aligned}
\frac{1}{1-1} &= 1 + 1 + 1 + \dots, \\
\frac{1}{(1-1)^2} &= 1 + 2 + 3 + \dots, \\
\frac{1}{(1-1)^3} &= 1 + 3 + 6 + \dots, \\
&\dots,
\end{aligned}
$$

which are the expansion of $\frac{1}{1-x}, \frac{1}{(1-x)^2}, \frac{1}{(1-x)^3}, \dots$, for $x = 1$, Euler obtained

$$
\begin{aligned}
\sum_{n=-\infty}^{0} a_n &= \frac{1}{1-1}a_1 + \frac{1}{(1-1)^2}(a_1 - a_2) + \frac{1}{(1-1)^3}(a_1 - 2a_2 + a_3) + \dots \\
&= a_1\left(\frac{1}{1-1} + \frac{1}{(1-1)^2} + \frac{1}{(1-1)^3} + \dots\right) \\
&\quad -a_2\left(\frac{1}{(1-1)^2} + \frac{2}{(1-1)^3} + \frac{3}{(1-1)^4} + \dots\dots\right) \\
&\quad +a_3\left(\frac{1}{(1-1)^3} + \frac{3}{(1-1)^4} + \frac{6}{(1-1)^5} + \dots\dots\right) + \dots \\
&= a_1\frac{1}{(1-1)-1} - a_2\frac{1}{((1-1)-1)^2} + a_3\frac{1}{((1-1)-1)^3} + \dots \\
&= -a_1 - a_2 - a_3 - \dots
\end{aligned}
$$

Hence,

$$
\sum_{n=-\infty}^{\infty} a_n = 0.
$$

Euler noted that this result sometimes differed significantly from the correct one. For instance, when n is negative, if we consider the series $\sum_{n=-\infty}^{\infty} \frac{1}{n^2}$, we obtain $\sum_{n=-\infty}^{\infty} \frac{1}{n^2} = 0$, whereas the sums of infinite positive numbers would be different from 0. Euler, however, stated that "this reasoning did not always fail"; indeed,

$$\sum_{n=-\infty}^{\infty} h^n = \sum_{n=-1}^{\infty} h^n + \sum_{n=-\infty}^{0} h^n = \frac{h}{1-h} + \frac{h}{h-1} = 0.$$

Euler ended the paper by observing that the exhibition of this example was no less useful than the exhibition of rigorously proved truths (see Euler [1739c, 360–363]).

In the *Consideratio*, Euler sensed the necessity for a deeper analysis for the concept of the sum. In effect, asymptotic series were puzzling to mathematicians' eyes. Until then mathematicians had thought that series could be used to approximate a quantity only if convergence held and that the sum of a series was its unique, true, ultimate value. Instead asymptotic series showed that certain divergent series could be used to derive numerically acceptable results, even if the sum of an asymptotic series was not the ultimate value. For instance, in his letter "on a logarithmetical mistake of some eminent mathematicians", Bayes explicitly observed that the series

$$\log x! = \frac{1}{2}\log 2\pi + \left(x + \frac{1}{2}\right)\log x - x + \frac{1}{12x}$$
$$-\frac{1}{360x^3} + \frac{1}{1260x^5} - \frac{1}{1680x^7} + \frac{1}{1188x^9}\cdots$$

"can never properly express any quantity at all" since "the whole series can have no ultimate value whatsoever" (Euler [1763, 279–271]).

However, when Bayes wrote these words, Euler had overcome the initial scepticism, as one can note his *Institutiones calculi differentialis*, where asymptotic series were used in an unproblematic way. In this treatise he gave several examples of asymptotic series by deriving them from the Euler–Maclaurin summation formula [1755, 351–356]. In particular, he gave a slightly modified version of (99):

$$\pi = \frac{1}{n} + \sum_{k=1}^{n} \frac{4n}{n^2 + k^2} + \sum_{h=0}^{\infty} (-1)^h B_{4h+2} \frac{1}{2^{2h}(2h+1)n^{4h+2}} - \frac{4\pi}{e^{2n\pi}-1} \quad (102)$$

The addition of the constant $-\frac{4\pi}{e^{2n\pi}-1}$, which was derived from the application of the Euler–Maclaurin summation formula, does not suffice to justify the numerical discrepancy originating from (99) and does not essentially change the question that arose in the *Consideratio*. However, in the *Institutiones*, Euler no longer pointed out the strangeness of the behavior of such series: of importance to him now was the formal relation. Even if (102) failed as an exact numerical equality, it retained the formal validity.

In the second part of the century amd into the new century, mathematicians, following Euler, accepted the use of asymptotic series (see, e.g., Laplace [1812], Legendre [1811–17]. However, this occurred by means of a substantial shift in the conception of series. According to the formal-quantitative concept of series, the relations $f(x) = \sum_{i=0}^{\infty} f_i(x)$ or $S = \sum_{i=0}^{\infty} a_i$ had a meaning that was independent of the procedure used for finding $f(x)$ and S: Indeed, it was the convergence of the series

$$\sum_{i=0}^{\infty} f_i(x) \quad \text{and} \quad \sum_{i=0}^{\infty} a_i$$

that gave a meaning to the equalities $f(x) = \sum_{i=0}^{\infty} f_i(x)$ and $S = \sum_{i=0}^{\infty} a_i$. In the case of asymptotic series, these equalities could not be regarded as convergence, since the final value was not given. This poses the following question:

> *How could mathematicians assert that the sum of the series $\sum_{i=0}^{\infty} a_i$ was precisely S and not another number S'?*

The writings of Euler, Lagrange and Laplace show clearly that mathematicians resorted to the formal derivation of the equalities $S = \sum_{i=0}^{\infty} a_i$ and $f(x) = \sum_{i=0}^{\infty} f_i(x)$: They thought that S *(and not S') was the sum of* $\sum_{i=0}^{\infty} a_i$ *because S (and not S') was the result of a formal procedure* [in the sense of proposition[249] (IE)]. In this way, formal manipulation was no longer the tool for deriving convergent series, but the one and only justification and guarantee of the exactness of certain relations.

---

[249]See Chapter 8, p. 117.

16   Infinite products and continued fractions

The investigations concerning the interpolation of sequences and the expression and approximation of integrals led Euler to deal with infinite products. In *De productis ex infinitis factoribus ortis* [1739a, 262–264], Euler sought to determine the value of the sequence

$$a_0 = 1, \quad a_i = (f + ig)a_{i-1}, \quad i = 1, 2, \ldots, \tag{103}$$

for $i = 1/2, 3/2, \ldots$. In order to do this, he defined the sequence A_i by setting

$$\begin{cases} A_{2i} = a_i, \\ A_1 = z, \quad A_{2i+1} = (f + \tfrac{2i+1}{2}g)A_i, \end{cases}$$

where z is to be determined. He stated that, when n is infinite, the sequence becomes indistinguishable from a geometric progression and, therefore, the formula

$$\alpha_i = \sqrt{\alpha_{i-1}\alpha_{i+1}}, \tag{104}$$

which is valid for a geometrical progression $\alpha_i = h^i$, $i = 0, 1, 2, \ldots$, can be applied to the sequence A_i. Thus, he obtained the following approximate expressions of z :

$$z = \sqrt{f + g},$$

$$z = \sqrt{\frac{(f+g)^2(f+2g)}{1(f+\tfrac{3}{2}g)^2}},$$

$$z = \sqrt{\frac{(f+g)^2(f+2g)^2(f+3g)}{1(f+\tfrac{3}{2}g)^2(f+\tfrac{5}{2}g)^2}},$$

$$\cdots$$

and assumed that the value of z is

$$z = \frac{(f+g)(f+2g)(f+3g)\cdots}{(f+\tfrac{3}{2}g)(f+\tfrac{5}{2}g)\cdots}.$$

Then Euler considered formula (87) and changed x into y^g in this formula. Since

$$\int_0^1 (-\log x)^{1/2}dx = \frac{\sqrt{\pi}}{2},$$

he obtained

$$\frac{(2f+g)(2f+3g)(2f+3g)(2f+5g)(2f+5g)\cdots}{(2f+2g)(2f+2g)(2f+4g)(2f+4g)(2f+6g)(2f+6g)\cdots}$$

$$= \frac{2f^2(2f+g)}{\pi g}\left(\int_0^1 y^{f-1}\sqrt{1-y^g}\,dy\right)^2.$$

In the *De productis*, Euler observed that if a quantity X is transformed into an infinite product $\prod_{i=1}^{\infty} \frac{p_i}{q_i}$, then it can be computed by logarithms $\left(\text{namely, } \log X = \sum_{i=1}^{\infty} \log \frac{p_i}{q_i}\right)$ and the more rapidly this series converges, the more the factors $\frac{p_i}{q_i}$ go to 1. He also stated that infinite products could be used to illustrate the nature of the quantity X [1739a, 261]. In effect, Euler used infinite (or finite) products as an intermediate step to derive various relations between integrals. For instance, he found

$$\pi = 2ag \left(\int_0^1 \frac{x^{a-1}}{\sqrt{1-x^{2g}}} dx \right) \left(\int_0^1 \frac{x^{a+g-1}}{\sqrt{1-x^{2g}}} dx \right)$$

and

$$\left(\int_0^1 \frac{x^{a-1}}{(1-x^b)^{1-m}} dx \right) \left(\int_0^1 \frac{x^{a+mb-1}}{(1-x^b)^{1-n}} dx \right) \tag{105}$$
$$= \left(\int_0^1 \frac{x^{a-1}}{(1-x^b)^{1-n}} dx \right) \left(\int_0^1 \frac{x^{a+nb-1}}{(1-x^b)^{1-m}} dx \right)$$

(see Euler [1739a, 266 and 288]). These are *de facto* results concerning the beta function, which was, however, not recognized as such, namely an integral of the type $\int_0^1 x^m (1-x)^n dx$ was not recognized as a new function. Moreover, in *De productis*, the above integrals are still viewed as the expression of geometrical quantities (areas).

It is worth noting that Euler was often interested in determining convergent products so that they could be computed effectively. However, even in this case, infinite products were used without an *a priori* consideration of their convergence. Furthermore, he also used these products as formal instruments for deriving relations such as (105) and, in these cases, the convergence of products seems to be of no importance.

$$* \quad * \quad *$$

Euler systematically investigated continued fractions from the 1730s onwards. In *De fractionibus continuis dissertatio*, he observed that this species of infinite expressions was not as usual as the other two types (series and infinite products),[250] although they were highly appropriate for expressing the values of certain quantities and for calculating them approximately [1737a, 188].

In *De fractionibus continuis dissertatio* [1737a, 189–196] and *De fractionibus continuis observationes* [1739b, 292–293], Euler showed that the continued fraction

$$C = a_1 + \frac{b_1}{a_2+} \frac{b_2}{a_3+} \frac{b_3}{a_4+} \dots$$

---

[250] In practice, after the initial results —described in Chapter 1— mathematicians showed little interest in continued fractions.

can be transformed into

$$C = C_1 + \sum_{k=1}^{\infty} \Delta C_k = a_1 + \sum_{k=1}^{\infty} (-1)^{k+1} \frac{\prod_{i=1}^{k} b_i}{q_k q_{k+1}},$$

where $C_k = \frac{p_k}{q_k}$ and p_k and q_k are defined for all $k > 1$ by the recursive rule

$$\begin{cases} p_k = a_k p_{k-1} + b_{k-1} p_{k-2} \\ q_k = a_k q_{k-1} + b_{k-1} q_{k-2} \end{cases}$$

with

$$p_1 = a_1, \ p_0 = 1, \ q_1 = 1, \ q_0 = 0,$$

and $\Delta C_k = C_{k+1} - C_k$. Vice versa, given the series

$$\sum_{k=1}^{\infty} (-1)^{k+1} \frac{\prod_{i=1}^{k} b_i}{q_k q_{k+1}},$$

Euler obtained

$$a_1 + \frac{b_1}{a_2+} \frac{b_2}{a_3+} \frac{b_3}{a_4+} \dots$$

by setting

$$a_i = \frac{q_i - a_{i-1} q_{i-2}}{q_{i-1}}.$$

In *De fractionibus continuis dissertatio* and *De fractionibus continuis observationes*, Euler paid attention to the convergence of the infinite processes he used: Indeed, he mainly aimed to find the value of certain integrals and convergence was essential. However, the link between series and continued fractions was viewed in a merely formal way, independently of the convergence of the series and fractions. *A series and a continued fraction were considered to be equal simply if they were derived from each other by the formal transformation*

$$a_1 + \frac{b_1}{a_2+} \frac{b_2}{a_3+} \frac{b_3}{a_4+} \dots = a_1 + \sum_{k=1}^{\infty} (-1)^{k+1} \frac{\prod_{i=1}^{k} b_i}{q_k q_{k+1}}. \qquad (106)$$

For instance, in [1739b, 296], by using the binomial expansion and transformation (106), he derived

$$\int_0^1 \frac{x^{n-1}}{(1+x^m)^{\frac{h}{v}}} dx = \frac{1}{n} + \sum_{k=1}^{\infty} (-1)^k \frac{h(h+v)\dots(h+(k-1)v)}{k! v^k (km+n)}$$

$$= \frac{1}{n+} \frac{hn^2}{vm+(v-h)n+} \frac{v(h+v)(m+n)^2}{(3v-h)m+(v-h)n} \dots.$$

Euler observed that if $v = 1$ and h is an integer, one had the fraction

$$\cfrac{1}{n+}\cfrac{hn^2}{m+(1-h)n+}\cfrac{(h+1)(m+n)^2}{(3-h)m+(1-h)n}\cdots,$$

whose series expression was not convergent. However, this was not considered to be a good reason to reject the general validity of equality (106).

If one assumed (106) to be valid independently of the convergence of the series and of the continued fraction, then it was very natural to use (106) to give a value to divergent series. During the 1740s, neither Euler nor other mathematicians tackled this question. Subsequently, however, Euler drew the logical conclusions from this.[251] In his [1784, 41], for instance, he regarded the transformation of divergent series into fractions as a secure way (and perhaps the only way) of finding the approximate value of a divergent series.

Some examples are appropriate to clarify the use of continued fractions. In *De fractionibus continuis observationes* [1739b, 302–306], after haved interpolating the sequence (103), Euler argued that if the sequence

$$A_k = \prod_{i=1}^{2k} a_i$$

was given, then $A_{1/2}$ was equal to a_1, $A_{3/2}$ was equal to $a_1 a_2 a_3$, ... He applied this idea to the series

$$\sum_{k=1}^{\infty} A_k = \sum_{k=1}^{\infty} \frac{p(p+2r)\dots(p+2(k-1)r}{(p+2q)(p+2q+2r)\dots(p+2q+2((k-1)r)}. \qquad (107)$$

He took

$$a_1 a_2 = \frac{p}{p+2q},$$

$$a_3 a_4 = \frac{p+2r}{p+2q+2r},$$

$$\dots$$

$$a_{2k-1} a_{2k} = \frac{p+2(k-1)r}{p+2q+2(k-1)r},$$

$$\dots$$

and

$$a_k = \frac{b_k}{p+2q+2(k-1)r},$$

---

[251] See Euler [1754–55] and [1784].

where the numbers b_k are to be determined appropriately, and found

$$b_1 \;=\; m - r + \cfrac{P}{m+} \cfrac{p(p+2q-r)(P+Q)}{2rR+}$$
$$\cfrac{(p+r)(p+2q)P(P+2Q)}{2r(R+S)+} \cfrac{(p+2r)(p+2q+r)(P+Q)(P+3Q)}{2r(R+2S)+} , \tag{108}$$

where m is an arbitrary quantity and

$$
\begin{aligned}
P &= p^2 + 2pq - pr - m^2 + mr, \\
Q &= 2r(p+q-m), \\
R &= p^2 + 2pq - mp - mq + qr, \\
S &= pr + qr - mr.
\end{aligned}
$$

At this point Euler applied (88). Setting

$$
\begin{aligned}
f + g &= p, \\
h + k &= p + 2q, \\
g &= k = 2r,
\end{aligned}
$$

and substituting n with $n-1$, he deduced

$$\frac{p(p+2r)\dots(p+2(n-1)r)}{(p+2q)(p+2q+2r)\dots(p+2q+2((n-1)r)} = \frac{\int_0^1 x^{\frac{p+2q-2r}{2r}}(1-x)^{n-1}dx}{\int_0^1 x^{\frac{p-2r}{2r}}(1-x)^{n-1}dx},$$

i.e., in more modern notation,

$$\frac{p(p+2r)\dots(p+2(n-1)r)}{(p+2q)(p+2q+2r)\dots(p+2q+2((n-1)r)} = \frac{B\left(\frac{p+2q}{2r}, n\right)}{B\left(\frac{p}{2r}, n\right)},$$

where $B(x,y)$ is the beta function. It follows, through a simple substitution in the integral, that the general term of (107) is

$$\frac{\int_0^1 y^{p+2q-1}(1-y^{2r})^{n-1}dy}{\int_0^1 y^{p-1}(1-y^{2r})^{n-1}dy}.$$

Hence,

$$b_1 = A_{\frac{1}{2}}(p+2q+r) = (p+2q-r)\frac{\int_0^1 y^{p+2q-1}\sqrt{(1-y^{2r})^{-1}}dy}{\int_0^1 y^{p-1}\sqrt{(1-y^{2r})^{-1}}dy}. \tag{109}$$

Therefore, the continued fraction (108) is expressed by the ratio of two integrals, and the crucial step for arriving at (109) is the series (107).

More generally, Euler obtained[252]

$$
\begin{aligned}
b_k &= p + q + (k-2)r + \cfrac{pr - q^2}{2(p+q+(k-2)r)+} \; \cfrac{2r^2 + pr - q^2}{2(p+q+(k-2)r)+} \\
&\quad \cfrac{6r^2 + pr - q^2}{2(p+q+(k-2)r)+} \; \cfrac{12r^2 + pr - q^2}{2(p+q+(k-2)r)+} \cdots \\
&= (p + 2q + (k-2)r) \frac{\int_0^1 y^{p+2q+(k-1)r-1}\sqrt{(1-y^{2r})^{-1}}dy}{\int_0^1 y^{p+k(-1)r-1}\sqrt{(1-y^{2r})^{-1}}dy}.
\end{aligned}
$$

It should be noted that series were merely conceived as a mere instrument relating continued fractions and integrals. It was thus of no importance whether series on their own had a meaning as quantities (i.e., they were convergent).

In [1739b, 324–325], Euler also considered the functions P and R satisfying the recursive relation

$$
(a + n\alpha) \int_0^1 P(x)R^n(x)dx = (b + n\beta) \int_0^1 P(x)R^{n+1}(x)dx \qquad (110)
$$
$$
+ (c + n\gamma) \int_0^1 P(x)R^{n+2}(x)dx;
$$

for $n = 0, 1, 2, \ldots$, he showed that

$$
\frac{\int_0^1 P(x)R(x)dx}{\int_0^1 P(x)dx} = \frac{a}{b+} \frac{(a+\alpha)c}{b+\beta+} \frac{(a+2c)(c+\gamma)}{b+2\beta+} \cdots. \qquad (111)
$$

The proof consisted in dividing the relation (110) by

$$
\int_0^1 P(x)R^{n+1}(x)dx, \qquad n = 0, 1, 2, \ldots
$$

to obtain the terms of the fraction step by step. Various interesting expansions follow from (111). For instance,

$$
\frac{\int_0^1 xe^x dx}{\int_0^1 e^x dx} = \frac{1}{1-e} = \frac{1}{1+} \frac{2}{2+} \frac{3}{3+} \cdots.
$$

(see Euler [1739b, 343–344]).

Another example concerns the use of continued fractions to solve differential equations. In [1739a, 345–347], Euler considered the Riccati equation

$$
ax^m dx + bx^{m-1}y dx + cy^2 dx + dy = 0 \qquad (112)
$$

---

[252] See Euler [1739b, 319–322].

and sought a solution of the like kind

$$y = \frac{1}{cx} - \frac{1}{zx^2},$$

where z is to be determined. By replacing $y = \frac{1}{cx} - \frac{1}{zx^2}$ in (112) and setting $x^{m+3} = t$, he obtained the equation

$$\frac{-c}{m+3}t^{-\frac{m+4}{m+3}}dt - \frac{b}{m+3}t^{\frac{1}{m+3}}zdt - \frac{ac+b}{(m+3)c}z^2dt + dz = 0,$$

which is of the same type as the previous. The reasoning can be repeated. He set

$$z = \frac{-(m+3)c}{(ac+b)t} - \frac{1}{vt^2},$$
$$t^{\frac{2m+5}{m+3}} = u$$

and replaced into the last equation and thus obtained a new equation of the same type. And so on.

Euler stated the equation had the solution

$$a_1x^{-1} + \frac{1}{-a_2x^{-1}+}\frac{1}{a_3x^{-1}+}\frac{1}{-a_4x^{-1}+}\cdots$$

where

$$a_1 = c^{-1}, \quad a_2 = \left(-\frac{ac+b}{(m+3)c}\right)^{-1}, \quad \text{etc.}$$

He preferred to write the solution in the form

$$cxy = 1 + \frac{(ac+b)x^{m+2}}{-(m+3)+}\frac{(ac-(m+2)b)x^{m+2}}{(2m+5)+}\frac{(ac+(m+3)b)x^{m+2}}{-(3m+7)+}\cdots$$

He also noted that the solution satisfied the condition $cxy = 1$ for $x = 0$ if $m+2 > 0$, and the condition $cxy = 1$ for $x = \infty$ if $m+2 < 0$.

The solution of a differential equation by continued fractions highlighted that continued fractions could be connected with *totally divergent series* (series with a radius of convergence equal to 0).[253] For instance, in *De seriebus divergentibus* [1754–55b, 615], Euler observed that the solution of the differential equations

$$x^m dx = x^{q+1}dy + (p-m)x^q ydx + ydx$$

was given by

$$y = \frac{x^m}{1+}\frac{px^q}{1+}\frac{qx^q}{1+}\frac{p(p+q)x^q}{1+}\frac{2qx^q}{1+}\frac{p(p+2q)x^q}{1+}\frac{3qx^q}{1+}\frac{p(p+3q)x^q}{1+}\cdots$$

---

[253] On series solutions to differential equations, see Chapter 26.

under the condition $y(0) = 0$. However, the solution can also given by the integral

$$e^{\frac{1}{qx^q}} x^{m-q} \int_0^x e^{-\frac{1}{qx^q}} \frac{dx}{x^{p-q-1}}$$

and by the series

$$y = x^m - px^{m+q} + p(m+q)x^{m+2q} - p(p+q)(p+2q)x^{m+3q} + \dots . \quad (113)$$

which, in general, has the radius of convergence equal to zero.

   These three solutions were conceived of as equivalent and it seemed natural to assume the value of a continued fraction is the value of the series. Indeed, setting $p = m = 1$, $q = 2$, $x = 1$, Euler found

$$1 - 1 + 1 \cdot 3 - 1 \cdot 3 \cdot 5 + 1 \cdot 3 \cdot 5 \cdot 7 - \dots = \frac{1}{1+} \frac{1}{1+} \frac{2}{1+} \frac{3}{1+} \frac{4}{1+} \frac{5}{1+} \dots = 0.65568;$$

hence, he assumed that the sum of $1 - 1 + 1 \cdot 3 - 1 \cdot 3 \cdot 5 + 1 \cdot 3 \cdot 5 \cdot 7 - \dots$ was 0.65568.

   I stress that this type of power series (totally divergent power series) was different from the type of power series upon which the formal-quantitative use of series was based. As we saw in Chapter 8, the formal-quantitative concept of series meant that a power series always had a positive radius of convergence and had a meaning that was independent of the formal transformation by which the series was derived: They represented quantities independently of the modality of derivation (I shall refer to these series as ordinary power series). Instead, *totally divergent series have no quantitative meaning by themselves*: Their meaning derived only by a formal transformation that linked them to a differential equation and a continued fraction or another expression of quantity.

17 Series and number theory

In *Variae observationes circa series infinitas*, Euler dealt with series that he argued were different from other series. In his opinion, mathematicians had hitherto mainly considered either series whose general term was given (by an analytical expression) or series whose terms could be computed by a recurrence formula. Instead, in *Variae observationes*, he considered series where neither their general term (in Euler's sense[254]) nor the rule of the continuation of terms was known, but whose nature was determined by means of other conditions. Moreover, their sum had to be computed by methods that differed from the customary ones, which required knowledge of the general term or the rule of continuation [1737b, 217].

Euler refers to series such as

$$\frac{1}{15} + \frac{1}{63} + \frac{1}{80} + \frac{1}{255} + \frac{1}{624} + \dots$$

and

$$\frac{1}{2} + \frac{1}{3} + \frac{1}{5} + \frac{1}{7} + \frac{1}{11} + \frac{1}{13} + \dots.$$

The general term of the first series is the fraction $\frac{1}{a}$ "whose denominators are one less all perfect squares which simultaneously are other powers" (Euler [1737b, 226]). The general term of the second series is $\frac{1}{p}$, where p is a prime number. I note that the series are, in reality, given by the general term, although these general terms do not satisfy the Eulerian definition by which the general term a_n was given by one single formula composed of the commonly used analytical symbols. This prevented Euler from investigating such numerical series according to his usual procedures, which consisted of reducing the treatment of a numerical series to that of an appropriate function series, even transforming a_n into a function $a(n)$ of a continuous variable.

The first series that Euler summed in *Variae observationes* was

$$\frac{1}{3} + \frac{1}{7} + \frac{1}{8} + \frac{1}{15} + \dots = \sum_{m,n>1} \frac{1}{m^n - 1}.$$

He used a sophisticated application of Leibniz's method described on p. 30. He set

$$x = 1 + \frac{1}{2} + \frac{1}{3} + \frac{1}{4} + \frac{1}{5} + \dots \tag{114}$$

and subtracted

$$1 = \frac{1}{2} + \frac{1}{4} + \frac{1}{8} + \frac{1}{16} + \dots$$

---

[254]See Section 13.1, p. 157.

from (114). Hence,

$$x - 1 = 1 + \frac{1}{3} + \frac{1}{5} + \frac{1}{6} + \frac{1}{7} + \frac{1}{9} + \dots \tag{115}$$

Then he subtracted

$$\frac{1}{2} = \frac{1}{3} + \frac{1}{3^2} + \frac{1}{3^3} + \dots$$

from (115) and obtained

$$x - 1 - \frac{1}{2} = 1 + \frac{1}{5} + \frac{1}{6} + \frac{1}{7} + \frac{1}{10} + \dots \tag{116}$$

He repeated the reasoning. By subtracting

$$\frac{1}{4} = \frac{1}{5} + \frac{1}{5^2} + \frac{1}{5^3} + \dots$$

from (116), he obtained

$$x - 1 - \frac{1}{2} - \frac{1}{4} = 1 + \frac{1}{6} + \frac{1}{7} + \frac{1}{10} + \dots$$

Euler stated that, in a similar way, all the remaining terms could be eliminated, leaving $x - 1 - \frac{1}{2} - \frac{1}{4} - \frac{1}{5} - \frac{1}{6} - \frac{1}{9} - \frac{1}{10} - \dots = 1$ or

$$x - 1 = 1 + \frac{1}{2} + \frac{1}{4} + \frac{1}{5} + \frac{1}{6} + \frac{1}{9} + \frac{1}{10} + \dots$$

Finally, he subtracted this series from (114) and obtained

$$\frac{1}{3} + \frac{1}{7} + \frac{1}{8} + \frac{1}{15} + \frac{1}{24} + \dots = 1.$$

The summation of

$$\frac{1}{15} + \frac{1}{63} + \frac{1}{80} + \frac{1}{255} + \frac{1}{624} + \dots$$

is similar but a little less hard to understand for the modern reader since the starting point is $\sum_{n=1}^{\infty} \frac{1}{n^2} = \frac{\pi^2}{6}$ rather than $\sum_{n=1}^{\infty} \frac{1}{n} = x$. Euler then considered the geometrical series

$$\frac{1}{n^2 - 1} = \frac{1}{n^2} + \frac{1}{(n^2)^2} + \frac{1}{(n^2)^3} + \dots$$

and observed that

$$1 + \frac{1}{4} + \frac{1}{9} + \frac{1}{16} + \frac{1}{25} + \dots$$
$$= 1 + \left(\frac{1}{4} + \frac{1}{16} + \frac{1}{64} + \dots \right) + \left(\frac{1}{9} + \frac{1}{81} + \frac{1}{729} + \dots \right) + \left(\frac{1}{25} + \frac{1}{625} + \dots \right) + \dots$$
$$= 1 + \frac{1}{3} + \frac{1}{8} + \frac{1}{24} + \frac{1}{35} + \dots.$$

The series $\frac{1}{15} + \frac{1}{63} + \frac{1}{80} + \frac{1}{255} + \frac{1}{624} + \dots$ is the difference between

$$1 + \sum_{n=2}^{\infty} \frac{1}{n^2 - 1} = 1 + \frac{1}{3} + \frac{1}{8} + \frac{1}{15} + \frac{1}{24} + \frac{1}{35} + \dots$$

and

$$\sum_{n=1}^{\infty} \frac{1}{n^2} = 1 + \frac{1}{4} + \frac{1}{9} + \frac{1}{16} + \frac{1}{25} + \dots = 1 + \frac{1}{3} + \frac{1}{8} + \frac{1}{24} + \dots$$

Since

$$\sum_{n=2}^{\infty} \frac{1}{n^2 - 1} = \frac{4}{3},$$

he obtained

$$\frac{1}{15} + \frac{1}{63} + \frac{1}{80} + \frac{1}{255} + \frac{1}{624} + \dots = \frac{7}{4} - \frac{\pi^2}{6}.$$

Then Euler derived various results concerning prime numbers. For example,

$$1 + \frac{1}{2} + \frac{1}{3} + \frac{1}{4} + \frac{1}{5} + \dots = \frac{2 \cdot 3 \cdot 5 \cdot 7 \cdot 11 \cdot 13 \cdot 17 \cdot 19 \cdot \dots}{1 \cdot 2 \cdot 4 \cdot 6 \cdot 10 \cdot 12 \cdot 16 \cdot 18 \cdot \dots}, \qquad (117)$$

where the numerator is the product of all the primes and the denominator is the product of all numbers one less than the primes [1737b, 228], and

$$\frac{\pi}{4} = \frac{3 \cdot 5 \cdot 7 \cdot 11 \cdot 13 \cdot 17 \cdot 19 \cdot \dots}{4 \cdot 4 \cdot 8 \cdot 12 \cdot 12 \cdot 16 \cdot 20 \cdot \dots},$$

where the denominator is the product of the numbers one greater or less than the corresponding numerator [1737b, 233].

In *Variae observationes* [1737b, 242–244], Euler also proved that the sum of the reciprocal of all the primes $\frac{1}{2} + \frac{1}{3} + \frac{1}{5} + \frac{1}{7} + \frac{1}{11} + \frac{1}{13} + \dots$ is infinite. He set

$$A = \frac{1}{2} + \frac{1}{3} + \frac{1}{5} + \frac{1}{7} + \frac{1}{11} + \dots, \qquad (118)$$

$$B = \frac{1}{2^2} + \frac{1}{3^2} + \frac{1}{5^2} + \frac{1}{7^2} + \frac{1}{11^2} + \dots,$$

$$C = \frac{1}{2^3} + \frac{1}{3^3} + \frac{1}{5^3} + \frac{1}{7^3} + \frac{1}{11^3} + \dots,$$

$$\dots.$$

Euler summed (118) column by column after multiplying the first row by 1, the second by $\frac{1}{2}$, the third by $\frac{1}{3}$, the fourth by $\frac{1}{4}$, etc. Since

$$\log(1 - x)^{-1} = x + \frac{x^2}{2} + \frac{x^3}{3} + \frac{x^4}{4} + \dots,$$

he obtained

$$A + \frac{1}{2}B + \frac{1}{3}C + \frac{1}{4}D + \ldots = \log\frac{2}{1} + \log\frac{3}{2} + \log\frac{5}{4} + \log\frac{7}{6} + \ldots$$

Hence,

$$e^{A+\frac{1}{2}B+\frac{1}{3}C+\frac{1}{4}D+\cdots} = \frac{2 \cdot 3 \cdot 5 \cdot 7 \cdot 11 \cdot 13 \cdot 17 \cdot 19 \cdot \ldots}{1 \cdot 2 \cdot 4 \cdot 6 \cdot 10 \cdot 12 \cdot 16 \cdot 18 \cdot \ldots}.$$

Euler applied (117) and wrote

$$e^{A+\frac{1}{2}B+\frac{1}{3}C+\frac{1}{4}D+\cdots} = 1 + \frac{1}{2} + \frac{1}{3} + \frac{1}{4} + \frac{1}{5} + \ldots \qquad (119)$$

Since $\frac{1}{2}B + \frac{1}{3}C + \frac{1}{4}D + \ldots$ has a finite value[255] and

$$e^{A+\frac{1}{2}B+\frac{1}{3}C+\frac{1}{4}D+\cdots} = \infty,$$

it follows that

$$A = \frac{1}{2} + \frac{1}{3} + \frac{1}{5} + \frac{1}{7} + \frac{1}{11} + \frac{1}{13} + \ldots = \infty.$$

Moreover, $\frac{1}{2}B + \frac{1}{3}C + \frac{1}{4}D + \ldots$ can be neglected with respect to $A = \infty$; therefore, (119) can be written as

$$e^{A} = 1 + \frac{1}{2} + \frac{1}{3} + \frac{1}{4} + \frac{1}{5} + \ldots$$

By passing to logarithms,

$$\begin{aligned} A &= \frac{1}{2} + \frac{1}{3} + \frac{1}{5} + \frac{1}{7} + \frac{1}{11} + \frac{1}{13} + \ldots \\ &= \log\left(1 + \frac{1}{2} + \frac{1}{3} + \frac{1}{4} + \frac{1}{5} + \ldots\right) = \log\log\infty. \end{aligned}$$

Euler concluded that

$$\frac{1}{2} + \frac{1}{3} + \frac{1}{5} + \frac{1}{7} + \frac{1}{11} + \frac{1}{13} + \ldots$$

is infinite but is infinitely less than

$$1 + \frac{1}{2} + \frac{1}{3} + \frac{1}{4} + \frac{1}{5} + \ldots$$

255

$$\frac{1}{2}B + \frac{1}{3}C + \frac{1}{4}D + \ldots < \prod_{n=1}^{\infty}\log\frac{n+1}{n} - \sum_{k=2}^{\infty}\frac{1}{k} = \log(n+1) - \sum_{k=2}^{\infty}\frac{1}{k} = C + 1,$$

where C is the Euler–Mascheroni constant.

Some years later Euler used series in tackling the problem of the partitions of numbers; this consists of determining the number of ways of writing the integer n as a sum of a fixed number ν of addends or any number of addends (the addends, belonging to a fixed set of positive integers, can all differ or not, as the case may be, but their order is not considered significant). The technique used in this case differed considerably from that of *Variae observationes circa series infinitas*. He now sought the power series whose coefficients provided the numbers of partitions (in modern terms, he was trying to find the generating function, though in this case the term "function" is not quite appropriate for Euler). Here, Euler made *a purely combinatorial use of series; they were needed for counting objects and he could disregard convergence completely.*

In his *Introductio in analysin infinitorum*[256] [1748a, 313–337], given a sequence n_j of integers, Euler sets

$$1 + \sum_{i=1}^{\infty} P_i z^i = \prod_{j=1}^{\infty} (1 + x^{n_j} z). \tag{120}$$

By the method of indeterminate coefficients, he obtains

$$P_1 = \sum_{j=1}^{\infty} x^{n_j},$$
$$P_2 = \sum_{j=1}^{\infty} \sum_{k>j}^{\infty} x^{n_j+n_k},$$
$$P_3 = \sum_{j=1}^{\infty} \sum_{k>j}^{\infty} \sum_{r>k}^{\infty} x^{n_j+n_k+n_r},$$
$$\cdots$$

It is clear that P_i is the sum of the infinite numbers x^{α}, where α is equal to the sum of i different terms of the sequence n_s. Since some sums $n_j + n_k + n_r + \ldots$ might be equal, Euler sets

$$P_v = \sum_{j=1}^{\infty} N_{v,m_j} x^{m_j},$$

where N_{v,m_j} gives the number of ways of writing the integer m_j as the sum of v different addends chosen between the terms of the sequence n_j. If one now denotes the set of the integers m_j (which are equal to the sum of i different terms of sequences n_s) by M, one can write

$$P_v = \sum_{m \in M}^{\infty} N_{v,m} x^m.$$

Then the number of ways of writing the integer m as the sum of v different addends chosen between the terms of the sequence n_j can be calculated by

---

[256]See also Euler [1750–51d].

determining the coefficients of $x^m z^n$ in the series

$$1 + \sum_{n=1}^{\infty} \left(\sum_{m \in M} N_{\nu,m} x^m \right) z^n = \prod_{j=1}^{\infty} (1 + x^{n_j} z). \qquad (121)$$

As an example, Euler considered the sequence of natural numbers $1, 2, 3, 4, \ldots$. In this case, formula (121) can be written as

$$(1 + xz)(1 + x^2 z)(1 + x^3 z)(1 + x^4 z)(1 + x^5 z) \ldots \qquad (122)$$
$$= 1 + z \left(x + x^2 + x^3 + x^4 + x^5 + x^6 + x^7 + x^8 + \ldots \right)$$
$$+ z^2 \left(x^3 + x^4 + 2x^5 + 2x^6 + 3x^7 + 3x^8 + 4x^9 + 4x^{10} + \ldots \right)$$
$$+ z^3 \left(x^6 + x^7 + 2x^8 + 3x^9 + 4x^{10} + 5x^{11} + 7x^{12} + 8x^{13} + \ldots \right)$$
$$+ z^4 \left(x^{10} + x^{11} + 2x^{12} + 3x^{13} + 5x^{14} + 6x^{15} + 9x^{16} + 11x^{17} + \ldots \right)$$
$$+ z^5 \left(x^{15} + x^{16} + 2x^{17} + 3x^{18} + 5x^{19} + 7x^{20} + 10x^{21} + 13x^{22} + \ldots \right)$$
$$+ z^6 \left(x^{21} + x^{22} + 2x^{23} + 3x^{24} + 5x^{25} + 7x^{26} + 11x^{27} + 14x^{28} + \ldots \right)$$
$$+ z^7 \left(x^{28} + x^{29} + 2x^{30} + 3x^{31} + 5x^{32} + 7x^{33} + 11x^{34} + 15x^{35} + \ldots \right)$$
$$\ldots$$

He thus found that the number of ways of writing the number 35 as the sum of 7 addends taken from the sequence of the natural number $1, 2, 3, 4, \ldots$ is 15, since 15 is the coefficient of $z^7 x^{35}$.

If we set $z = 1$ and rearrange the series (121), we have

$$1 + \sum_{m \in M}^{\infty} R_m x^m = \prod_{j=1}^{\infty} (1 + x^{n_j}),$$

where

$$R_m = N_{1,m} + N_{2,m} + \ldots.$$

The coefficient R_m gives the numbers of ways of writing the integer m as the sum of any number of different addends belonging to the sequence n_j. For example, for $z = 1$, we have

$$(1 + x)(1 + x^2)(1 + x^3)(1 + x^4)(1 + x^5) \ldots.$$
$$= 1 + x + x^2 + 2x^3 + 2x^4 + 3x^5 + 4x^6 + 5x^7 + 6x^8 + \ldots.$$

We thus found that the number of ways of writing the number 8 as the sum of any number of different addends taken from the sequence $1, 2, 3, 4, 5, \ldots$ is 6.

Euler continued with the case whereby the addends can equal each other. Given a sequence n_j of integers, he considered

$$\frac{1}{\prod_{j=1}^{\infty} (1 + x^{n_j} z)}$$

and expanded this product into a power series

$$1 + \sum_{n=1}^{\infty} \sum_{m \in M}^{\infty} P_{n,m} x^m z^n,$$

where M is the set of the integers that are equal to the sum of i terms of sequences n_j. Reasoning in the way described above, Euler showed that the number of ways of writing an appropriate integer m_j as the sum of v addends (not necessarily different), chosen from the terms of the sequence n_j, could be calculated by determining the coefficients of $x^m z^n$ of the series $1 + \sum_{n=1}^{\infty} \sum_{m \in M}^{\infty} P_{n,m} x^m z^n$. In particular, when $n_j = j$, $j = 1, 2, 3, \ldots$ he obtained

$$\frac{1}{(1+xz)(1+x^2 z)(1+x^3 z)(1+x^4 z)(1+x^5 z) \ \ldots.}$$
$$= \ 1 + z\left(x + x^2 + x^3 + x^4 + x^5 + x^6 + x^7 + x^8 + x^9 + \ldots\right)$$
$$+ z^2\left(x^2 + x^3 + 2x^4 + 2x^5 + 3x^6 + 3x^7 + 4x^8 + \ldots\right)$$
$$+ z^3\left(x^3 + x^4 + 2x^5 + 3x^6 + 4x^7 + 5x^8 + 7x^9 + \ldots\ldots\right)$$
$$+ z^4\left(x^4 + x^5 + 2x^6 + 3x^7 + 5x^8 + 6x^9 + 9x^{10} + \ldots\right)$$
$$+ z^5\left(x^5 + x^6 + 2x^7 + 3x^8 + 5x^9 + 7x^{10} + 10x^{11} + \ldots\right)$$
$$\ldots$$

For example, he found that the number of ways of writing the number 11 as the sum of 5 integer positive numbers is 10. By setting $z = 1$, Euler obtained

$$\frac{1}{(1+x)(1+x^2)(1+x^3)(1+x^4)(1+x^5) \ \ldots}$$
$$= \ 1 + x + 2x^2 + 3x^3 + 5x^4 + 7x^5 + 11x^6 + 15x^7 + 22x^8 + \ldots$$

The coefficients of the latter series (which today is considered as defining the partition function) give the number of ways of writing an integer n as the sum of any number of (both different or equal) positive integers. For instance, 6 can be written in 11 ways as the sum of positive integers.

18 Analysis after the 1740s

In the previous chapters of this second part, I described the growth of the theory of series from the 1720s to the 1750s. However, this evolution was part of a more general change in analysis which, during the 18th century, became an autonomous discipline, independent of geometry and arithmetic.[257] This change matured in the 1730s and 1740s and was made manifest by the publication of Euler's *Introductio in analysin infinitorum* in 1748.

In this chapter, I shall discuss the basic principles of the 18th-century concept of analysis, which lasted through to the first decades of the 19th century. The success and decline of the formal theory of series would be unintelligible if it was not considered within this context.

18.1 Eighteenth-century analysis as nonfigural and symbolic investigation of the real

In the preface to the *Institutiones calculi differentialis*, Euler made two remarkable observations about the nature of the differential calculus. First, he explicitly rejected geometrical confirmation as a means of testing the validity of the calculus, namely, he refused to accept proofs of the calculus' correctness based solely on the fact that the calculus reached the same conclusions as elementary geometry: The calculus cannot have its own foundation in a geometrical reference [1755, 6]. He then observed:

> I mention nothing of the use of this calculus in the geometry of curved lines: that will be least felt, since this part has been investigated so comprehensively that even the first principles of the differential calculus are, so to speak, derived from geometry and, as soon as they had been sufficiently developed, were applied with extreme care to this science. Here, instead, everything is contained within the limits of pure analysis so that *no figure is necessary to explain the rules of this calculus.* (Euler [1755, 9; my emphasis])

Similar statements can be found in Lagrange's writings. Indeed, in 1773, he wrote:

> I hope that the solutions I shall give will interest geometers both in terms of the methods and the results. These solutions are purely analytical and can be understood without figures. (Lagrange [1773, 661])

And, in his *Traité de mécanique analytique*, he stated:

---

[257]On 18th-century analysis, see Fraser [1989] and [1997].

One will find no figures in this work. The methods that
I present require neither constructions nor geometrical or mechanical reasonings, but only algebraic operations, subject to a
regular and uniform course. Those who admire analysis will with
pleasure see mechanics become a new branch of it and will be
grateful to me for having extended its domain (Lagrange [1788,
2]).

The insistence on figures can be easily understood if one thinks of the
role that figures played in geometry (I refer to Chapter 7).[258] In effect, when
Euler and Lagrange claimed that figures were absent from their treatises,
they were claiming the absence of inference derived by the mere inspection
of a figure and therefore the independence of analysis from geometry, understood as a figural study of curves. This gives rise to a crucial question:
*What basic principles and instruments were used by 18th-century analysts
to make analysis truly independent of geometry?*

To answer this question, I shall begin by observing that d'Alembert
considered the principles of analysis to be "based upon merely intellectual
notions, upon ideas that we ourselves shaped by abstraction, by simplifying
and generalising the 'first' ideas".[259] In other terms, analysis was considered
as a system of merely intellectual notions, where the term "intellectual"
referred to a form of knowledge that was not based on material awareness
but was conceptual and mediated; it functioned in a discursive way along
abstract notions. Whereas geometry was entrusted, to a certain extent, to
the intuitive immediacy of an inspection of the figure and the perception
of the relationships shown in the diagram, analysis was understood as a
conceptual system where deduction was merely linguistic and mediated, or
to put it another way, proceeded from one proposition to another discur-
sively. Eighteenth-century analysis was not simply the linear continuation
of Leibniz's or Newton's analysis but was based on a new way of doing
mathematics. This new concept of analysis is undoubtedly closer to modern
analysis than the previous one, even though it presents some aspects that
significantly distinguish it from the modern concept. One of these aspects
was the very notion of mathematical theory, which I shall examine in the
remainder of this section.

$$* \quad \overset{*}{} \quad *$$

In Chapter 7, we already saw that the decisive aspect of analytical symbolism was the fact that signs were the concrete objects of a calculation,

---

[258]In Chapter 14, we also saw that this conception of the function of figures was still
true for Euler.

[259]See d'Alembert [1773, 5:154]. By contrast, geometry and mechanics were "material
and sensible" science; in particular, geometry was "the science of the properties of extension as it is considered as merely extended and figured" (d'Alembert [1773, 5:158]).

namely of a manipulation performed according to certain rules, independently of the meaning of the symbols (syntactically, in modern terms). Of
course, analysis cannot be reduced to the mechanical or blind manipulation of letters. It is not only a matter of the inventiveness necessary to
derive formulas that are not reduced to a simple exercise; rather, the point
is that doing mathematics does not merely consist of deriving formulas but
of deriving formulas that have an interest or a sense in a certain context.

This is also true for modern formal theories. A theorem T of a formal
theory is the last proposition of a sequence of propositions P_i, $i = 1, \ldots, n$,
where $P_n = T$ and P_i, $i = 1, \ldots, n - 1$, is an axiom or is deduced by a rule
of inference from the preceding propositions. While all derivable propositions in the given theory are theorems in this sense, in mathematical praxis,
only some propositions (significant for whatever reason) are theorems. The
decision that P_n is a theorem, while P_{n-1} is not, is not part of the formal
structure of theory. However, the goal of a formal theory is to yield theorems
in this more restricted sense.[260]

I would argue that the nature of analytical or algebraic derivations is
necessarily syntactical and, as such, one handles signs associated with certain rules regardless of the meaning of the objects of calculation; however,
the syntactical rules that govern analytical signs must make sense for the
mathematician and must yield results that make sense or have some interest.

Eighteenth-century analysis was symbolical in the sense that it dealt
with quantities that were reified into concrete signs and were manipulated
according to certain fixed transformations. However, the way in which the
syntactical structure was constructed differed profoundly from the way it is
conceived today. Today the rules[261] used in a theory are explicit axioms,
which in principle are freely chosen, or, to use a widely employed term, arbitrary.[262] Within the limits of the given system of axioms, mathematical
objects can freely be created by arbitrary definitions. In this way, the development of a theory is entirely syntactical and it is possible to make a
distinction between syntactical correctness and semantic truth.

This is not the case for 18th-century mathematicians. The idea of the free
creation of mathematical objects was lacking. Mathematical objects did not
exist in virtue of implicit or explicit definitions. They were always connected

---

[260] On this, see Panza [1997, 366–367].

[261] It is clear that by "rules of manipulation", I do not intend rules of inference, but rules
of the type $ab = ba$, which in modern formal theory are axioms (or theorems derived from
axioms).

[262] Here freedom and arbitrariness do not mean that one chooses the system of axioms and
gives definitions without reason; rather, it means that axioms and definitions are fixed by
an act of will determined by the targets that one wants to achieve, without other restraint
to the achievement of such targets. Axioms and definitions have no intrinsic necessity,
neither do they consist of a description of physical or geometrical reality; however, they
must have the capability of representing certain concepts adequately.

with reality, directly or indirectly.[263] The rules of manipulation were not arbitrary: They were derived from the notion of quantity and expressed properties of quantities (or of numbers). For instance, $a + b = b + a$ was not an arbitrary axiom associated with the operation $+$ (which we may or may not choose, according to the objectives of our theory); it was a mere consequence of the concept of joining two quantities.

A system of explicit axioms in the modern sense and an accurate construction of certain mathematical objects (e.g., the construction of the different species of numbers) were lacking. In their place, 18th-century mathematics admitted the reference to the intuitive knowledge of the mathematical notions drawn from pre-mathematical experiences.

This depended on the 18th-century concept of mathematics as a "science of nature".[264] Analysis was considered as a mirror of reality; its objects were idealizations derived from the physical world and had an intrinsic existence before and independently of their definition. Mathematical propositions were not merely hypothetical but concerned reality, and were true or false accordingly to whether or not they corresponded to the facts. For instance, d'Alembert stated: "The physicist ignorant of mathematics considers the truths of geometry as if they were grounded upon arbitrary hypotheses and as mere whims (jeux d'esprit) that entirely lack any applications."[265]

This led to a lack of distinction between syntax and semantics and to the impossibility of distinguishing a syntactically correct theory from semantically true theory. Today, stating that a proposition "p" of the mathematical theory T is syntactically correct is not the same as saying that it is semantically true. The truth can be predicated of "p" if and only if we specify what universes of objects constitute the models of the theory T. In this case, we say that "p" is true if the event p occurs in the model M where T has been interpreted. Given the theory L_1 containing the statement p and the theory L_2 containing the statement non-p, if one asks: "May L_1 and L_2 be correct simultaneously?", we today answer that L_1 and L_1 can be syntactically correct at the same time and, even, both true provided they are interpreted by two different models. In the 18th century, mathematicians thought that a theory was acceptable only if it conformed to the reality. Since reality is *unique*, two alternative theories based on alternative definitions of certain notions (e.g., the sum of a series and limit of a sequence) or different axioms could not be correct simultaneously.

---

[263]If no intuitive interpretation of them was known —e.g., imaginary numbers—, they were viewed as tools for improving the analytical theory of quantity, in the same manner as the sign 0 improves the notation of natural numbers that counts objects even though it denotes no object.

[264]See, for example, the preliminary discourse to the *Encyclopédie* of d'Alembert.

[265]See d'Alembert [1773, 5:121].

18.2   Functions, relations, and analytical expressions

The transformation of analysis into a system based on linguistic deduction was made possible by the notion of function.[266] By "function" Leibniz initially denoted a line that performs a special duty in a given figure (Youschkevitch [1976, 56]). Later, Leibniz used this term to denote a part of a straight line that is cut off by straight lines drawn solely by means of a fixed point and points of a given curve.[267] Therefore, functions were merely geometric variables.

The calculus, however, expressed geometric quantities analytically (by "indeterminates and constants") and, already during the first decades of the calculus, mathematicians felt the need to give a name to such analytical expressions of geometric quantities. Thus, while investigating the isoperimetric problem that consists of minimizing the area enclosed by a curve, Johann Bernoulli termed them "functions", with Leibniz's agreement (see Leibniz [GMS, 3:506–507 and 526]).[268] However, it was only as a result of Euler's work that the notion of a function assumed a crucial role in mathematics. According to Euler, "A function of a variable quantity is an analytical expression composed in any way of that variable and numbers or constant quantities" [1748a, 1:18].[269] At first glance, this definition seems to reduce a function to an analytical expression. In reality, the problem is considerably more complex. In order to make this point clear, let us examine Euler's definition for functions of more than one variable:

> 77. Even though we have so far examined more than one variable quantity, they were connected so that each of them was the function of only one variable and once the value of one variable was determined, the others would be simultaneously determined at the same time. We shall now consider certain variable quantities that do not depend on one another; if a determined value is given to one of these variables, the others remain indeterminate and variable. It would be convenient to denote such variables with x, y, z, because they comprise all determined values; if they are compared with each other, they will completely unconnected, since it is legitimate to replace any value of one of them such as z, and the others, x and y, remain entirely free as before. This is the difference between dependent variable quantities and independent variable quantities. In the first case, if we determine one, all the others are determined. In the second case, the de-

---

[266]On the 18th-century concept of a function, see Fraser [1989] and Panza [1996].

[267]For instance, see Leibniz [GMS, 5:268 and 316].

[268]In his [1718, 241], Bernoulli gave the following definition: "I call a function of a variable quantity, a quantity composed in whatever way of that variable quantity and constants."

[269]See the definition of the general term (Section 13.1, p. 157).

termination of a variable in no way restricts the meanings of the others.

78. Therefore a function of two or more variable quantities x, y, z is an expression composed of these quantities in whatever manner. (Euler [1748a, 1: 91])

Euler first, in Section 77, spoke of "dependence" among variables; he later, in Section 78, defined a function of more than one variable as an analytical expression. This seems to be a contradiction. This apparent contradiction can often be found in 18th-century texts (see Panza [1992, 695–696]). Thus, in his *Théorie des fonctions analytiques*, Lagrange first stated: "The term function of one or more quantities shall be given to every expression of calculus to which these quantities belong, with or without other quantities which are considered as given and invariable, so that the quantities of the function can have all possible values" (Lagrange [1797, 15]). However, he was later to assert: "In general, by the characteristic f or F placed before a variable, we shall denote any function of this variable, that is to say, any quantity dependent on this variable and that vary according to it following a given law" (Lagrange [1797, 21]).

In my opinion, the 18th-century concept of function effectively contained both the idea of dependence or relation among variables and the idea of analytical expression. A function was intended as the analytical expression of a relation between general quantities: *It was a pair consisting of a relation between quantities and of the formula that analytically expressed this relation.* Not only were the notions of analytical expressions and relations between quantities not contrasted with each others, but they were closely intertwined. An analytical expression was a function since it reifies a relation between quantities; conversely, a relation between quantities could be the object of study in analysis only insofar as it was expressed by an analytical expression or formula.

I also specify that 18th-century mathematicians often referred to a function as a quantity. For instance, Euler stated: "A function itself of a variable quantity is a variable" [1748a, 1:18]. Here the word "quantity" denoted a quantity depending on other quantities (and therefore the word "quantity" denoted what I have termed as a "relation between quantities"). By using this terminology, one can state that *a function was a pair consisting of a quantity —depending on other quantities— and the analytical expression of this quantity.*[270]

A crucial aspect of the 18th-century concept of function was that only the relationships that were analytically expressed by means of *certain determined formulas* were actually accepted as functions. The following excerpt

---

[270] Afterwards I shall often conform to this use and speak of a "quantity" in place of a "relation between quantities."

from Lagrange's *Leçons sur le calcul des fonctions* is useful to make this clear:

> The functions that we have considered in the last three lessons [they are: x^m, a^x, $\log_a x$, $\sin x$, $\cos x$, $\arcsin x$, $\arccos x$] are as the elements of which all functions, which can be formed by algebraic operations, are composed. For this reason I have thought to start by trying the derived functions of these simple functions; now I go on to illustrate how one can find the derived functions of the functions that are composed by means of the simple functions in any way (Lagrange [1806, 48]).[271]

Lagrange identified a set of functions that acted as the basic building blocks. For the sake of simplicity, I shall term them "basic functions". In his opinion, all other functions were constructed by means of basic functions using only the operations of addition, subtraction, multiplication, division and composition of functions. This concept was widely shared in the 18th century. Following Fraser,[272] one can therefore state that

(F) a function was given by one analytical expression constructed from variables in a finite number of steps using basic functions, algebraic operations and composition of functions.

In the remainder of this book I shall refer to the functions included in this notion as "elementary functions".

I specify that functions could also be given in an implicit form. For example, in the first chapter of [1748a], Euler wrote:

> [A]lgebraic functions can not often be exhibited explicitly, a function of z of this type is Z if it is defined by an equation such as $Z^5 = azzZ^3 - bz^4Z^2 + cz^3Z - 1$. Indeed, although this equation cannot be solved, it is however known that Z is equal to any expression composed of the variable z and constants and, therefore, Z is a certain function of z (Euler [1748a, 19–20]).

If $F(x,y)$ was an elementary function of x and y, then the equation $F(x,y) = 0$ expressed a function y of x, even when the explicit form of the function $y = y(x)$ was not known. Similarly, if $y = f(x)$ was a function of x, it was assumed that x was a function of y, even though the analytical expression $x = g(y)$ was unknown. In both cases, it was considered sufficient to have an analytical expression, $F(x,y)$ or $f(x)$, upon which one could operate.

---

[271]The same concept is expressed in Lagrange [1813, 25–26].
[272]See Fraser [1989, 325].

There were historical reasons for the choice of the above-mentioned set of basic functions. At the end of the 17th century, only algebraic relations between quantities were expressed using formulas. Transcendental relations "were expressed by means of certain circumlocutions in prose" (see Bos [1974, 5]). Subsequently, logarithmic and exponential relations were expressed analytically by formulas involving numbers, letters, and abbreviations; trigonometric functions followed around 1740.[273] In the 1740s, the set of basic functions was created and was explicitly described in Euler [1748a]. In the second half of the century, it remained substantially unchanged.

The specific set of basic functions was also connected with the widespread notion of analysis as a unitary theory based upon the step-by-step extensions of arithmetical rules.[274] For instance, Lagrange stated that analysis considered the functions that resulted from the generalization and symbolic representation of arithmetical operations (Lagrange [1806, 10]).[275] This was due to the fact that functions were not really defined if the term "definition" is taken to mean a free act of will by which we create the *definiendum*. Thus, in the case of the exponential function a^x, mathematicians did not define it but assumed the existence of a quantity with the following properties:

(1) It interpolated a^n;

(2) it possessed the same properties as the arithmetical operation of raising to a power;

(3) it could be represented by a "nice" curve.

The function a^x was merely the expression of this quantity in an abstract form using symbolic notation.

It should be noted that basic functions were thought to satisfy the following conditions, which made them different from other relations between quantities:

(C1) *A special calculus concerning these functions existed (i.e., a group of algorithmic rules related to the analytical expression, such as the rule of the calculus of trigonometric functions).*[276]

---

[273]They were introduced later, when their link with the exponential function had been established and it had been highlighted that they occurred as solutions to certain differential equations (see Katz [1987]).

[274]See Panza [1992, 701–702] and Jahnke [1993, 281].

[275]This idea even made the introduction of trigonometric functions problematic (see Panza [1992, 701]).

[276]In [1754–55a], Euler wrote: "The different kinds of quantities, which Analysis deals with, generate different types of calculus, where rules had to be adapted to any kind of quantities. Thus one teaches the special algorithm of both fractions and irrational quantities in elementary Analysis. The same use occurs in higher Analysis. There, since logarithmic and exponential quantities, which formed a new kind of transcendental quantities,

(C2) *The values of basic functions were considered as given since they could be calculated by performing algebraic operations and using tables of values.*

Conditions (C1) and (C2) were considered to be preserved by algebraic operations and compositions of functions and to be shared by all elementary functions. Eighteenth-century mathematicians regarded elementary functions as satisfying these conditions.

It should be noted that (C1) and (C2) were precisely those conditions that allowed the object "function" to be accepted as the solution to a problem. In general, it is necessary to exhibit a known object in order to solve a problem. During the 18th century, only elementary functions (in the above sense) were thought to be known objects[277] to the point that they could be accepted as the final solution to a problem.

Conditions (C1) and (C2) were vague and did not specify when a relation was to be considered as known; they did not imply that the set of accepted functions was definitively fixed. In effect, 18th-century mathematicians were prepared to extend it by introducing new functions, once certain (relations between) quantities were considered as known, but the introduction of new functions only took place very slowly and with great uncertainty. Only around 1800 did many mathematicians, such as Legendre,[278] accept new functions, but Lagrange did not consider them in any edition of his treatises published from 1797 to 1813. I will return on this question in Chapter 29.

Moreover, 18th-century functions were characterized in an essential way by the use of a formal methodology that it made it possible to operate upon analytical expressions, independently of their meaning. In Chapter 7, we saw that formal methodology was based upon two closely connected principles and examined one of these principles, the extension of rules and procedures from the finite to the infinite. I shall now illustrate the other principle, the generality of algebra. It consisted of the following assumption:

(GA) *If an analytical formula was derived by using the rules of algebra,*[279]

---

enter in computations, one usually teaches a special type of algorithm concerning both symbols and rules. It was termed exponential calculus by the inventor Joh. Bernoulli and also treats the theory of logarithm and their differentiation and integration. In addition to the logarithmic and exponential quantities there occurs in analysis a very important type of transcendental quantity, namely the sine, cosine and tangent of angles, whose use is certainly the most frequent. Therefore this type rightly merits, or rather demands, that a special calculus be given, whose invention in so far as the special signs and rules are comprised, the celebrated author of this dissertation [Euler] is able rightly to claim all for himself, and of which he gave examples in his Introduction to Analysis and Institutions of Differential Calculus" [1754–55a, 542–543].

[277]In analysis, an object was considered as known if it had an analytical expression on which one could operate and if one could at least partially calculate its values.

[278]See Chapter 29.

[279]Here the expression "the rules of algebra" is meant in a general sense. It includes not only the rules of the algebra of finites but also the rules of analysis of infinite quantities.

then it was thought to be valid in general.

In his *Calculus as Algebraic Analysis*, Fraser expressed this principle by stating "The existence of an equation among variables implies the global validity of the relation in question" (Fraser [1989, 329]). The generality of algebra made it possible to view a function as a whole. Its behavior became a global matter that could not be reduced to the sum of the behavior of the points of its domain: It could not have a property P here, and a different property there. This does not mean that 18th-century mathematicians merely considered functions that had the property P in every point: Rather they assumed rules that were valid over an interval I_x (or, more precisely, for certain values that this variable x assumed moving with continuity) as globally valid. For this reason, if one proved that a function $f(x)$ had the property P in the interval I_x, then one could extend this property beyond the interval I_x. For instance, the rules concerning the function $\log x$ were derived for positive values of quantity x; however, it was assumed that the properties of the expression "$\log x$" lasted beyond the original interval of definition, even when x is negative or even imaginary.[280]

Of course, if what was valid in an interval was generally valid, not only did a function possess the same properties everywhere but also it maintained the same form everywhere since the form embodied all its properties.[281] Therefore, one function necessarily consisted of one single formula[282] and a relation such as

$$f(x) = \begin{cases} x^2 & \text{for } x \geq 0 \\ x^3 & \text{for } x < 0 \end{cases}$$

was never considered a function.[283]

In this context, it is necessary to emphasise the fact that the equality $f(x) = \sum_{i=0}^{\infty} a_i x^i$ was not restricted to the values of x where the series was convergent. The generality of algebra implied that the relation $f(x) = \sum_{i=0}^{\infty} a_i x^i$ could not be regarded as a local relation, only valid for an interval. Vice versa, a statement of the type

---

[280]Euler stated: "For, as this calculus concerns variable quantities, that is quantities considered in general, if it were not generally true that $d(\log x) = dx/x$, whatever value we give to x, either positive, negative or even imaginary, we would never able to make use of this rule, the truth of the differential calculus being founded on the generality of the rules it contains" [1749b, 143–144]. For two examples of the generality of algebra from Legendre, see Chapter 29.

[281]One may ask to what extent this conception also belonged to pre-Eulerian analysis. The generality of algebra was also part of pre-Eulerian analysis insofar as it used analytical expressions, but its impact proved to be somewhat restricted by the vicinity of geometric reference, the disappearance of which led to the explosion of formalism.

[282]See Fraser [1989, 325].

[283]I note that an analytical expression of the type $y^2 = f(x)$ was considered as a many-valued implicit function (see p. 207).

$$f(x) = \sum_{i=0}^{\infty} a_i x^i \text{ for } x \in (a,b) \text{ and } f(x) \neq \sum_{i=0}^{\infty} a_i x^i \text{ for } x \notin (a,b)$$

involved a rejection of the generality of algebra.

A heavy reliance on formal methods prevented 18th-century mathematicians from appreciating the difference between complex and real variables and, therefore, between complex and real analysis. The transition from real to imaginary values usually occurred only by applying the generality of algebra, without proving specific rules for imaginary values of variables. Complex functions were not an autonomous object of study but, instead, were useful tools for the theory of real functions, and their use was restricted to exceptional circumstances.

Finally, it is also worthwhile noting that the generality of algebra was restricted to analysis, where functions were studied without *a priori* restrictions concerning variables. In arithmetic, geometry, and mechanics, functions and variables have a natural range; therefore, mathematicians were obliged to take into consideration the restrictions that the nature of the specific problem under examination imposed. When the results derived from the use of the generality of algebra were applied to other sciences, they had to be subjected to appropriate reinterpretations that adapted them to concrete circumstances. This approach is an aspect of the mathematical method for studying natural science in the 18th century, which Dhombres [1988] referred to as the "functional method." By solving a problem mathematically, appropriate symbols replaced concrete quantities and their relations come to be conceived as formulas and equations. The solutions to these equations were to be interpreted in relation to the specific problem and by eliminating anything that was meaningless for this particular problem. The whole theory of series is an example of this conception, since the convergence was studied *a posteriori* as a condition for the applicability of the formally derived results.

18.3   On the continuity of curves and functions

When referring to 18th-century functions or curves, the term "continuity" can be understood in two different (though connected) ways.

First, continuity can be understood as the absence of jumps or the assumption of any intermediate state between two given states or gradual change. I shall refer to this sense of continuity as local[284] continuity, or L-continuity for short. In the 18th century, functions were thought to be intrinsically continuous in this sense. This concept depended on the fact that a function $y = f(x)$ was a relation between the general quantities y and x, or in other words, the general quantity y was considered as depending on the

---

[284]Of course, local continuity does not mean pointwise continuity. According to the description on p. 102, the continuous was not reducible to points and a function —in the sense of a relation between general quantities— was not defined pointwise.

quantity x. Since a general quantity was continuous, any function $y = f(x)$ was considered as such. In analytical terms, this implied that the following property held for any function:[285]

(LC) $\Delta f = f(x + \omega) - f(x)$ is infinitesimal if ω is infinitesimal.[286]

I explicitly emphasize that this property was not the definition of the continuity of a function but merely a trivial consequence of the notion of continuous quantity.[287]

Second, continuity can be thought of as coinciding with uniqueness. The basic idea behind this concept is that an object is continuous if it is an unbroken object, i.e., if it is not broken in two objects and is therefore *one* object.[288] Even though 18th-century mathematicians always considered functions and curves as locally continuous, the usually accepted definition of the continuity of a function or curve was based on the property of uniqueness. I shall refer to this way of understanding the continuous as global continuity, or G-continuity or Euler's continuity for short. I shall investigate global continuity in this section, as concerns continuous curves, and in Chapter 23, as concerns G-continuous functions.

First of all, I observe that if continuity is the same thing as uniqueness, *one* curved line was G-continuous merely because it was one. Therefore, the notion of a continuous curve may appear superfluous and useless. However, in analytical geometry, a curve is represented by an analytical expression and one analytical expression does not necessarily correspond to an unbroken curve. For instance, the function $y = \frac{k}{x}$ is G-continuous since it is one, but its geometrical counterpart, the hyperbola of the equation $y = \frac{k}{x}$, is broken into two pieces: It is then very natural to ask whether the hyperbola is continuous, i.e., whether its two pieces form an unique curve. Put in more general terms, *how does one recognize that an object is one?* The most obvious answer is that an object is one if it retains its properties. Now, if we study a curve analytically, its properties are included within its analytic expression. If we accepted this view, then it is entirely natural that the criterion of uniqueness must be applied to the analytical expression, as Euler did in classifying curves.[289] Indeed, he stated that although some curves could be described mechanically, he aimed to study curves insofar as they were originated by functions because this method was the most general and best suited to calculation. According to Euler, from such an idea about curved lines, it immediately follows that they should be divided

---

[285]See, for instance, Euler [1755, 82] and Lagrange [1797, 28].

[286]The property (LC) held at least for over one interval. However, the generality of algebra made it possible to consider L-continuity as a property of the whole function.

[287]See Chapter 7.

[288]This is precisely the Aristotelian concept of continuity. On the notion of continuity in Aristotle, see Panza [1989, 39–65].

[289]See Euler [1748a, 2: Section 8].

into *continuous* and *discontinuous* or mixed. A curve was continuous if its nature was determined by only one function, and discontinuous or mixed if it was described piecewise by more than one function and, consequently, was not formed according to an unique law. Uniqueness did not apply to the course of a curve, which was seen as an outward manifestation, but to the function itself as a primary object. The number of the branches of a curve was therefore of no importance.

Euler also subdivided curves into complex and not complex ones using a similar criterion. He noted that the equation of certain algebraic curves could be broken down into rational factors:

> Such equations include not one but many continuous curves, each of which can be expressed by a particular equation. They are connected with each other only because their equations are multiplied mutually. Since their link depends upon our discretion, such curved lines cannot be classified as constituting a single continuous line. Such equations (referred to above as complex) do not give rise to continuous curves, although they are composed of continuous lines. For this reason, we shall call these curves complex. (Euler [1748a, 2: Section 61])

The complex curves (like mixed ones) were discontinuous because their equation was characterized by arbitrariness; in other words, they are not determined by exactly one analytical law. Their difference is that the complex curves were composed of more than one whole curve, whereas mixed curves were composed of pieces of more than one curve.[290]

In his *Introductio*, Euler only considered G-discontinuous curves rather than G-discontinuous functions. This is entirely coherent with the 18th-century principle being developed by Euler. Indeed, the application of global continuity to functions required that an essential property of functions did not hold anywhere, whereas the principle of the generality of algebra made it possible to consider a function as a unitary object: The properties valid on an interval were considered valid anywhere. In this context it seemed impossible to attribute a meaning to the term "discontinuous" when referring to functions. Nevertheless, when the controversy about the vibrating string arose, the existence of noncontinuous functions was admitted and the relationship between quantities and analytical expressions appeared to be

---

[290]On the basis of these subdivisions, the curve of equation $y = \sqrt{x^2}$ is not continuous. Although it appears to be a G-continuous curve since it derives from one two-valued function, it is, in reality, the complex curve corresponding to the (implicit function) equation $y^2 - x^2 = 0$. According to Euler, uniqueness did not refer to the "apparent", complex form, but to the essential, irreducible form. In the light of this observation, Cauchy's objection to Euler's classification in [1844] should also be considered.

problematic. The way this occurred is dealt with subsequently when I illustrate the emergence of certain difficulties in the structure of 18th-century analysis.

19 The formal concept of series

The evolution of series theory after 1720 revealed tension between the formal and quantitative aspects. This led to the formulation of a different concept of series, where the previously existing balance was upset. I term this new concept *formal*. In the present chapter, I illustrate how Euler made the formal concept of series explicit.

19.1 Criticisms to the infinite extension of finite rules

The sum of $\sum_{n=1}^{\infty} \frac{1}{n^{2k}}$ in *De summis serierum reciprocarum* was one of Euler's most important successes,[291] though the method he had used also became the object of severe criticism. Some mathematicians[292] noted that such a method could be applied to other quantities and not only to $\sin s$: In this case the series

$$\sum_{n=1}^{\infty} \frac{1}{n^2}$$

would have sums that were different from those derived by Euler. Indeed, given an ellipse of axes m, l ($m > l$), let s be the length of the arc AP, where A has coordinates $\left(\frac{m}{2}, 0\right)$ and P is an arbitrary point (Fig. 20). If one considered the ordinate of P as a function $y(s)$ of s and applied Euler's method to the equation

$$y(s) = s - \frac{1}{6c^2}s^3 + \left(-\frac{1}{10rc^3} + \frac{13}{120c^4}\right)s^5$$
$$+ \left(-\frac{1}{14r^2c^4} + \frac{71}{420rc^5} - \frac{493}{5040rc^6}\right)s^7 + \ldots = 0,$$

where $r = \frac{m}{2}$ and $c = \frac{l^2}{2m}$, one obtained

$$\sum_{n=1}^{\infty} \frac{1}{n^2} = \frac{1}{6c^2} \cdot \frac{S^2}{4} = \frac{S^2 m^2}{6l^4},$$

where S is the circumference of the ellipse, in contradiction with $\sum_{n=1}^{\infty} \frac{1}{n^2} = \frac{\pi^2}{6}$.

In *De summis serierum reciprocarum dissertatio altera*, Euler[293] replied to this criticism by stating that the equation $y(s) = 0$ had imaginary roots, unlike $\sin s = 0$, and it was therefore not possible to deduce the sum of series

---

[291] See Section 13.3.

[292] See Fuss [1843, 2:477 and 683].

[293] It is worth observing that Euler also responded to criticisms by seeking new proofs of the results of [1734–35a] (see, e.g., *Démonstration de la somme de cette suite* $1 + \frac{1}{4} + \frac{1}{9} + \frac{1}{16} + \frac{1}{25} + \frac{1}{36} + \ldots$ [1743b]).

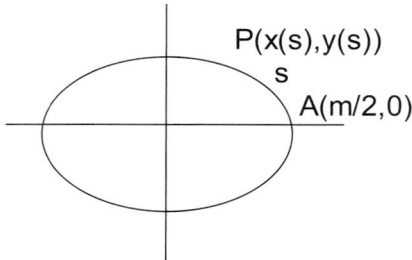

Fig. 20

from this equation (Euler [1743a, 139–140]). In a letter to Euler, however, Nikolaus Bernoulli II observed that, even if $\sin s = 0$ did not have imaginary roots, his reasoning was illegitimate unless "it had been demonstrated that the series $s - \frac{1}{6}s^3 + \frac{1}{120}s^5$—etc. was convergent and gave the sinus of the arc s accurately, whatever value was assigned to s" (Fuss [1843, 2:683]). In Bernoulli's opinion, if one was not sure that series were convergent, one might be wrong, and this fact occurred when one reasoned on $y(s)$ instead of $\sin s$. In effect, the validity of the factorization

$$1 - \frac{s}{y} + \frac{1}{3!y}s^3 - \frac{1}{5!y}s^5 + \ldots = \left(1 - \frac{s}{A_1}\right)\left(1 - \frac{s}{A_2}\right)\left(1 - \frac{s}{A_3}\right)\ldots$$

depended on the convergence of $s - \frac{1}{6}s^3 + \frac{1}{120}s^5 - \ldots$:

> and for this reason the objection concerning the series of the sinus of the elliptic arc $s - \frac{s^3}{6c^2}$+etc. is solved. This series is divergent for increasing s; from which it cannot be deduced, as in the circle where the series is not divergent, that the coefficient, taken as a negative, of the second term in the infinite equation $0 = s - \frac{s^3}{6c^2} + etc$, i.d.. $\frac{1}{6c^2}$ expresses the sum of all $\frac{1}{ss}$, namely it is $= \frac{1}{\pi\pi} + \frac{1}{4\pi\pi} + \frac{1}{9\pi\pi}$+etc. (Fuss [1843, 2: 691])

On April 6, 1743, Bernoulli explicitly wrote to Euler:

> I cannot persuade myself that you [Euler] think that a divergent series ... provides the exact value of a quantity which is expanded into the series. Indeed, e.g., $\frac{1}{1-x}$ is not $= 1 + x + xx + x^3 + \ldots + x^\infty$, but $= 1 + x + xx + x^3 + \ldots + x^\infty + \frac{x^{\infty+1}}{1-x}$. (Fuss [1843, 2:701–702])

Later, Bernoulli clarified: "I think the idea of the sum or aggregation of more terms cannot be associated with the idea of terms advancing endlessly, and I regard these two ideas as contradictory" and argued that the properties applied to a finite equation could not be applicable to an infinite equation (Fuss [1843, 2:708–709]).

Although N. Bernoulli's concept might seem close to modern ideas, in reality he reproposed the traditional formal-quantitative concept of series. Indeed, in his *Inquisitio in summa seriei* $1 + \frac{1}{4} + \frac{1}{9} + \frac{1}{16} + \frac{1}{25} +$ etc. [1738], in order to sum the recurrent series $\sum_{n=1}^{\infty} a_n$, with

$$a_{n+1} = \frac{An(n-1) + Bn + C}{E(n+1)n + F(n+1) + G} a_n$$

(A, B, C, E, F, and G are constants), N. Bernoulli considered

$$y(x) = \sum_{n=1}^{\infty} a_n x^n$$

and derived the expression of dy and d^2y by differentiating term by term. By using principle[294] (IE), he obtained

$$Gy dx^2 + Fx dy dx + Ex^2 d^2 y$$
$$= \left(\sum_{n=1}^{\infty} G a_n x^n \right) dx^2 + \left(\sum_{n=1}^{\infty} Fn a_n x^n \right) dx^2 + \left(\sum_{n=2}^{\infty} En(n-1) a_n x^n \right) dx^2$$
$$= \left(G a_1 x + \sum_{n=1}^{\infty} C a_n x^{n+1} \right) dx^2 + \left(F a_1 x + \sum_{n=1}^{\infty} Bn a_n x^{n+1} \right) dx^2$$
$$+ \left(\sum_{n=2}^{\infty} An(n-1) a_n x^{n+1} \right) dx^2$$
$$= G a_1 x dx^2 + F a_1 x dx^2 + Cxy dx^2 + Bx^2 dy dx + Ax^3 d^2 y.$$

The differential equation

$$Gy dx^2 + Fx dy dx + Ex^2 d^2 y$$
$$= G a_1 x dx^2 + F a_1 x dx^2 + Cxy dx^2 + Bx^2 dy dx + Ax^3 d^2 y$$

provided the sum $y(x)$. Surely N. Bernoulli intended to apply this procedure to a convergent series, but any *a priori* discussion of the convergence and of the legitimacy of the steps of the proof is lacking. He differentiated term by term and freely rearranged the series. However, by differentiating and integrating term by term and freely rearranging, a convergent series can be turned into a divergent one, and vice versa. This had already been noted by

---

[294] See p. 117.

Leibniz in his *Epistola ad V. Cl. Christianum Wolfium*[295] [1713] and was to be one of the reasons for Daniel Bernoulli in his *De summationibus serierum* [1771, 76] to use of the sum of divergent series. D. Bernoulli observed that, by integrating

$$\sum_{n=0}^{\infty}(-1)^n(n+1)x^n = \frac{1}{(1+x)^2},$$

one had

$$\sum_{n=0}^{\infty}(-1)^n x^{n+1} = \frac{x}{1+x}.$$

If one divided this equality by x and integrated again, one obtained

$$\sum_{n=0}^{\infty}(-1)^n \frac{x^{n+1}}{n+1} = log(1+x).$$

Setting $x = 1$, the three series yielded

$$\sum_{n=0}^{\infty}(-1)^n(n+1) = \frac{1}{4},$$

$$\sum_{n=0}^{\infty}(-1)^n = \frac{1}{2},$$

$$\sum_{n=0}^{\infty}(-1)^n \frac{1}{n+1} = log\,2,$$

respectively. Daniel Bernoulli thought that the validity of the last relation, in a sense, guaranteed the validity of the two others. Of course, Nikolaus Bernoulli could have replied that in the first two cases the series were not convergent; however, it had to be observed that he never analyzed the condition of convergence *a priori* and, above all, did not provide the conditions of applicability of certain procedure. In his *Inquisitio* he limited himself to observing *a posteriori* that a formal theorem was applied to a particular, convergent series. The formal study of a theorem and consideration of the conditions of its applicability in concrete cases *a posteriori* is just one of the characteristics of a concept that contains formal components. For this reason, his criticisms of the infinite extension of finite laws were weak: He used the same principle (IE) that Euler had, even though Euler used it in a stronger form. N. Bernoulli's ideas could be developed in a more modern sense, but this did not occur during the 18th century. People who criticised Euler were unable to avoid formal methods entirely. When considered seriously, N. Bernoulli's arguments involved a rethinking of the whole analysis,

---

[295]See footnote no. 185.

which went beyond the intention of the mathematicians of the time and the state of the art.

There was a second reason why N. Bernoulli's standpoint was weak. It did not produce results of wide interest; by contrast, divergent series gave rise to a number of significant findings and were to prove fertile ground for further investigations in later decades. I would like to emphasize this point: *The formal point of view contributed to the development of mathematical knowledge*, whereas the approach of their opponents was substantially sterile during the middle of 18th century. I think that this was the heart of question. *The formal approach was triumphant because it was capable of producing new mathematics.*[296]

19.2   The impossibility of the quantitative approach

In the 1750s, the criticisms of his opponents led Euler to make the formal concept of series explicit and to go beyond the formal-quantitative approach. He sought to give a definition of the sum that generalized the old notion and provided a basis for new findings. Before examining the formal concept, however, it is necessary to make some preliminary remarks.

In writing to Euler, N. Bernoulli asserted that the "idea of the sum" was to be clarified in order to avoid the situation whereby "the dispute becomes a logomachy" (Fuss [1843, 2:708]). Euler observed that the problem lay in the word "sum" and thought it was appropriate to provide an adequate definition of this term [1754–55b, 589]. These statements should not give the idea that the disputants gave a "definition" in the modern sense of the term, i.e., in the sense of the numerous definitions of the sum of a divergent series given after 1890. As we saw, in the 18th century, a mathematical theory was not the set of the logical deduction of axioms and definitions, subject only to the principle of contradiction; rather, it was an idealization derived from physical reality. In this context, definitions were clarifications of the unique and necessary "truth" that already existed in nature. For this reason, Euler believed that $1 - 1 + 1 - 1 + \ldots$ was equal to $1/2$ not because of an arbitrary definition, but because this equality was the unique possible equality that could be derived from the laws of analysis. In a similar way, Euler thought that

$$1 + 1 + 1 + \ldots = \infty$$

and could not assert

$$1 + 1 + 1 + \ldots = -1/2,$$

unlike Ramanujan's modern definition of the sum.[297] In Euler's opinion, setting $\sum_{n=0}^{\infty} a_n = C$ without a justification but only by   definition was

---

[296]See Part III.

[297]On this definition of the sum of a divergent series sum, see Hardy [1949, 327 and 333].

unacceptable. By defining the sum of a series, Euler wanted to indicate there was a "really existent" link between that series and a certain quantity; he did not want to create an arbitrary notion whose logical value would be independent from any verification of utility or correspondence to external truth.

Since Euler was seeking the "true" notion of the sum, in *De seriebus divergentibus* [1754–55b] and *Institutiones calculi differentialis* [1755], he provided a long analysis that allowed him to exclude the possibility of a quantitative interpretation of the sum of divergent series. In *De seriebus divergentibus*, he subdivided divergent series into four classes according to the sign and behavior of the nth term:

$$\sum_{n=0}^{\infty} a_n \text{ and } a_n < M, \tag{123}$$

$$\sum_{n=0}^{\infty} (-1)^n a_n \text{ and } a_n < M,$$

$$\sum_{n=0}^{\infty} a_n \text{ and } a_\infty = \infty, \tag{124}$$

$$\sum_{n=0}^{\infty} (-1)^n a_n \text{ and } a_\infty = \infty,$$

where M denotes a positive constant and $a_n > 0$ is a nondecreasing sequence.[298]

Divergent series with positive terms gave rise to the most serious problems: Unlike alternating divergent series, it would be natural to assign the value ∞ as their sum. For Euler, only the series of type (123) actually had an infinite sum.[299] The series of the type (124) could have a finite sum; however, the real meaning of equalities such as

$$1 + 2 + 4 + \ldots = -1$$

and

$$1 + 3 + 9 + \ldots = -\frac{1}{2}$$

was problematic. These series had negative sums but, at the same time, had to have sums greater than

$$1 + 1 + 1 + \ldots = \infty;$$

---

[298]According to the usual terminology of 18th-century, Euler termed a series $\sum_{n=0}^{\infty} a_n$ convergent if $a_n > 0$ is a decreasing sequence and $a_\infty = 0$. Euler did not discuss series of this kind, whose sum can be infinite. He probably thought that "convergent" series (in this sense) had always a quantitative meaning even when their sum was infinite.

[299]He stated: "There is no doubt that the sum of these series can be shown by means of expressions of the type $\frac{a}{0}$" [1754–55b, 589]

hence, negative numbers should be greater than infinity. Euler believed that the concept according to which negative numbers are both greater than infinite and less than zero is not far from the truth; using a modern image, we could say that the set of real numbers is closed by the infinite. In *Institutiones calculi differentialis*, Euler justified this idea by examining several sequences of the kind

$$\ldots, -\frac{1}{4}, \ -\frac{1}{3}, -\frac{1}{2}, -\frac{1}{1}, \ +\frac{1}{0}, \ +\frac{1}{1}, +\frac{1}{2}, +\frac{1}{3}, \ldots,$$
$$\ldots, -4, \ -3, \ -2, \ -1, \ +0, \ +1, \ +2, \ +3, \ldots.$$

He highlighted analogies between the zero and the infinite, showing that the "transition" from the negative to the positive occurred via both the zero and the infinite, and appealed to the law of continuity and geometry [1755, 2: Sections 98–101].

This idea was compatible with a quantitative interpretation of the sum; however, it did not prevent difficulties. Further, in his *Institutiones calculi differentialis*, Euler observed that

$$\frac{1}{(1-x)^2} = 1 + 2x + 3x^2 + 4x^3 + \ldots$$

yielded

$$\infty = \frac{1}{(1-1)^2} = 1 + 2 + 3 + 4 + \ldots \quad \text{for } x = 1$$

and

$$1 = \frac{1}{(1-2)^2} = 1 + 4 + 12 + 32 + \ldots \quad \text{for } x = 2;$$

therefore, even 1 should be greater than ∞ [1755, 2: Section 104]. He was thus obliged to reject the quantitative concept of the sum of divergent series, as it was unsuitable for any relation of order between numbers and the infinite.

With respect to alternating divergent series, the main objection to their use (a valid objection for positive term series as well) concerned the possibility of neglecting the remainder. Euler criticized mathematicians such as Nikolaus Bernoulli II, who thought that the remainder $\frac{a^{n+1}}{1+a}$ of

$$\frac{1}{1+a} = 1 - a + a^2 - a^3 + \ldots \pm a^n \mp \frac{a^{n+1}}{1+a}$$

had not to be neglected for $n = \infty$, except for $(0 <)a < 1$. However, he did not deny that a divergent series approached no limit and, therefore, the remainder could not be eliminated if one considered the relation

$$\frac{1}{1+a} = 1 - a + a^2 - a^3 + \ldots$$

as a quantitative relation (Euler [1755, 1: Section 109]). That objection proved only that divergent series could not be summed in a quantitative sense and that the quantitative notion of the sum was not acceptable if one wished to consider the relation $\frac{1}{1+a} = 1 - a + a^2 - a^3 + \ldots$ as valid independently of the condition $|a| < 1$, which was required by the principles of 18th-century analysis.

19.3   Euler's definition of the sum

In the *Institutiones calculi differentialis*, Euler ended the lengthy discussion on the sum of series by giving the well-known definition:

> I term the sum of an infinite series to be the finite expression, from the expansion of which the series is generated (Euler [1755, 1: Section 111]).

This definition, which is a generalization of the definition of the sum of recurrent series,[300] made the formal concept of series theory explicit. It was given by Euler in a letter[301] to Goldbach on August 7, 1745, and was later published on many occasions. I point out some consequences of this concept:

1. every series was conceived to have its own generating closed expression[302], which was identified with the series;

2. the reciprocal substitution between a series and its generating closed expression was always possible;[303]

3. a series was not an autonomous object of study.[304] While the quantity expressed by means of the closed expression (in particular, a function) existed independently of its series expansion, a series was merely a transformed form. Consequently, a function could never be defined by a series.

These three aspects cannot be considered as entirely novel. In the formal-quantitative approach to the theory of power series, it was obvious that the summing of series consisted of inverting the operation of development, that a function could be replaced by its expansion, and that a series was simply

---

[300]See Chapter 10, p. 137.

[301]See Fuss [1843, 2:323–326]. In this letter Euler distinguished between the sum and the value of a series. The former was the sum in the quantitative sense while the latter was the sum in the formal sense.

[302]It could be a function but also a variable quantity expressed in closed form and even a constant quantity expressed by a number.

[303]See, e.g., Euler [1754–55b, 593–594].

[304]On this subject, see also Fraser [1989, 322].

a tool for investigating analytical quantities; however, in this approach the sum of a power series was also a function to which the series converges, at least over an interval of value of x. According to Euler's definition, the fact that a series was the result of a formal transformation of a function, according to principle[305] (IE), was the *only* justification for stating that the function was the sum of the series.[306] This definition made it possible to generalize the notion of series and could be applied to series that differed from ordinary power series (for instance, to totally divergent series[307]). It was later applied to symbols of operations.[308]

I also observe explicitly that the more formal notion of series did not imply a rejection of the formal-quantitative approach, but rather that the former was the necessary presupposition of the latter. Even though new types of series were admitted, mathematicians continued to use the same procedures to expand functions and always based these procedures upon principle (IE). The possibility of inventing new procedures that were not merely an infinite extension of finite procedures was never considered.

$$* \quad * \quad *$$

In order to clarify how Euler's definition was applied in practice, I shall examine one of the methods he used: the method of finite differences.[309] In *Institutiones calculi differentialis*, he first associated the power series $\sum_{n=1}^{\infty} a_n x^n$ with the numerical series $\sum_{n=1}^{\infty} a_n$. By applying the transformation

$$x = \frac{y}{1-y}$$

to the power series, he derived

$$\sum_{n=1}^{\infty} a_n x^n = \sum_{n=1}^{\infty} \Delta^{n-1} a_1 y^n = \sum_{n=1}^{\infty} \Delta^{n-1} a_1 \left(\frac{x}{1-x}\right)^n, \qquad (125)$$

where $\Delta a_n = a_{n+1} - a_n$ and $\Delta^n a_n = \Delta^{n-1} a_{n+1} - \Delta^{n-1} a_n$. If the differences $\Delta^n a_1$ became equal to 0, (125) transformed an infinite series into a finite expression and, therefore, exhibited its sum. For instance, the sum of

$$\sum_{n=0}^{\infty} (2n+1) x^{n+1}$$

---

[305] See p. 117.

[306] In early theory a series represented a quantity insofar as it converged to that quantity: It was assumed that a formal transformation yielded a convergent power series at least on one interval. A series was now considered as being able to represent, a quantity even when it was not convergent and not only convergent power series were considered. The assumption that the expansion of any quantity yielded a convergent series continued to hold.

[307] See p. 191.

[308] See Chapters 21 and 28.

[309] It had old roots. See Goldbach [1720].

is

$$\frac{x}{1-x} + \frac{2x^2}{(1-x)^2},$$

as $a_1 = 1$, $\Delta a_1 = 2$, and $\Delta^n a_1 = 0$ for $n > 1$ (see [1755, 2: Chapter 1]).

If there is no number k such that $\Delta^n a_1 = 0$ for $n > k$, (125) is unsuitable for approximating $\sum_{n=1}^{\infty} a_n x^n$. Indeed, if $x < 1$ ("in which case only summation in the proper sense of the word can take place" [1755, 221]), then

$$\frac{x}{1-x} > x,$$

and (125) does not improve the convergence of $\sum_{n=1}^{\infty} a_n x^n$. Furthermore, for $x = 1$, (125) does not yield a numerical value.

However, by changing x into $-x$ in (125), Euler derived

$$\sum_{n=1}^{\infty} (-1)^{n-1} a_n x^n = \sum_{n=1}^{\infty} (-1)^n \Delta^{n-1} a_1 \left(\frac{x}{1+x}\right)^n.$$

Since $\frac{x}{1+x} < 1$ and $\frac{x}{1+x} < x$ for $0 < x < 1$,

$$\sum_{n=1}^{\infty} (-1)^n \Delta^{n-1} a_1 \left(\frac{x}{1+x}\right)^n$$

sped up

$$\sum_{n=1}^{\infty} (-1)^{n-1} a_n x^n$$

and provided an appropriate "approximation of the value". Setting $x = 1$, he obtained

$$\sum_{n=1}^{\infty} (-1)^{n-1} a_n = \sum_{n=1}^{\infty} (-1)^{n-1} 2^{-n} \Delta^{n-1} a_1. \qquad (126)$$

Euler stated that, in certain cases, (126) could provide the sum of divergent series. Of course, it did not provide the sum in the quantitative sense but in the formal sense or, to use his words, the value of the finite expression whose expansion generates the given series (Euler [1755, 222–223]). This could occur either because (126) changed the divergent series into a finite one or because it transformed a divergent series into a convergent one.[310]

---

[310]By applying (126) repeatedly, Euler found that $1! - 2! + 3! - 4! + 5! - \ldots$ was approximately equal to $0,40082038$: He said that it was a value not much different from $0,4036524077$, the more precise value he determined elsewhere [1755, 222–223]. Euler probably referred to the last of the values of $1 - (1! - 2! + 3! - 4! + 5! - \ldots)$, which he had determined in his [1754–55b].

For instance,

$$1 - 2 + 4 - 8 + \ldots = \frac{1}{2} - \frac{1}{4} + \frac{1}{8} - \frac{1}{16} = \frac{1}{3}.$$

Euler also used (126) to transform slowly convergent series into rapidly convergent series, such as

$$1 - \frac{1}{2} + \frac{1}{3} - \frac{1}{4} + \frac{1}{5} - \ldots = \frac{1}{2} + \frac{1}{2 \cdot 4} + \frac{1}{3 \cdot 8} + \frac{1}{5 \cdot 32} + \ldots,$$

without appreciating the crucial difference (from a modern viewpoint) between the application of (126) to a divergent or convergent series. Euler thought that the essence of the method was to derive a series

$$\sum_{n=1}^{\infty} b_n$$

from a given series

$$\sum_{n=1}^{\infty} (-1)^{n-1} a_n$$

such that

$$|b_n| < |a_n|$$

independently of the fact that we speed up the convergence or go from the divergence to the convergence or decelerate the divergence. The similarity to the modern summation method $(E, 1)$ is therefore only apparent.[311] Nowadays, if

$$\sum_{n=1}^{\infty} (-1)^{n-1} a_n$$

is divergent, (126) conventionally defines its sum, which exists by means of this definition; on the other hand, if the series is convergent, it allows us to calculate its ordinary sum more easily, which exists by virtue of another definition. By contrast, for Euler, it externalized the true sum but did not "create" it.

$$* \quad * \quad *$$

In *Institutiones calculi differentialis*, Euler argued that divergent series never led to an error being made. This was one of the reasons in favo of divergent series, because if their sums were false, they would not always lead us

---

[311] On this method of summation, see Hardy [1949].

to the truth.[312] Such a justification must have already been communicated to N. Bernoulli, who, in response, pointed out some sums of divergent series that led to an error being made (see Fuss [1843, 2:709]). On the other hand, in *Consideratio progressionis cuiusdam ad circuli quadraturam inveniendam idoneae* [1739a], Euler himself had pointed out that divergent series could cause discrepancies. As we saw in Chapter 15, he had noted that

(a) the theorem "given any series $\sum_{n=1}^{\infty} a_n$, then $\sum_{n=-\infty}^{\infty} a_n = 0$, where $a_0, a_{-1}, \dots$ are defined by (100)" was not always true;

(b) Equation (99) provided a good approximation of π but was not a quantitative equality in the sense of convergent series.

In *De seriebus divergentibus* and *Institutiones calculi differentialis*, Euler did not mention the above difficulties[313] and limited himself to inviting mathematicians to correct any accidental mistake as the absence of systematic errors was certain [1755, 81].[314] Euler therefore interpreted the results of the *Consideratio* differently.

I think that Euler merely considered the counterexample $\sum_{n=-\infty}^{\infty} \frac{1}{n^2} \neq 0$ to the theorem $\sum_{n=-\infty}^{\infty} a_n = 0$ as an exception. This approach was already implicit in the *Consideratio* when he stated that $\sum_{n=-\infty}^{\infty} a_n = 0$ did not fail in any case but led to the truth in numerous cases [1739a, 362]. From a modern viewpoint, only one counterexample is sufficient to reject a proposition. According to Euler, the existence of a counterexample on its own did not mean that the proof of $\sum_{n=-\infty}^{\infty} a_n = 0$ is incorrect. Euler conceived counterexamples (if they were sporadic) as exceptions to the rule, which had to be examined and explained, but did not invalidate a formal result.

This approach was widespread in 18th-century analysis. Mathematicians thought that a theorem concerning a generic object X could admit exceptional cases in which it fails. In other words, the theorem

the generic object X has the property P

did not imply that all the propositions

the object X_1 has the property P,

the object X_2 has the property P,

the object X_3 has the property P,

. . .

---

[312]See Euler [1755, 2: Section109].

[313]It is appropriate to note that the results of *Consideratio* were based upon the same principles Euler used in *De seriebus divergentibus* and *Institutiones calculi differentialis*: the interpretation of a_n as a function of n (in Euler's sense), the extrapolation by means of (100); the replacement of $\frac{1}{1-1}$, $\frac{1}{(1-1)^2}$, $\frac{1}{(1-1)^3}$ with $1 + 1 + 1 + \dots$, $1 + 2 + 3 + \dots$, $1 + 3 + 6 + \dots$, $\dots$, the possibility of rearranging the terms of a series.

[314]See Chapter 15, p. 183.

where X_1, X_2, X_3, ..., are specific cases of X, was true.

In the above example, the theorem concerns the behavior of a generic series $\sum_{n=1}^{\infty} a_n$ and the exceptional case was the specific series $\sum_{n=-\infty}^{\infty} \frac{1}{n^2}$. More frequently, the theorem had the form

a function $f(x)$ has the property P.

In this case, the theorem was thought to be valid and rigorous as long as the variable x remained indeterminate; but this was no longer the case if one gave a determinate value to x. As Fraser[315] puts it, "isolated exceptional values at which the relation fails are not significant". For instance, in his *Leçon sur le calcul des functions*, Lagrange proved that, given a generic function $f(x)$, the development

$$f(x) + ai + bi^2 + ci^3 + \ldots + qi^n + \ldots$$

of $f(x + i)$ included no fractional power of i.[316] When referring to this theorem, Lagrange asserted:

> This demonstration is general and rigorous as long as x and i remain indeterminate; but this is no longer the case if one gives a determinate value to x. (Lagrange [1797, 23])

In his [1771], Daniel Bernoulli explained such a concept by stating that the formal sum may exclude certain points:

> whose existence and location cannot be indicated by abstract analysis. Thus the tangent method cannot indicate cuspidal points if they are in the given curve. As a consequence, however, neither can the tangent method be disproved nor can one be convinced of its falsity. (D. Bernoulli [1771, 84])

I explicitly emphasize that, in this context, no specific counterexample could make the derivation of a theorem (and in particular a theorem on series) invalid.

$$*\quad*\quad*$$

In the early theory of series, we saw that the investigation of the effective convergence of a numerical series or the determination of the interval of convergence of a power series was an *a posteriori* question; it concerned the moment of the application of the results found using analytical techniques.[317]

---

[315]Cf. Fraser [1989, 331].
[316]See Chapter 28.
[317]See Chapter 8, p. 119.

This is also true for the formal concept, but with differences regarding the mathematical context in which series theory was included. Previously, the context had been geometric: Series served to investigate geometric quantities and the moments of determination of a certain expansion and that of its application were two moments of the same field of mathematics. The context was now analytical and series served to investigate general quantities. Geometric and numerical applications of series were not part of analysis; they belonged to different fields of mathematics and this increasingly separated the two moments and made the formal more evident. Analysis on its own did not deal with convergence explicitly,[318] but when series were applied to geometry, mechanics, and to what today is termed numerical analysis, they were treated quantitatively, i.e., one observed whether, and within what limits, a formal, general equality became a quantitative, special equality. In these applications,[319] convergence was important.

This concept explains the observations of historians such as Golland and Golland, according to whom Euler was more concerned with convergence issues than he has traditionally been credited.[320] The Gollands' analysis is based on the papers concerning the application of trigonometric series to mechanics: This is one of the cases in which 18th-century mathematicians paid attention to questions of convergence, approximations, and the evaluations of error. I recall the words of Lagrange written in a letter regarding astronomy:

> It is not difficult to reduce the problem to an equation but since this equation was differential, it required some integrations. Such integrations can only be obtained through series. The entire question came down to whether they were convergent. (Lagrange [Oeuvres, 14: 280])

For instance, in *Institutiones calculi differentialis* [1755], Euler first derived the Taylor theorem formally, without regard to the interval of convergence, and only subsequently took care to consider convergence when he applied the Taylor series to the calculation of the sth roots of a number c. He[321] considered the Taylor expansion of $y = x^n$:

$$(x + \omega)^n = x^n + nx^{n-1}\omega + \frac{n(n-1)}{2!}x^{n-2}\omega^2 \qquad (127)$$
$$+ \frac{n(n-1)(n-2)}{3!}x^{n-3}\omega^3 + \dots$$

---

[318]It should be recalled that the quantitative was masked within the accepted procedures and the notion of variable quantity (see Chapter 8).

[319]Applications (especially numerical applications) were sometimes contained in the treatises of the analysis of the infinite, but they were conceptually separate (the clearest example is Lagrange [1797]).

[320]See Golland and Golland [1993, 55].

[321]See Euler [1755, 277–279].

Setting $\omega = u$, $n = r/s$, $x = a^s$ in (127), he derived

$$(a^s + u)^{r/s} = a^r \left(1 + \frac{ru}{sa^s} + \frac{r\,(r-s)\,u^2}{s\cdot 2s \cdot a^{2s}} \right.$$
$$\left. + \frac{r\,(r-s)\,(r-2s)u^3}{s\cdot 2s\cdot 3s\cdot a^{3s}} + \ldots \right). \qquad (128)$$

This formula could be applied to the calculation of $c^{1/s}$ by an appropriate decomposition of c of the type $a^s + u$. However, Euler was not satisfied and improved the convergence by changing ω into $-\frac{a^s u}{a^s + u}$ and n into $-\frac{r}{s}$ in formula (127). He obtained

$$(a^s + u)^{r/s} = a^r \left(1 + \frac{ru}{s(a^s + u)} + \frac{r\,(r+s)\,u^2}{s\cdot 2s\cdot (a^s + u)^2} \right. \qquad (129)$$
$$\left. + \frac{r\,(r+s)\,(r+2s)u^3}{s\cdot 2s\cdot 3s\cdot (a^s + u)^3} + \ldots \right).$$

Euler observed:

> the latter series [i.e., (129)] converges more than the former [i.e., (128)], since its terms still decrease if $u > a^s$, in which case the latter instead series diverges. (Euler [1755, 277])

Then he showed that one could further improve the convergence of (129) by means of an appropriate decomposition of c in $a^s + u$ or by calculating the roots

$$hc^{1/s} = (h^s c)^{1/s}$$

in place of $c^{1/s}$ and decomposing $h^s c$ in $b^s + u$, with $b > a$.

In the context of the actual use of series for numerical calculations, Euler also provided two criteria that series had to satisfy in order to be suitable for numerical calculations.

First, the terms of series should not be complex.

Second, "the series must be vehemently convergent[322] or of such a kind that any term is much less than the preceding one." In this way not many terms provided a close enough approximation to the sought sum (see Euler [1737c, 247]).

Finally, I note that when analysis (abstract analysis, to use Bernoulli's words [1771, 71 and 84]) was applied to geometry, mechanics, and numerical analysis, convergence criteria were even used: They served to avoid useless calculations when one had to compute the value of a series by approximation.[323]

---

[322] As usual in the 18th century, he was interested in rapid convergence rather than the convergence on its own: A slowly convergent series is of little use for calculating its values in almost the same way as a divergent series.

[323] For instance, see Euler [1794a].

Part III
The theory of series after 1760: Successes and problems of the triumphant formalism

In the years after 1760, the theory of series developed in the ways indicated by Euler. The formal approach produced many remarkable results, such as the Lagrange series and Laplace's theory of generating functions. Further findings led to the discovery of the calculus of operations, while series was largely employed in the solution to differential equations, and even trigonometric series became the subject of research. However, the use of series solutions for differential equations and of trigonometric series shows signs of certain difficulties in the theory of series and, more generally, of 18th-century analysis. At the end of the century, series theory appeared incapable of making a contribution to the growth of mathematics or becoming part of a coherent analytical theory. Nevertheless, the mathematicians of the period did not appreciate this or, at least, were unable to offer a solution.

This third part is divided into 11 chapters. The first three chapters are devoted to illustrating some of the greatest successes of triumphant formalism: the Lagrange inversion theorem (Chapter 20), Laplace's theory of generating functions (Chapter 22), the first steps toward the calculus of operations (Chapter 21). From Chapters 23 to 25, I examine issues related to the representation of transcendental quantities and their analytical investigation. I then discuss problems concerning the series solution to differential equations (Chapter 26) and the emergence of trigonometric series (Chapter 27). In Chapter 28, I deal with the attempts to prove the binomial theorem, Lagrange's view of the Taylor theorem, and other significant developments that took place between the end of the 18th century and the beginning of the 19th century. Finally, the problematic attempt of Legendre to enlarge the realm of accepted functions and the emergence of techniques of inequalities in d'Alembert's and Lagrange's work are the subject matter of Chapters 29 and 30.

20 Lagrange inversion theorem

During the second part of the 18th century, the bulk of research on series placed increasing emphasis on the formal aspect. One of the most remarkable results was the "Lagrange series", which I examine in this chapter.[324] In *Nouvelle méthode pour résoudre le équations littérales par le moyen des séries* [1768, 14–25] Lagrange proved the following theorem:

Given the equation $\alpha - x + \varphi(x) = 0$, where $\varphi(x)$ is any function, if p is a root of this equation and $f(p)$ is a function of p, then

$$f(p) = f(x) + \varphi(x)f'(x) + \frac{1}{2!}\frac{d}{dx}\left(f'(x)\varphi^2(x)\right) + \frac{1}{3!}\frac{d^2}{dx^2}\left(f'(x)\varphi^3(x)\right) + \ldots \tag{130}$$

by changing x into α after differentiation.

In order to prove this theorem, Lagrange considered the equation $a_1 - a_2 x + a_3 x^2 - \ldots + (-1)^n a_{n+1} x^n = 0$, where $a_1, a_2, a_3, \ldots, a_{n+1}$ are unspecified coefficients, and set

$$a_1 - a_2 x + a_3 x^2 - \ldots + (-1)^n a_{n+1} x^n = a_1\left(1 - \frac{x}{p_1}\right)\left(1 - \frac{x}{p_2}\right)\ldots\left(1 - \frac{x}{p_n}\right).$$

He divided by $a_2 x$ and obtained

$$
\begin{aligned}
&1 - \frac{a_1}{a_2 x} - \frac{a_3 x - \ldots + (-1)^n a_{n+1} x^{n-1}}{a_2}\\
={}& -\frac{a_1}{a_2 x}\left(1 - \frac{x}{p_1}\right)\left(1 - \frac{x}{p_2}\right)\ldots\left(1 - \frac{x}{p_n}\right)\\
={}& \frac{a_1}{a_2 p_1}\left(1 - \frac{p_1}{x}\right)\left(1 - \frac{x}{p_2}\right)\ldots\left(1 - \frac{x}{p_n}\right).
\end{aligned}
$$

Then he set

$$\xi = \frac{a_3 x - \ldots + (-1)^n a_{n+1} x^{n-1}}{a_2}$$

and wrote

$$
\begin{aligned}
\log\left(1 - \frac{a_1}{a_2 x} - \xi\right) ={}& \log\frac{a_1}{a_2 p_1} + \log\left(1 - \frac{p_1}{x}\right)\\
& + \log\left(1 - \frac{x}{p_2}\right) + \ldots + \log\left(1 - \frac{x}{p_n}\right), \tag{131}
\end{aligned}
$$

---

[324]In the reconstruction of the topics described in Chapter 20, 21, and 22, I follow Panza [1992].

where $\log N$ denotes the natural logarithm of the numbers N. Lagrange expanded the right-hand side of (131):

$$\log \frac{a_1}{a_2 p_1} + \log \left(1 - \frac{p_1}{x}\right) + \log \left(1 - \frac{x}{p_2}\right) + \log \left(1 - \frac{x}{p_3}\right)$$

$$+ \ldots + \log \left(1 - \frac{x}{p_n}\right)$$

$$= \log \frac{a_1}{a_2 p_1} - \sum_{r=1}^{\infty} \frac{1}{r} \left(\frac{p_1}{x}\right)^r - \sum_{r=1}^{\infty} \frac{1}{r} \left(\frac{x}{p_2}\right)^r - \sum_{r=1}^{\infty} \frac{1}{r} \left(\frac{x}{p_3}\right)^r$$

$$- \ldots - \sum_{r=1}^{\infty} \frac{1}{r} \left(\frac{x}{p_n}\right)^r .$$

By rearranging, he had

$$\log \frac{a_1}{a_2 p_1} + \log \left(1 - \frac{p_1}{x}\right) + \log \left(1 - \frac{x}{p_2}\right) + \ldots + \log \left(1 - \frac{x}{p_n}\right)$$

$$= \log \frac{a_1}{a_2 p_1} - \sum_{r=1}^{\infty} \frac{1}{r} \left(\frac{p_1}{x}\right)^r - \sum_{r=1}^{\infty} \frac{1}{r} \left(\sum_{s=2}^{n} \left(\frac{1}{p_s}\right)\right)^r x^r . \tag{132}$$

Then Lagrange wrote the left-hand side of (131) in the form

$$\log \left(1 - \frac{a_1}{a_2 x} - \xi\right) = \log \left(1 - \frac{a_1}{a_2 x}\right) + \log \left(1 - \frac{\xi}{1 - \frac{a_1}{a_2 x}}\right)$$

$$= - \sum_{r=1}^{\infty} \frac{1}{r} \left(\frac{a_1}{a_2 x}\right)^r - \sum_{r=1}^{\infty} \frac{1}{r} \left(\frac{\xi}{1 - \frac{a_1}{a_2 x}}\right)^r . \tag{133}$$

He set $\xi^r = \sum_{i=0}^{\infty} A_{r,i} x^i$ $(r > 1)$. Consequently,

$$\frac{1}{r} \left(\frac{\xi}{1 - \frac{a_1}{a_2 x}}\right)^r = \frac{1}{r} \xi^r \left(1 - \frac{a_1}{a_2 x}\right)^{-r}$$

$$= \frac{1}{r} \left[\sum_{i=0}^{\infty} A_{r,i} x^i\right] \left[\sum_{j=0}^{\infty} \binom{r+j-1}{j} \left(\frac{a_1}{a_2 x}\right)^j\right]$$

$$= \frac{1}{r} \left[\sum_{i=0}^{\infty} \sum_{j=0}^{\infty} \binom{r+j-1}{j} A_{r,i} \left(\frac{a_1}{a_2}\right)^j x^{i-j}\right] \tag{134}$$

and

$$\log \left(1 - \frac{a_1}{a_2 x} - \xi\right) = - \sum_{r=1}^{\infty} \frac{1}{r} \left(\frac{a_1}{a_2 x}\right)^r \tag{135}$$

$$- \sum_{r=1}^{\infty} \frac{1}{r} \left[\sum_{i=0}^{\infty} \sum_{j=0}^{\infty} \binom{r+j-1}{j} A_{r,i} \left(\frac{a_1}{a_2}\right)^j x\right] ,$$

By comparing (135) and (132), he obtained

$$\sum_{r=1}^{\infty} \frac{1}{r}\left(\frac{a_1}{a_2 x}\right)^r + \sum_{n=1}^{\infty}\frac{1}{n}\left[\sum_{i=0}^{\infty}\sum_{j=0}^{\infty}\binom{r+j-1}{j}A_{n,i}\left(\frac{a_1}{a_2}\right)^j x^{i-j}\right]$$

$$= \log\frac{a_2 p_1}{a_1} + \sum_{r=1}^{\infty}\frac{1}{r}\left(\frac{p_1}{x}\right)^r + \sum_{r=1}^{\infty}\frac{1}{r}\left(\sum_{s=2}^{n}\left(\frac{1}{p_s}\right)\right)^r x^r.$$

Lagrange assumed that the coefficients of x^r had to be equal. In particular, he considered the coefficient of x^0, x^{-1}, x^{-2}, ... and obtained

$$\sum_{n=1}^{\infty}\frac{1}{n}\left[\sum_{i=0}^{\infty}\binom{n+i-1}{i}A_{n,i}\left(\frac{a_1}{a_2}\right)^i\right] = \log\frac{a_2 p_1}{a_1},$$

$$\frac{a_1}{a_2} + \sum_{n=1}^{\infty}\frac{1}{n}\left[\sum_{i=0}^{\infty}\binom{n+i}{i+1}A_{n,i}\left(\frac{a_1}{a_2}\right)^{i+1}\right] = p_1,$$

$$\left(\frac{a_1}{a_2}\right)^2 + \sum_{n=1}^{\infty}\frac{1}{n}\left[\sum_{i=0}^{\infty}\binom{n+i+1}{i+2}A_{n,i}\left(\frac{a_1}{a_2}\right)^{i+2}\right] = \frac{1}{2}\left(p_1\right)^2,$$

$$\dots$$

In this way, Lagrange obtained formulas for $\log p_1 \left(= \log\frac{a_2 p_1}{a_1} + \log\frac{a_1}{a_2}\right)$ and every power of p_1. Finally, Lagrange wrote these formulas by using differentials. He set $p = p_1$, $x = \frac{a_1}{a_2}$ and had

$$\log p = \log x + \sum_{r=1}^{\infty}\frac{1}{r}\left[\sum_{i=0}^{\infty}\binom{r+i-1}{i}A_{r,i}x^i\right]$$

$$= \log x + \left[\sum_{i=0}^{\infty}A_{0,i}x^i\right] + \frac{1}{2}\left[\sum_{i=0}^{\infty}(i+1)A_{1,i}x^i\right]$$

$$+ \frac{1}{3}\left[\sum_{i=0}^{\infty}\frac{(i+2)(i+1)}{2!}A_{2,i}x^i\right] + \dots$$

$$= \log x + \xi + \frac{1}{2!}\frac{d}{dx}\left(x\xi^2\right) + \frac{1}{3!}\frac{d^2}{dx^2}\left(x^2\xi^3\right) + \dots.$$

Similarly,

$$p^k = x^k + k\left[x^k\xi + \frac{1}{2!}\frac{d}{dx}\left(x^{k+1}\xi^2\right) + \frac{1}{3!}\frac{d^2}{dx^2}\left(x^{k+2}\xi^3\right) + \dots\right]. \qquad (136)$$

Since

$$a_1 - a_2 x + a_3 x^2 - \dots + (-1)^n a_{n+1}x^n = a_1 - a_2 x + \xi a_2 x,$$

by setting $\alpha = \frac{a_1}{a_2}$ and $\varphi(x) = \frac{\xi(x)}{x}$, one had

$$\alpha - x + \varphi(x) = 0. \qquad (137)$$

Lagrange [1768, 24] considered this as a generic equation, where $\varphi(x)$ is any function, and stated that if p is a root of Equation (137), then p^k is given by (136) by setting $x = \alpha$ after the differentiation. He also stated that it is easy to see that any function $f(p)$ of the root p can be expressed in the form (130).[325]

An important aspect of Lagrange's proof is that he used the properties of $\frac{d^k}{dx^k}(x^i y^j)$ to represent the terms of certain series in a compact form. Lagrange was not interested in the meaning of $\frac{d^k}{dx^k}\left(x^i y^j\right)$ but in how x and y combined after the application of $\frac{d^k}{dx^k}$, which is understood as an operator.[326] In this sense, it resembles Euler's derivation of the Euler–Maclaurin formula in *Institutiones calculi differentialis* [1755].

Lagrange devoted a section of the *Nouvelle méthode* to convergence. In this section he clearly shows his adherence to the view commonly held by 18th-century mathematicians. Convergence was not considered in the proof. However, the actual application of the series to the solution of the equation required series to be convergent; only in this case it represented actually the values of (finite) quantity [1768, 60–61]. For this reason, Lagrange made some *a posteriori* considerations on convergence and proved that

$$\left| \frac{1}{n!} \frac{d^n}{dx^n} \left(f'(x) \varphi^n(x) \right) \right|_{x=\alpha} = 0$$

for $n = \infty$ (under certain conditions concerning the function φ).

In *Sur le problème de Kepler* [1769], Lagrange used his inversion theorem to provide an approximate solution to Kepler's equation

$$M = E - e \sin E. \tag{138}$$

Kepler's equation arose from the description of the motion of celestial bodies. It gives the relation between the polar coordinates of a planet and the time elapsed from a given initial point. In Equation (138), M denotes the mean anomaly (a parameterization of time) and E the eccentric anomaly of a body orbiting on an ellipse with eccentricity e.

Lagrange showed that the solution E (E is a function of M) is given by

$$E = M - e \sin M + \frac{1}{2!} \frac{d}{dM} \left(e^2 \sin^2 M \right) - \frac{1}{3!} \frac{d^2}{dM^2} \left(e^3 \sin^3 M \right) + \dots$$

In order to compute E, Lagrange wrote the power $\sin^n M$ as functions of $\cos kM$ and $\sin kM$ and obtained

$$E = M - e \sin M + \frac{e^2}{2 \cdot 2!} \left[2 \sin 2M \right] - \frac{e^3}{2^2 3!} \left[3 \sin M - 3^2 \sin 3M \right] + \dots$$

---

[325]Lagrange explicitly proved this only in his [1798, 252–253]. The proof depended on the assumption that any function could be expanded into a power series.
[326]See Panza [1992, 556–557].

Lagrange did not discuss the interval of convergence. The radius of convergence was determined by Laplace in his [1799–1825, 5:479]).

According to Lacroix, the Lagrange inversion theorem "was an epoch-making discovery in the history of analysis with regard to the expansion of functions into series" (Lacroix [1810–19, 1:286]). By 1770 it had become the subject matter of several studies. In 1772, Lambert had already published a new proof of the theorem by deriving it from an application of Taylor series (see Lambert [1770]). Laplace generalized it to several variables in *Mémoire sur l'usage du calcul aux différences partielles dans la théorie des suites* [1777] and *Théorie du mouvement et de la figure elliptique des planètes* [1784]. In *Traité de mècanique céleste* [1799–1825], he applied the theorem to the calculation of the orbits of planets. In his *Traitè de la résolution des équations numériques de tous les degrés* [1798], Lagrange returned to the inversion theorem and gave a new proof of it. It was also subject of investigation by the Combinatorial School[327] (see p. 286) and many other studies followed in the next century.

---

[327]In his [1793], Rothe dealt with the problem of reversion of series, which I discussed in Chapter 4. Given the equality $\sum_{n=1}^{\infty} b_n z^{\beta+nk} = \sum_{n=1}^{\infty} a_n x^{\alpha+nh}$ $(a_1, \alpha > 0)$, Rothe [1793] represented an arbitrary power x^λ as a power series in y by using the polynomial formula (see Chapter 28) to calculate the coefficients of this power series. In 1795, Pfaff and Rothe showed that Rothe's results and the Lagrange inversion theorem are equivalent if the Taylor series is presupposed (see Jahnke [1993, 273]).

21 Toward the calculus of operations

Leibniz's analogy,[328] which Lagrange rediscovered in 1754,[329] was the starting point of numerous significant results. In his *Sur une nouvelle espèce de calcul*,[330] Lagrange stated that

$$\Delta u = e^{\frac{du}{dx}\xi + \frac{du}{dy}\varphi + \frac{du}{dz}\zeta + \cdots} - 1, \tag{139}$$

where $u(x, y, z, \ldots)$ is a function of the variables $x, y, z, \ldots$, ξ, φ, ζ, $\ldots$ are increments of these variables, and $\Delta u = u(x + \xi) - u(x)$. Lagrange's reasoning can be so summarized [For the sake of simplicity, I consider the case of a function $u(x)$ of the sole variable x]. By changing x into $\frac{du}{dx}\xi$ in the expansion $e^x - 1 = \frac{x}{1} + \frac{x^2}{2!} + \frac{x^3}{3!} + \ldots$, one can write

$$\frac{du}{dx}\xi + \frac{1}{2!}\left(\frac{du}{dx}\right)^2\xi^2 + \frac{1}{3!}\left(\frac{du}{dx}\right)^3\xi^3 + \ldots = e^{\frac{du}{dx}\xi} - 1.$$

If one now identifies $\left(\frac{du}{du}\right)^k$ with $\frac{d^k \cdot u}{du.}$, then one obtains

$$\frac{du}{dx}\xi + \frac{1}{2!}\frac{d^2}{dx^2}\xi^2 + \frac{1}{3!}\frac{d^3 u}{dx^3}\xi^3 + \ldots = u(x + \xi) - u(x) = \Delta u.$$

Hence,

$$\Delta u = e^{\frac{du}{dx}\xi} - 1. \tag{140}$$

From (139) Lagrange derived

$$\Delta^\lambda u = \left(e^{\frac{du}{dx}\xi + \frac{du}{dy}\varphi + \frac{du}{dz} + \cdots} - 1\right)^\lambda \tag{141}$$

if λ is a positive integer, and

$$\Sigma^\lambda u = \frac{1}{\left(e^{\frac{du}{dx}\xi + \frac{du}{dy}\varphi + \frac{du}{dz} + \cdots} - 1\right)^{-\lambda}} = \Delta^{-\lambda} u \tag{142}$$

if λ is a negative integer.

Lagrange stated that the operation by which one goes from Δu to $\Sigma^\lambda u$ and $\Delta^\lambda u$ was not based on clear and rigorous principles; nevertheless, it is exact, as can be ensured *a posteriori*. However, it would have probably been extremely difficult to provide a direct and analytical demonstration of it (see Lagrange [1772, 451]).

---

[328] See Chapter 3, p. 50.
[329] See *Lettera a Giulio Carlo di Fagnano* (Lagrange [1754]).
[330] See Lagrange [1772, 448–452].

In his paper, Lagrange showed some applications of these formulas. He considered the expansion

$$(e^w - 1)^\lambda = w^\lambda (1 + Aw + Bw^2 + Cw^3 + \ldots), \qquad (143)$$

where A, B, C, ... are to be determined, and calculated the logarithmic derivative of both sides. He obtained

$$\lambda \left(\frac{e^w}{e^w - 1} - \frac{1}{w} \right) = \frac{Aw + 2Bw^2 + 3Cw^3 + \ldots}{1 + Aw + Bw^2 + Cw^3 + \ldots}.$$

Since

$$\frac{e^w}{e^w - 1} = \frac{1}{1 - e^{-w}} = \left(\sum_{i=1}^{\infty} (-1)^{i+1} \frac{w^i}{i!} \right)^{-1},$$

Lagrange substituted this expression into the last equation and compared the coefficients. He thus found

$$A = \frac{\lambda}{2!},$$

$$B = \frac{1}{2} \left(\frac{(\lambda+1)A}{2!} - \frac{\lambda}{3!} \right) = \frac{\lambda}{24} + \frac{\lambda^2}{8},$$

$$C = \frac{1}{3} \left(\frac{(\lambda+2)B}{2!} - \frac{(\lambda+1)A}{3!} + \frac{\lambda}{4!} \right) = \frac{\lambda^2}{48} + \frac{\lambda^3}{48},$$

$$\ldots\ldots$$

By a comparison between $\Delta^\lambda u = \left(e^{\frac{du}{dx}\xi} - 1 \right)^\lambda$ and (143), he obtained

$$\Delta^\lambda u = \frac{d^\lambda u}{dx^\lambda} \xi^\lambda + \frac{\lambda}{2} \frac{d^{\lambda+1} u}{dx^{\lambda+1}} \xi^{\lambda+1} + \left(\frac{\lambda}{24} + \frac{\lambda^2}{8} \right) \frac{d^{\lambda+2} u}{dx^{\lambda+2}} \xi^{\lambda+2} \qquad (144)$$

$$+ \left(\frac{\lambda^2}{48} + \frac{\lambda^3}{48} \right) \frac{d^{\lambda+3} u}{dx^{\lambda+3}} \xi^{\lambda+3} + \ldots$$

Finally, Lagrange replaced λ by $-\lambda$ and obtained the analogous expression for $\Sigma^\lambda u$.

For $\lambda = -1$, (144) provided the Euler–Maclaurin summation formula. Lagrange did not go into detail and referred to Euler and Maclaurin. He probably considered the coincidence of the results as a proof of the correctness of the method (see Lagrange [1772, 456-457]).

Lagrange went on to obtain new formulas. In [1772, 457-458], he derived

$$\frac{du}{dx} \xi = \log(1 + \Delta u) = \Delta u - \frac{\Delta^2 u}{2} + \frac{\Delta^3 u}{3} - \frac{\Delta^4 u}{4} + \ldots \qquad (145)$$

from (140). This equation is to be interpreted as a series in powers of Δu where the symbols $(\Delta u)^\lambda$ are replaced by $\Delta^\lambda u$. Then Lagrange assumed

$$\frac{d^\lambda u}{dx^\lambda}\xi^\lambda = \log^\lambda(1+\Delta u) \tag{146}$$

and obtained

$$\frac{d^\lambda u}{dx^\lambda}\xi^\lambda = \Delta^\lambda u + M\Delta^{\lambda+1}u + N\Delta^{\lambda+2}u + P\Delta^{\lambda+3}u + \dots,$$

where

$$M = -\frac{\lambda}{2!},$$
$$N = -\frac{1}{2}\left(\frac{(\lambda+1)M}{2} - \frac{\lambda}{2\cdot3}\right),$$
$$P = -\frac{1}{3}\left(\frac{(\lambda+2)N}{2} - \frac{(\lambda+1)M}{2\cdot3} + \frac{\lambda}{3\cdot4}\right),$$
$$\dots$$

In a similar way, he derived

$$\frac{\int^\lambda u\,dx}{\xi^\lambda} = {\sum}^\lambda u + \mu{\sum}^{\lambda-1}u + \nu{\sum}^{\lambda-2}u + \dots,$$

where

$$\mu = \frac{\lambda}{2!},$$
$$\nu = \frac{1}{2}\left(\frac{(\lambda-1)\mu}{2} - \frac{\lambda}{2\cdot3}\right),$$
$$\dots$$

For $\lambda=1$, he obtained Gregory's formula:

$$\frac{\int u\,dx}{\xi} = \sum u + \mu u + \nu\Delta u + \dots.$$

In [1792, 663–684], Lagrange also applied similar procedures to obtain some results concerning interpolation. He considered the differences

$$\Delta^m = \sum_{r=0}^m (-1)^r\binom{m}{r}T_{m-r}, \quad m>1, \text{ and } \Delta^0 = T_0$$

of a given sequence T_r. He noted that $T_1 = \Delta^0 + \Delta^1$ and

$$T_n = (T_1)^n = (\Delta^0+\Delta^1)^n = \Delta^0 + n\Delta^1 + \frac{n(n-1)}{1\cdot2}\Delta^2$$
$$+\frac{n(n-1)(n-2)}{1\cdot2\cdot3}\Delta^3 + \dots,$$

which is Newton's interpolation formula. Then he stated that

$$\Delta^0 \Delta^1 = \Delta^{0+1} = \Delta^1$$

and $T_1 = \Delta^0 + \Delta^1 = \Delta^0(1 + \Delta^1)$. Since $1 + \Delta^1 = e^{\log(1+\Delta^1)}$, he derived

$$T_n = (T_1)^n = \Delta^0 e^{n \log(1+\Delta^1)}.$$

Since the law of exponents implied that

$$
\begin{aligned}
\Delta^0 [\log(1 + \Delta^1)]^m &= \Delta^0 [\log(1 + \Delta^1)]^m \\
&= [\Delta^0 [\Delta^1 - \frac{1}{2}(\Delta^1)^2 + \ldots]]^m \\
&= [\Delta^1 - \frac{1}{2}(\Delta^1)^2 + \ldots]^m = [\log(1 + \Delta^1)]^m,
\end{aligned}
$$

Lagrange expanded $e^{n \log(1+\Delta^1)}$ and obtained:

$$
\begin{aligned}
T_n &= \Delta^0 e^{n \log(1+\Delta^1)} \\
&= \Delta^0 + n\Delta^0 \log(1 + \Delta^1) + \frac{n^2}{2}\Delta^0[\log(1 + \Delta^1)]^2 + \frac{n^3}{6}\Delta^0[\log(1 + \Delta^1)]^3 + \ldots \\
&= \Delta^0 + n \log(1 + \Delta^1) + \frac{n^2}{2}[\log(1 + \Delta^1)]^2 + \frac{n^3}{6}[\log(1 + \Delta^1)]^3 + \ldots.
\end{aligned}
$$

Setting

$$P_0 = \Delta_0 = T_0 \quad \text{and} \quad P_r = [\log(1 + \Delta^1)]^r,$$

he derived

$$T_n = P_0 + nP_1 + \frac{n^2}{2}(P_2)^2 + \frac{n^3}{6}(P_3)^3 + \ldots.$$

and

$$1 + \Delta^1 = e^{P_1} = 1 + P_1 + \frac{1}{2}P_2 + \frac{1}{6}P_3 + \ldots$$

In this way, Lagrange had various relationships between T_n, P_n, and Δ^n.

Lagrange's results were the starting point for some of Laplace's research. In his *Mémoire sur l'inclinaison moyenne des orbites des comètes, sur la fig-ure de la Terre et sur les fonctions*, Laplace recalled Lagrange's words on the difficulty of justifying the analogy between positive powers and differences, and between negative powers and sums; however, he thought he had found a method that was as direct and simple as possible and that showed the reason of the analogy *a priori* (see Laplace [1773, 314]). For this purpose, Laplace wrote

$$u(x + \xi) = u(x) + \alpha(x, \xi) \tag{147}$$

[with $\alpha(x, 0) = 0$, for every x]. By differentiating with respect to variable ξ and applying the theorem of mixed partial differentials, he obtained

$$\frac{\partial u(x + \xi)}{\partial x} = \frac{\partial \alpha(x, \xi)}{\partial \xi}.$$

Hence,

$$\alpha(x, \xi) = \int_0^\xi \frac{\partial u(x + \xi)}{\partial x} d\xi.$$

By replacing in (147), he had

$$u(x + \xi) = u(x) + \int_0^\xi \frac{\partial u(x + \xi)}{\partial x} d\xi. \tag{148}$$

Hence,

$$\frac{\partial u(x + \xi)}{\partial x} = \frac{du(x)}{dx} + \int_0^\xi \frac{\partial^2 u(x + \xi)}{\partial x^2} d\xi. \tag{149}$$

He substituted

$$\frac{\partial u(x + \xi)}{\partial x} d\xi$$

by

$$\frac{du(x)}{dx} + \int_0^\xi \frac{\partial^2 u(x + \xi)}{\partial x^2} d\xi$$

into (148), resulting in

$$\begin{aligned} u(x + \xi) &= u(x) + \int_0^\xi \left(\frac{du(x)}{dx} + \int_0^\xi \frac{\partial^2 u(x + \xi)}{\partial x^2} d\xi \right) d\xi \tag{150} \\ &= u(x) + \frac{du(x)}{dx} \xi + \int_0^\xi \left(\int_0^\xi \frac{\partial^2 u(x + \xi)}{\partial x^2} d\xi \right) d\xi. \end{aligned}$$

Then Laplace differentiated (149) with respect to x and obtained an expression of $\frac{\partial^2 u(x+\xi)}{\partial x^2}$, which he replaced into (150). By repeating the procedure, he derived

$$\Delta u(x) = \frac{du(x)}{dx} \xi + A_1 \frac{d^2 u(x)}{dx^2} \xi^2 + A_2 \frac{d^3 u(x)}{dx^3} \xi^3 + \dots . \tag{151}$$

The coefficients A_i depended neither on x, u, nor ξ. Laplace did not determine the coefficients (since it did not serve his purpose); however, he *de facto* provided a proof of the Taylor series by repeated integration. In the following steps, he even provided a generalization of (151). Indeed, he substituted x by $x + \xi$ into (151). Since $d(x + \xi) = dx$, he wrote

$$\Delta^2 u(x) = \left(\frac{\partial u(x + \xi)}{\partial x} - \frac{du(x)}{dx} \right) \xi \tag{152}$$

$$+ A_1 \left(\frac{\partial^2 u(x + \xi)}{\partial x^2} - \frac{d^2 u(x)}{dx^2} \right) \xi^2$$

$$+ A_2 \left(\frac{\partial^3 u(x + \xi)}{\partial x^3} - \frac{d^3 u(x)}{dx^3} \right) \xi^3 + \dots$$

From

$$\frac{\partial^h u(x + \xi)}{\partial x^h} = \frac{d^h u(x)}{dx^h} + \int_0^\xi \frac{\partial^{h+1} u(x + \xi)}{\partial x^{h+1}} d\xi,$$

it follows that

$$\Delta^2 u(x) = \left(\int_0^\xi \frac{\partial^2 u(x + \xi)}{\partial x^2} d\xi \right) \xi + A_1 \left(\int_0^\xi \frac{\partial^3 u(x + \xi)}{\partial x^3} d\xi \right) \xi^2$$

$$+ A_2 \left(\int_0^\xi \frac{\partial^3 u(x + \xi)}{\partial x^3} d\xi \right) \xi^3 + \dots$$

After various calculations, he obtained

$$\Delta^2 u(x) = \frac{d^2 u(x)}{dx^2} \xi^2 + B_1 \frac{d^3 u(x)}{dx^3} \xi^3 + A_2 \frac{d^4 u(x)}{dx^4} \xi^4 + \dots,$$

where the numerical coefficients depend neither on x, u, nor ξ. Laplace [1773, 317] stated that, in general, the following formula held:

$$\Delta^k u(x) = \frac{d^k u(x)}{dx^k} \xi^k + K_1 \frac{d^{k+1} u(x)}{dx^{k+1}} \xi^{k+1} + K_2 \frac{d^{k+2} u(x)}{dx^{k+2}} \xi^{k+2} + \dots, \tag{153}$$

where the numerical coefficients K_i depended neither on x, u, nor ξ.

Laplace observed that, if one took a particular function, such as $u = e^x$, one could determine the coefficients K_i and thus reobtain Lagrange's results (see Laplace [1773, 318–321]). In this way, Laplace thought he had provided an explanation of Leibniz's analogy. In effect, he was still very far off (see Panza [1992, 595–604]).

In the years that followed, however, the subject underwent a remarkable development that gave rise to the calculus of operations and to a real explanation of the analogy.

22 Laplace's calculus of generating functions

The research into Leibniz's analogy and the calculus of finite differences led Laplace to the calculus of generating functions.[331] Laplace formulated the calculus of generating functions in his *Mémoire sur les suites* [1779], and later he returned to it in on several occasions, in particular in *Mémoire sur diverses points d'analyse* [1809] and *Théorie analytique des probabilités* [1812], without modifying the basic principles expounded in the first paper.

In *Mémoire sur les suites*, Laplace considered a function y_x of t and formed the infinite series

$$y_0 + y_1 t + y_2 t^2 + \ldots + y_x t^x + \ldots + y_\infty t^\infty \qquad (154)$$

He denoted the sum of the series by u and termed the function u as the generating function of the sequence y_x. Laplace specified that he was using the term "sum" to mean the function whose expansion generates the series (see Laplace [1779, 5]).[332] He then stated:

> A generating function of any variable y_x is thus generally a function of t, which, expanded according to the powers of t, has this variable y_x for the coefficient of t^x; and, reciprocally, the corresponding variable of a generating function is the coefficient of t^x in the expansion of this function according to the powers of t. (Laplace [1779, 5–6])

In modern terms, Laplace identified an operation $O(y_x) = u$ that associates the function u to the sequence y_x such that the expansion of u is y_x.

It follows from the definition that

- *if $u = u(t)$ is the generating function of y_x, then $z = t^r u(t)$ is the generating function of y_{x-r}, where r is any positive or negative integer [namely, if $O(y_x) = u(t)$ then $O(y_{x-r}) = t^r u(t)$].*

Laplace justified this theorem by observing that the coefficient of t^x in the expansion of $z = t^r u(t)$ is equal to the coefficient of t^{x-r} in the expansion of $u(t)$ and, therefore, it is equal to y_{x-r} (see Laplace [1779, 6]).

He then observed that the general term of $(t^{-1} - 1)u(t)$ is the difference of the general term of $t^{-1}u(t)$ —which is equal to y_{x+1} for the previous theorem— and $u(t)$. Therefore,

---

[331] For a slightly different reconstruction, see Panza [1992, 550–584].

[332] The explicit reference to the Eulerian definition is lacking in *Théorie analytique des probabilités*, where Laplace stated that (154) "can always be conceived as a function of t, which is developed according to the power of t" (Laplace [1812, 7]). However, there are no further substantial changes, and the basic concepts on which Laplace's calculus is grounded remain the same.

- *the generating function of $\triangle y_x = y_{x+1} - y_x$ is $(t^{-1} - 1)u(t)$.*

More generally, Laplace stated that

- *if $u = u(t)$ is the generating function of y_x, then $(t^{-1} - 1)^n u(t)$ is the generating function of $\triangle^n y_x$, where $\triangle^n y_x = \triangle(\triangle^{n-1} y_x)$ (see Laplace [1779, 6]).*

At this point, Laplace set $\nabla y_x = a_0 y_x + a_1 y_{x+1} + a_2 y_{x+2} + \ldots + a_n y_{x+n}$, where $a_0, a_1, a_2, \ldots, a_n$ are constants, and $\nabla^k y_x = \nabla(\nabla^{k-1} y_x)$ and showed that

- *the generating function of ∇y_x is $u(t)(a_0 + a_1 t + a_2 t^{-2} + \ldots + a_n t^{-n})$,*

- *the generating function of $\nabla^k y_x = \nabla(\nabla^{k-1} y_x)$ is $u(t)(a_0 + a_1 t + a_2 t^{-2} + \ldots + a_n t^{-n})^k$,*

- *the generating function of $\triangle^s \nabla^k y_{x+r}$ is $u(t)(a_0 + a_1 t + a_2 t^{-2} + \ldots + a_n t^{-n})^k (t^{-1} - 1)^s t^r$ (see Lapace [1779, 6–7]).[333]*

Finally, Laplace denoted the inverse operation of $\triangle$ by Σ. He proved that if $u = u(t)$ is the generating function of y_x, then

$$\frac{u(t) + \sum_{i=1}^{n} A_i t^{-i}}{(t^{-1} - 1)^n} = \frac{u(t)t^n + \sum_{i=1}^{n} A_i t^{n-i}}{(1 - t)^n},$$

where A_i are arbitrary constants, is the generating function of $\Sigma^n y_x$ (Laplace [1779, 8]). Laplace stated that by setting aside the arbitrary constants, one could obtain the generating function of $\Sigma^n y_x$ by changing n into $-n$ in the generating function of $\triangle^n y_x$ (Laplace [1779, 8]).

In *Mémoire sur les suites* and *Théorie analytique des probabilités,* Laplace used the above theorems in a extremely powerful way: He showed how to derive a vast number of formal identities and provided several applications. In *Mémoire sur les suites,*[334] he observed that

$$\frac{u(t)}{t^i} = u(t) \left(1 + (\frac{1}{t} - 1)\right)^i$$

$$= u\left(1 + i(t^{-1} - 1) + \frac{i(i-1)}{1 \cdot 2}(t^{-1} - 1)^2\right.$$

$$\left. + \frac{i(i-1)(i-2)}{1 \cdot 2 \cdot 3}(t^{-1} - 1)^3 + \ldots\right).$$

---

[333]Laplace also generalized these results for case where $\nabla y_x = \sum_{i=0}^{\infty} a_i y_{x+i}$ and the exponents are rational or irrational numbers (see Laplace [1779, 7] and [1812, 9]).
[334]See Laplace [1779, 9].

By passing from the generating functions to the corresponding variables, he obtained[335] the well-known formula[336]

$$y_{x+i} = y_x + i\Delta y_x + \frac{i(i-1)}{1\cdot 2}\Delta^2 y_x + \frac{i(i-1)(i-2)}{1\cdot 2\cdot 3}\Delta^3 y_x + \dots$$

According to Laplace, the various different ways of expanding the power $\frac{1}{t^i}$ give a corresponding number of methods for interpolating the series. For example, in *Mémoire sur les suites*, he set $t^{-1} = 1+\alpha t^{-r}$ and expanded t^{-i} as a function of α. He obtained:

$$\begin{aligned} \frac{u}{t^i} &= u + i\alpha + \frac{i(i+2r-1)}{2!}\alpha^2 + \frac{i(i+3r-1)(i-3r-2)}{3!}\alpha^3 \\ &+ \frac{i(i+4r-1)(i-4r-2)(i-4r-3)}{4!}\alpha^4 + \dots \end{aligned}$$

Hence, by applying the rules of calculus, he derived the interpolation formula

$$\begin{aligned} y_{x+i} &= y_x + i\Delta y_{x-r} + \frac{i(i+2r-1)}{2!}\Delta^2 y_{x-2r} \\ &+ \frac{i(i+3r-1)(i-3r-2)}{3!}\Delta^3 y_{x-3r} \\ &+ \frac{i(i+4r-1)(i-4r-2)(i-4r-3)}{4!}\Delta^4 y_{x-4r} + \dots \end{aligned}$$

(see Laplace [1779, 9–10]).

Another interesting instance of Laplace's methodology is the following. Laplace[337] set

$$z = t(t^{-1}-1)^2$$

and sought the expansion of t^{-i} as a power series in z. He observed that t^{-i} is equal to the coefficient of θ^i in the expansion of

$$\frac{1}{1-\frac{\theta}{t}}.$$

Since $\frac{1}{t}+t = 2+z$, one had

$$\frac{1}{1-\frac{\theta}{t}} = \frac{1-\theta t}{1-\theta(\frac{1}{t}+t)+\theta^2} = \frac{1-\theta t}{(1-\theta)^2-z\theta} = \sum_{i=0}^{\infty}\frac{z^i\theta^i}{(1-\theta)^{2i+2}}.$$

---

[335] Laplace briefly mentioned convergence: "This equation, having place whatever be i, will serve to interpolate the series of which the differences of the terms go by decreasing" [1772, 10].
[336] See p. 22 and p. 88.
[337] See Laplace [1779, 11–12].

He named Z the coefficient of θ^i in the expansion

$$\sum_{i=0}^{\infty} \frac{z^i \theta^i}{(1-\theta)^{2i+2}}$$

of $\frac{1-\theta t}{(1-\theta)^2 - z\theta}$. By expanding the fractions $\frac{1}{(1-\theta)^{2i+2}}$ and rearranging, Lagrange obtained

$$Z = \sum_{k=0}^{\infty} \frac{(i+k+1)_{2k+1}}{(2k+1)!} z^k,$$

where $(\alpha)_n = \alpha(\alpha - 1) \ldots (\alpha - n + 1)$ or

$$Z = 1 + i + \sum_{k=1}^{\infty} (1+i) \frac{\prod_{j=1}^{k} [(i+1)^2 - j^2]}{(2k+1)!} z^k. \tag{155}$$

After that Laplace named Z' the coefficient of θ^i in the expansion of $\frac{\theta}{(1-\theta)^2 - z\theta}$. He remarked that it can be obtained by Z by changing i into $i-1$ in (155) and, thus, the sought coefficient of θ^i in the expansion of $\frac{1-\theta t}{(1-\theta)^2 - z\theta}$ is $Z - tZ'$. Hence,

$$u(t)t^{-i} = u(t)(Z - tZ').$$

By applying the principles of his calculus, Laplace obtained the formula today named after Everett:

$$
\begin{aligned}
y_{x+i} &= (1+i)y_x + \sum_{k=1}^{\infty} (1+i) \frac{\prod_{j=1}^{k}[(i+1)^2 - j^2]}{(2k+1)!} \Delta^{2k} y_{x-k} \tag{156} \\
&\quad - iy_{x-1} + \sum_{k=1}^{\infty} i \frac{\prod_{j=1}^{k}[i^2 - j^2]}{(2k+1)!} \Delta^{2k} y_{x-k-1}.
\end{aligned}
$$

Laplace subsequently transformed it and obtained the so-called Newton–Stirling formula (see Laplace [1779, 13]). In a similar way Laplace derived many interpolation formulas, such as the Laplace summation formula:

$$
\begin{aligned}
\int_0^n y_x dx &= \frac{1}{2}y_0 + y_1 + y_2 + \ldots + y_{n-1} + \frac{1}{2}y_n - \frac{1}{2}[\Delta y_{n-1} - \Delta y_0] \\
&\quad - \frac{1}{24}[\Delta^2 y_{n-2} - \Delta^2 y_0] - \frac{19}{720}[\Delta^3 y_{n-3} - \Delta^3 y_0] + \ldots
\end{aligned}
$$

(see [1799–1825, 4:205–208]).

To end this chapter, I mention Laplace's treatment of difference equations by means of generating functions. In [1812, 80], Laplace showed that if the sequence y_x is given by the finite difference equation $\sum_{n=0}^{m} K_n y_{x+n} = 0$, then its generating function has the form

$$\frac{\sum\limits_{n=0}^{m-1} R_n t^n}{\sum\limits_{n=0}^{m} K_n t^{m-n}},$$

where R_n are arbitrary constants. He then[338] stated that if the generating functions of difference equations are known, one can obtain the solutions to these equations in terms of definite integrals. Indeed, he set

$$
\begin{aligned}
u(t,z) &= \sum_{n=0}^{m} y_x(z)t^x, \\
t &= e^{iw}, \\
U(w,z) &= u(e^{iw}, z),
\end{aligned}
$$

where i is the imaginary unity, and integrated $\int_{-\pi}^{\pi} U(w,z)e^{-iwx} dw$. Hence,

$$y_x = \frac{1}{2\pi} \int_{-\pi}^{\pi} U(w,z)(\cos xw - i \, \sin xw)dw.$$

This formula has the disadvantage of introducing complex numbers. To eliminate them, Laplace supposed that

$$y_x = \int_a^b t^{-x-1} T(t)dt,$$

where the function $T(t)$ is to be determined, as are the limits of integration. He assumed that y_x was given by the equation

$$\sum_{n=0}^{m} a_n y_{x+n} + \sum_{n=0}^{m} b_n y_{x+n} = 0. \tag{157}$$

On substituting $\int_a^b t^{-x-1} T(t)dt$ for y_x in (157), he obtained

$$\left| -T(t)t^{-x} \sum_{n=0}^{m} b_n t^{-n} \right|_a^b + \int_a^b t^{-x-1} \left(T(t) \sum_{n=0}^{m} a_n t^{-n} + t\frac{d}{dt} \left\{ T(t) \sum_{n=0}^{m} b_n t^{-n} \right\} \right) dt = 0.$$

Lagrange set

$$\int_a^b t^{-x-1} \left(T(t) \sum_{n=0}^{m} a_n t^{-n} + t\frac{d}{dt} \left\{ T(t) \sum_{n=0}^{m} b_n t^{-n} \right\} \right) dt = 0$$

---

[338] See Laplace [1812, 83].

and considered the differential equation

$$T(t) \sum_{n=0}^{m} a_n t^{-n} + t \frac{d}{dt} \left\{ T(t) \sum_{n=0}^{m} b_n t^{-n} \right\} = 0,$$

which determines $T(t)$. In order to obtain the limits of the integral, Lagrange set

$$\left| -T(t) t^{-x} \sum_{n=0}^{m} b_n t^{-n} \right|_a^b = 0.$$

Therefore, the limits are $b = \infty$ and $a = t_\alpha$, where t_α is one of the solutions to the equation $\sum_{n=0}^{m} b_n t^{-n} = 0$ (see Laplace [1812, 83–85]).

23 The problem of analytical representation of nonelementary quantities

In the previous chapters, I illustrated some of the most remarkable results obtained by formal methodology. In this chapter and the following three chapters, I shall examine a different group of results derived during the 18th century. They were also based on a formal methodology, although they contain within them the germ of future difficulties. Such difficulties, which emerged for the most part at the end of the historical period considered here, mainly stemmed from the fact that series theory (in the Eulerian and Lagrangian sense) was not suitable for representing and dealing with certain quantities, especially quantities linked to the investigation of certain physical phenomena. The complications involved the notion of a function and other basic concepts related to it. Therefore, I shall first return to the concept of a function.

We saw that analysis investigated quantities insofar as they were expressed by functions and that the set of functions was restricted. During the second part of the 18th century, mathematicians felt the need to investigate certain (relations between) quantities that could not be expressed using elementary functions (later I will refer to them as nonelementary transcendental functions) and sometimes —though not always[339]— termed them "functions".[340] In this way the term "function" underwent various terminological shifts. This shift did not affect the concept of analysis. The real subject of analysis remained elementary functions, which were to be treated by formal methodology: nonelementary transcendental functions were not considered as functions in the strict sense of term and had a status different from the former.

To justify this statement, I note that 18th-century evidence shows that the term "function" was used in three different senses (apart from the sense of elementary functions).

First, the term "function", in certain cases, denoted certain quantities that did not have an analytical expression or that were analytically expressed by means of more than one analytical expression.

---

[339]The expression "(transcendental) quantity" or the words "formula" and "expression" (with reference to integral functions) were often used.

[340]To distinguish between different senses of the term, the adjective "analytical" was at times added to the noun "function" in order to denote the functions that had an appropriate analytical expression and were the true object of analysis. According to Youschkevitch [1976, 75–76], Condorcet was the first to use it in his unpublished *Traité du calcul*. On the notion of a function and of an analytical function in Condorcet, see Gilain [1988], especially p. 103, where Gilain stated that Condorcet's analytical functions correspond to explicit elementary functions. For Lagrange, analytical functions were all the functions studied in analysis, namely elementary functions; indeed, there are only elementary functions in his two foundational treatises, which are entitled *Théorie des fonctions analytiques* and *Leçon sur le calcul des fonctions*.

Second, the term "function" was associated with certain quantities derived by Wallis's interpolation (inexplicable functions).

Third, the term "function" was associated with certain quantities that were analytically expressed by integrals or differential equations.

I refer to Chapters 24 and 25 for a discussion of the second and third meaning. I shall now dwell upon the first meaning.

The word "function" was employed for the first time to denote a relation without an analytical expression in the controversy regarding the vibrating string. In his *Recherches sur la courbe que forme une corde tendüe mise en vibration* [1747], d'Alembert described the motion of a stretched elastic string by an equation equivalent to a partial differential equation

$$\frac{\partial^2 y}{\partial x^2} = a^2 \frac{\partial^2 y}{\partial t^2}.$$

He solved this equation and found $y = f(t + x) + F(t - x)$, for $a = 1$, f and F being two arbitrary functions. D'Alembert thought that the solution to the problem had to be interpreted only by means of relations that have one only analytical expression, because the rules of the calculus had been formulated with reference to such functions. In contrast, Euler tried to eliminate this restriction in the geometric or mechanical applications but without prejudicing the nature of the calculus. In the summary of *De usu functionum discontinarum in Analysi*, Euler explained:

> The solutions that Geometers gave to the problem of the vibrating motion of strings include nothing but the assumption that the figure, which is given to the string to the beginning of the motion, is regular and can be represented by a certain equation. Instead they denied that the other case (if this figure is discontinuous or irregular) was of relevance for analysis or that the motion that originated from this configuration might be reasonably defined. (Euler [1765, 7])

He thought that similar problems involved the use of discontinuous functions necessarily but merely added the new G-discontinuous functions to old continuous functions, without changing the concept of the latter. Euler obtained this result by a change in terminology and a peculiar interpretation of the constants resulting from the integration of partial differential equations.

In *Introductio in analysin infinitorum* [1748a], the term "function" always denoted an analytical written expression (embodying a relation between quantities), and the word "curve" had an obvious geometrical meaning; any function could be represented geometrically by a curve; the converse was not true, since some curves were not analytically expressible. For this reason a function had to be continuous and a curve could be discontinuous. In *De usu functionum discontinarum*, every curve was instead viewed as analytically expressible by a function[341] and Euler denoted the analytical expression by the term "equation", while he indicated the relation between quantities by the terms "curve" and "function" (the one was often used in place of the other in the paper). In this way, Euler could introduce the notion of discontinuous functions:

Curves or functions were said to be discontinuous if they were unions of more than one equation (see Euler [1765, 4–5]).

After having defined discontinuous functions, Euler had to explain how these new functions entered in the calculus (he indeed agreed that the calculus concerned single analytical expressions, i.e., continuous functions). He resorted to a special interpretation of the constants produced by integration. He observed that these new functions, absolutely indefinite and dependent upon our discretion, originate from the integration of a function of two variables, a new and little-developed field of the integral calculus, which "differs very much from the common integral calculus, where functions of a variable only occur. It demands entirely special rules, even if it also used the devices of the first part [of the calculus]" (Euler [1765, 20]).

In his *Institutiones calculi integralis*,[342] Euler explained that if one integrates a function $X(x)$ of one variable x, one obtains

$$\int X(x) = F(x) + C,$$

where $F(x)$ is a function such that

$$\frac{dF(x)}{dx} = X(x),$$

and the constant C is determined by the nature of problem whose integration gives the solution. In the same way, if one integrates a function $Z(x, y)$ of the variables x and y with respect to x, one obtains

$$\int Z(x, y)dx = F(x, y) + f(y),$$

where $F(x, y)$ was a function such that

$$\frac{dF(x, y)}{dx} = Z(x, y)dx$$

and $f(y)$ as an arbitrary quantity dependent on y.[343] The character of the quantity $f(y)$ is determined by the nature of the problem and could even be

---

[341] See Euler [1765, 3].
[342] See Euler [1768–70, 3:35–37].
[343] On this question, see also Euler [1765, 20].

a quantity that is not expressible by a formula but can be thought of as the ordinate of a curve whose abscissa is y.

Thus, if one considers the wave equation $\frac{\partial^2 z}{\partial y^2} = a^2 \frac{\partial^2 z}{\partial x^2}$, by a change of variable $t = x + ay$, $u = x - ay$, one obtains

$$\frac{\partial^2 z}{\partial t \partial u} = 0.$$

By integrating with respect t, one has a function

$$\frac{\partial z}{\partial u} = f(u);$$

hence,

$$z = \int h(u)du + f(t) = F(u) + f(t)$$

and

$$z = f(x + ay) + F(x - ay). \tag{158}$$

According to Euler, the functions f and F could be discontinuous. Since integration naturally contains an element of arbitrariness, Euler believed that the integral calculus of functions of more than one variable could directly provide a relation, without the intermediate step of the formula. Of course, in order to give a sense to this interpretation of integration, it was necessary to explain what the differential ratio of a G-discontinuous function is. Euler merely used the geometric meaning of a function and stated that if $f(x)$ represented a curve, then $f'(x)$ was the slope of the tangent, whereas if $f(x)$ was interpreted as an area, then $f'(x)$ was a curve (see Euler [1768–70, 3:69]).

In his [1768–70, 3:192–193], Euler was, however, obliged to admit that the use of an immediately geometrical notion in an analytical context gave rise to a remarkable lack. He indeed observed that if one applied (158) to the equation

$$\frac{\partial^2 z}{\partial y^2} + a^2 \frac{\partial^2 z}{\partial x^2} = 0,$$

then one obtained the complex solution

$$z = f(x + ay\sqrt{-1}) + F(x - ay\sqrt{-1}).$$

Euler passes to an equation having a complex coefficient $\pm a\sqrt{-1}$ without any special hypothesis. As was customary in the 18th century, he did not appreciate the difference between complex and real analysis. An interpretation of this solution, which was obviously influenced by a weak knowledge of the conditions of differentiability of a function of a complex variable, is beyond the scope of this book. I limit myself to illustrate how Euler derived "real solutions" from $z = f(x + ay\sqrt{-1}) + F(x - ay\sqrt{-1})$ provided f and F were continuous.

He indeed observed that if f and F are continuous, then they can be reduced to the form $P \pm Q\sqrt{-1}$. Hence, it is easy, he said, to obtain solutions in the real form[344]

$$z = \frac{1}{2}[f(x+ay\sqrt{-1})+f(x-ay\sqrt{-1})]+\frac{1}{2\sqrt{-1}}[F(x-ay\sqrt{-1})-F(x-ay\sqrt{-1})].$$
(159)

He probably realized that if $P \pm Q\sqrt{-1}$ satisfies $\frac{\partial^2 z}{\partial y^2} + a^2\frac{\partial^2 z}{\partial x^2} = 0$, then

$$
\begin{aligned}
P + Q &= \text{Re}[f(x + ay\sqrt{-1}) + F(x - ay\sqrt{-1})] \\
&\quad + \text{Im}[f(x + ay\sqrt{-1}) + F(x - ay\sqrt{-1})] \\
&= \frac{1}{2}\left[f(\omega) + \overline{f(\omega)} + F(\overline{\omega}) + \overline{F(\overline{\omega})}\right] \\
&\quad + \frac{1}{2\sqrt{-1}}\left[f(\omega) - \overline{f(\omega)} + F(\overline{\omega}) - \overline{F(\overline{\omega})}\right]
\end{aligned}
$$

also does (here, I take $\omega = x + ay\sqrt{-1}$ and denote the conjugate, real part and imaginary part of the complex number ω by $\overline{\omega}$, $\text{Re}(\omega), \text{Im}(\omega)$, respectively).

Euler assumed that, for every continuous function, $\overline{h(\omega)} = h(\overline{\omega})$; therefore,

$$
\begin{aligned}
P + Q &= \frac{1}{2}[f(\omega) + f(\overline{\omega}) + F(\overline{\omega}) + F(\omega)] \\
&\quad + \frac{1}{2\sqrt{-1}}[f(\omega) - f(\overline{\omega}) + F(\overline{\omega}) - F(\omega)].
\end{aligned}
$$

Since f and F are two generic continuous functions, the latter expression furnished (159). Euler justified the equality $\overline{h(\omega)} = h(\overline{\omega})$ as follows. Setting $x = s\cos\phi$ and $ay = s\sin\phi$, one has

$$(x \pm ay\sqrt{-1})^n = s^n(\cos n\phi \pm \sqrt{-1}\sin n\phi),$$

and since h is a continuous functions, namely it is composed of analytical (algebraic or elementary transcendental) operations, its values can be exhibited by means of the sine and cosine (Euler assumed that every continuous function can be expanded in power series with real coefficients).

If f and F are discontinuous, then they cannot be reduced to a real form:

---

[344]See Euler [1768-70, 3:192].

In any curve traced by a free stroke of the hand, what meaning will one give the ordinates corresponding to the abscissas

$$x + ay\sqrt{-1} \quad \text{and} \quad x - ay\sqrt{-1}$$

according to the nature of imaginaries and their real sums [the real part of their sums] or the difference which will also be real if it is divided by $\sqrt{-1}$? Therefore we note this not slight lack of calculus, for which one can make up in no way yet. (Euler [1768–1770, 3:193])

Despite this fact, Euler's solution to the problem of the vibrating string was substantially accepted in the 18th century. G-discontinuous functions were considered as tools that made up for a local insufficiency of analysis, just as imaginary quantities made up for local insufficiencies of real quantities. With hindsight, the controversy of the vibrating string posed the question of the lack of analytical tools for describing certain more complicated phenomena: *It actually showed the restricted nature of 18th-century analysis and its overall inadequacy for more sophisticated investigations.* Mathematicians had to resort to a direct geometrical interpretation in order to give meaning to discontinuous functions, so one of the presuppositions of Euler's and Lagrange's analysis (analysis as an autonomous theory) failed. To avoid a "return to geometry"[345] and to make G-discontinuous functions true analytical objects, it was necessary to restructure analysis, but this did not happen. The result was that the discontinuous functions were on the margins of analysis, which remained grounded upon single analytical expressions.

---

[345]See Grattan-Guiness [1970, 11].

24 Inexplicable functions

The second type of quantities to which the name "function" was given was inexplicable functions. Euler termed inexplicable those functions having neither a determinate expression nor an expression by means of an equation (namely, an implicit algebraic expression). He actually considered the sums $\sum_{n=1}^{x} a_n$ and the products $\prod_{n=1}^{x} a_n$ and studied the functions obtained by giving nonintegral values to x. This was effectively a new form of the old problem of Wallis's interpolation.

In *Institutiones calculi differentialis*, Euler attempted to establish a theory of inexplicable functions and mainly aimed to determine the differentials of these functions. He first operated in general — in other words, he determined the form of differentials of the general function of the kinds[346] $\sum_{n=1}^{x} a_n$ and $\prod_{n=1}^{x} a_n$; he then applied the obtained results to particular cases. The two most important particular cases were

$$1 + \frac{1}{2} + \frac{1}{3} + \frac{1}{4} + \ldots + \frac{1}{x}$$

and

$$1 \cdot 2 \cdot 3 \cdot \ldots \cdot x,$$

and, while investigating these two cases, Euler derived some interesting series expansions of digamma[347] and gamma functions [such as (161), (162) and (168)].

Euler[348] set $S(x) = \sum_{n=1}^{x} a_x$ and $T(x) = S(x + \omega)$ and considered the sequence

$$S(x), \quad S(x+1), \quad S(x+2), \quad S(x+3), \quad \ldots$$

---

[346] See formulas (160), (164), and (165).

[347] The inexpicable function $H(x) = 1 + \frac{1}{2} + \frac{1}{3} + \frac{1}{4} + \ldots + \frac{1}{x}$ differs from the digamma function for a constant. The digamma function is defined as

$$\Psi(z) = \frac{d}{dz} \log \Gamma(z),$$

where $\Gamma(z)$ is the gamma function, or as

$$F(z) = \frac{d}{dz} \log z!,$$

where $z!$ is the factorial function. It is well known that

$$F(z) = \Psi(z+1) = H(\omega) = -\gamma - \sum_{n=1}^{\infty} \frac{z}{n(n+z)} \ ,$$

where γ is the Euler constant. A comparison with (161) shows that

$$\Psi(n) = -\gamma - H(n-1).$$

[348] See Euler [1755, 2: Sections 369-374].

Since $S(x+n) - S(x+n-1) = a_{x+n}$, if one assumes that the numbers a_{x+n} converge to a number L, one has

$$S(\infty+1) - S(\infty) = a_{\infty+1} = L, \quad S(\infty+2) - S(\infty+1) = a_{\infty+2} = L, \quad \ldots,$$

i.e., $S(\infty)$, $S(\infty+1)$, $S(\infty+2)$, $S(\infty+3)$, ... are an arithmetic progression. Consequently,

$$S(\infty + \omega) - S(\infty) = L \cdot \omega = a_{\infty+1} \cdot \omega.$$

Since $T(\infty) = S(\infty + \omega) = S(\infty) + \omega a_{\infty+1}$, one can write

$$T(\infty) = S(x) + \sum_{n=1}^{\infty} a_{x+n} + \omega a_{\infty+1}.$$

As $T(\infty) = T(x) + \sum_{n=1}^{\infty} a_{x+n+\omega}$, one has

$$S(x + \omega) = S(x) + \omega a_{\infty+1} + \sum_{n=1}^{\infty} a_{x+n} - \sum_{n=1}^{\infty} a_{x+n+\omega}. \tag{160}$$

If ω is infinitesimal, $S(x + \omega) - S(x)$ is the expression of the differential dS of the function. This aspect of (160) attracted Euler's attention, not the possibility that one could obtain a representation of the function $S(x)$ by means of an infinite series from (160). The infinite expression of $S(x)$ was only viewed as a tool for deriving the differential of $S(x)$ or for applying the function to the problem of interpolation. In no case did Euler conceive of the possibility of defining $S(x)$ by means of (160) or (165). For instance, in Section 371, he considered $H(x) = 1 + \frac{1}{2} + \frac{1}{3} + \frac{1}{4} + \ldots + \frac{1}{x}$ and derived

$$
\begin{aligned}
H(x + \omega) &= H(x) + \sum_{n=1}^{\infty} \frac{1}{x+n} - \sum_{n=1}^{\infty} \frac{1}{x+n+\omega} \\
&= H(x) + \sum_{n=1}^{\infty} \frac{\omega}{(x+n)(x+n+\omega)} \\
&= H(x) + \sum_{n=1}^{\infty} \frac{\omega}{x+n} \sum_{h=1}^{\infty} (-1)^{h-1} \frac{\omega^{h-1}}{(x+n)^h} \\
&= H(x) + \sum_{h=1}^{\infty} (-1)^{h-1} \omega^h \sum_{n=1}^{\infty} \frac{1}{(x+n)^{h+1}}.
\end{aligned}
$$

For $\omega = dx$, he had

$$dH = \sum_{h=1}^{\infty} (-1)^{h-1} dx^h \sum_{n=1}^{\infty} \frac{1}{(x+n)^{h+1}}.$$

Only subsequently, in Section 372, did he derive the infinite expression of $H(x)$ in order to interpolate $1 + \frac{1}{2} + \frac{1}{3} + \frac{1}{4} + \ldots + \frac{1}{x}$. Since $H(0) = 0$, he obtained the expansions

$$H(\omega) = \sum_{n=1}^{\infty} \frac{\omega}{n(n+\omega)} \qquad (161)$$

and

$$H(\omega) = \sum_{h=1}^{\infty} (-1)^{h-1} \omega^h \sum_{n=1}^{\infty} \frac{1}{n^{h+1}}. \qquad (162)$$

These provide the sum H, even if ω is not an integer, but they do not define $H(x)$. [In Euler's opinion, the definition of $H(x)$ was: $H(x)$ is the quantity interpolating $1 + \frac{1}{2} + \frac{1}{3} + \frac{1}{4} + \ldots + \frac{1}{x}$.]

For a general inexplicable function $S(x)$, Euler considered the Taylor series of

$$a_{n+x+\omega} = a(n+x+\omega) = \sum_{h=0}^{\infty} \frac{\omega^h d^h a(x+n)}{h! dx^h}, \qquad \text{for } n = 0, 1, 2, 3, \ldots$$

(he applied the methodology already used to seek the summation formula[349]) and obtained

$$
\begin{aligned}
S(x+\omega) &= S(x) + \omega a_{\infty+1} + \sum_{n=1}^{\infty} a(x+n) - a(x+n+\omega) \\
&= S(x) + \omega a_{\infty+1} - \sum_{n=1}^{\infty} \sum_{h=1}^{\infty} \frac{\omega^h d^h a(x+n)}{h! dx^h} \\
&= S(x) + \omega a_{\infty+1} - \sum_{h=1}^{\infty} \frac{\omega^h}{h!} \sum_{n=1}^{\infty} \frac{d^h a(x+n)}{dx^h}.
\end{aligned}
$$

As $a_{\infty+1} = a(1) + \sum_{h=1}^{\infty} [a(h+1) - a(h)]$, he obtained

$$S(x+\omega) = S(x) + \omega a(1) + \omega \sum_{h=1}^{\infty} [a(h+1) - a(h)] - \sum_{h=1}^{\infty} \frac{\omega^h}{h!} \sum_{n=1}^{\infty} \frac{d^h a(x+n)}{dx^h},$$
$$(163)$$

which provides the "complete differential"

$$dS = a(1)dx + dx \sum_{h=1}^{\infty} [a(h+1) - a(h)] - \sum_{h=1}^{\infty} \sum_{n=1}^{\infty} \frac{d^h a(x+n)}{h!} \qquad (164)$$

---

[349]See Chapter 14.

Euler later derived another expression of dS using the power series of $S(x)$. Setting $x = 0$ and

$$G_h = \frac{1}{h!} \sum_{n=1}^{\infty} \frac{d^h a(x+n)}{dx^h},$$

we have

$$S(\omega) = \omega a(1) + \omega \sum_{h=1}^{\infty} [a(h+1) - a(h)] - \sum_{h=1}^{\infty} G_h \omega^h,$$

and, changing ω into x, we obtain

$$S(x) = xa(1) + \sum_{h=1}^{\infty} [a(h+1) - a(h)] x - \sum_{h=1}^{\infty} G_h x^h. \qquad (165)$$

Euler generalized these achievements (derived subject to the condition that, for $n = \infty$, the sequence a_n becomes a constant L), considering the case in which the second or third differences of $S(\infty), S(\infty+1), S(\infty+2), \ldots$ equal 0. He then reduced the products $S(x) = \prod_{n=1}^{x} a_x$ to sums by means of logarithms. Under the conditions $\log a_\infty = 0$ and $\log a_{\infty+1} - \log a_\infty = 0$, he derived

$$S(x+\omega) = S(x) \prod_{n=1}^{\infty} \frac{a_{n+x}}{a_{n+x+\omega}} \quad \text{and} \quad S(x+\omega) = S(x) a_{x+1}^{\omega} \prod_{n=1}^{\infty} \frac{a_{x+n+1}^{\omega} a_{x+n}^{1-\omega}}{a_{n+x+\omega}},$$

respectively. For $x = 0$, one has $S(0) = 1$; changing ω into x, one obtains

$$S(\omega) = \prod_{n=1}^{\infty} \frac{a_n}{a_{n+x}} \quad \text{and} \quad S(\omega) = a_1^x \prod_{n=1}^{\infty} \frac{a_{n+1}^x a_n^{1-x}}{a_{n+x}}$$

(see Euler [1755, 2: Sections 381–382]).

The latter may be applied to $G(x) = 1 \cdot 2 \cdot 3 \cdot \ldots \cdot x$ (in this case, the first difference of the logarithms of the terms whose index is infinite is 0) to obtain (82) (see Euler [1755, 2: Section 402]). For

$$\log S(x) = \log \prod_{n=1}^{\infty} a_n = \sum_{n=1}^{\infty} \log a_n$$

and $\log a_{\infty+1} - \log a_\infty = 0$, (163) is transformed into

$$\log S(x+\omega) = \log S(x) + \omega \log a(1) + \omega \sum_{h=1}^{\infty} \log \frac{a(x+h+1)}{a(x+h)} \qquad (166)$$

$$- \sum_{h=1}^{\infty} \frac{\omega^h}{h!} \sum_{n=1}^{\infty} \frac{d^h \log a(x+n)}{dx^h}$$

and then

$$\frac{dS}{S} = dx \log a(x+1) + dx \sum_{h=1}^{\infty} \log \frac{a(x+h+1)}{a(x+h)} - \sum_{h=1}^{\infty} \sum_{n=1}^{\infty} \frac{d^h \log a(x+n)}{h!}$$

(see Euler [1755, 2: Section 385]).

Euler applied this formula to $G(x) = 1 \cdot 2 \cdot 3 \cdot \ldots \cdot x$ to obtain

$$\frac{dG}{G} = dx \log(x+1) + dx \sum_{h=1}^{\infty} \log \frac{x+h+1}{x+h} + \sum_{h=1}^{\infty} (-1)^h \frac{dx^h}{h} \sum_{n=1}^{\infty} \frac{1}{(x+n)^h}.$$

(167)

He did not write the infinite expression of $\log G(x)$ by deriving it from (166); he merely sought the differential of $\log G(x)$ and did not consider the expression of $\log G(x)$ of importance.

Finally, from (165), Euler derived

$$\log S(x) = x \log a(1) + x \sum_{h=1}^{\infty} \log \left[a(h+1) - a(h) \right] - \sum_{h=1}^{\infty} G_h x^h,$$

where $G_h = \frac{1}{h!} \sum_{n=1}^{\infty} \frac{d^h \log a(x+n)}{dx^h}$ and $\log a_{\infty+1} - \log a_{\infty} = 0$, and then

$$\frac{dS}{S} = dx \log a(1) + dx \sum_{h=1}^{\infty} \log \left[a(h+1) - a(h) \right] - \sum_{h=1}^{\infty} h G_h x^{h-1} dx.$$

If $G(x) = 1 \cdot 2 \cdot 3 \cdot \ldots \cdot x$, he obtained $\log G(x) = \sum_{h=1}^{\infty} \frac{(-1)^h}{h} C_h x^h$ and

$$\frac{dG}{G} = \sum_{h=1}^{\infty} (-1)^h C_h x^{h-1},$$

(168)

where C_1 is Euler's constant and $C_h = \sum_{n=1}^{\infty} \frac{1}{n^h}$, for $h = 2, 3, 4, \ldots$.

However, even in this case, the power series of $\log G(x)$ is only an intermediate step for arriving at (168) and in particular for calculating C_1.[350]

---

[350]By applying the sum formula (98) to the inexplicable functions $S(x) = \sum_{n=1}^{x} a_n$, Euler [1755, 2: Sections 386–388] inferred

$$dS = a(x)dx + \frac{1}{2} da(x) + \sum_{n=1}^{\infty} (-1)^{n-1} \frac{B_{2n}}{(2n)!} \frac{d^{2n-1} a(x)}{dx^{2n-1}} da(x).$$

As a particular case, he considered $G = 1 \cdot 2 \cdot 3 \cdot \ldots \cdot x$ and, using the logarithms, derived

$$\frac{dG}{G} = \log x dx + \frac{dx}{2x} - \sum_{n=1}^{\infty} (-1)^{n-1} \frac{B_{2n}}{2n} \frac{dx}{x^{2n}}$$

The latter is more suitable than (167) or (168) for calculating dS/S when x is very large.

Euler was not entirely satisfied by this exposition of the theory of inexplicable functions and later tried to clarify it in his *Delucidationes in capita postrema calculi mei differentialis de functionibus inexplicabilibus* (see Euler [1787]). Despite his efforts, during the second half the 18th century, inexplicable functions did not actually enter into analysis in the same ways as the other known transcendental (logarithmic and trigonometric) functions.[351] The reasons for this are closely connected with the limits of the formal methodology, which I will discuss in Chapter 29.

---

[351]See Chapter 25.

25   Integration and functions

I shall now go on to examine the case in which the term "function" was associated with certain quantities that were analytically expressed in a closed form by integrals. The stimulus for the investigation of such quantities stemmed mainly from geometry and applied mathematics and from the attempt to mathematicize nature. For instance, probabilists ran up against gamma and beta functions, elliptic integrals arose from questions regarding physics and geometry, while astronomy in general required new instruments other than analytical expressions and their power series, etc. In the last 30 years of the century, nonelementary transcendental functions became an important field of research in analysis. Many results were derived; although nonelementary transcendental functions were not considered well known enough to be accepted as true functions. For example, in his *Institutiones calculi integralis*, Euler argued that the solution to an integral $\int f(x)dx$ could be a transcendental function, although only logarithmic and trigonometric functions — among all transcendental functions — were to be placed on the same plane as algebraic functions (see Euler [1768-70, 1:14]). He investigated various nonelementary transcendental functions in this treatise; the first was

$$\int_0^z \frac{dz}{\log z}.$$

According to him, if the integral $\int_0^z \frac{dz}{\log z}$ could be assigned, it should have been of a very wide use in analysis (Euler [1768-70, 1:122]. It merited a more careful investigation, however, for the time, "the nature of this function is not known enough" (Euler [1768-70, 1:128]).

   Similarly, in his *Sur l'intégration de quelques équations différentielles*, Lagrange suggested that investigation of integrability of such expressions as

$$\frac{1}{\sqrt{a + bx + cx^2 + cx^3 + dx^4 + \ldots}}$$

opened a vast field to research (see Lagrange [1766, 33]). However, as Fraser noted,[352] Lagrange limited himself to stating the existence of this possibility, which was not developed in his work and, in effect, such expressions did not enter into the list of his analytical functions.

   This approach can be better understood if we briefly examine the 18th-century concept of integration. Leibniz regarded integration as summation of infinitesimals. In the geometrical context of the first years of the calculus, integration was an operation that, given the analytical expression of a geometrical quantity, made it possible to determine the analytical expression of another quantity, the existence of which was geometrically self-evident (e.g.,

---

[352]See Fraser [1987, 40].

given the ordinate of a curve, integration made it possible to calculate the area or the length of the curve — the area and the length existed on the basis of geometrical evidence).

In the second part of 18th century, integration was usually defined as the inverse operation of differentiation or derivatio,[353] namely that the integral $\int f(x)dx$ of the function $f(x)$ was a function $F(x)$ such that $dF = f(x)dx$. For instance, Lagrange gave the name "direct analysis of functions" to that part of analysis that investigated the rules by which it was possible to move from a function to its derivatives of different orders; instead, the part of the calculus that scrutinized the return from derivatives to their primitive function was termed the "inverse analysis of functions" (see Lagrange [1797, 140–141]). Euler regarded differentiation as the direct operation while he considered integration to be the inverse operation by which one returned from the differential to the function generating the differential (see *Institutiones calculi integralis*, [1768–70, 1:5]). Similarly, Lacroix stated that the integral calculus was the inverse of the differential calculus and consisted of returning from differential coefficients to functions from which they were derived (see Lacroix [1797–1800, 2:1]).

Mathematicians were aware that many simple functions could not be integrated by means of elementary functions and that this concept of integration posed the problem of nonelementarily integrable functions. Euler and Lagrange briefly mentioned the question in their treatises. They compared integration with inverse arithmetical operations. According to Lagrange, there were certain cases in which the operations of division and extraction of a root could not be performed in an exact manner but only by approximation (Lagrange [1797, 141]). Similarly, inverse analysis did not always succeed in finding primitives, although it was still possible to find an approximate solution to differential equations (by series) when this occurred. According to Euler, integration led to new transcendental "quantities" in the same manner as the operations of subtraction, division, and extracting a root led to negative, rational, and irrational numbers (Euler [1768-1770, 1:13]).

This analogy was significant for the status of transcendental quantities. Indeed, we saw that only integers and fractions were numbers in the strict sense of the term.[354] The extension of the term "number" to incommensurable ratios was considered incorrect because "number" presupposed an exact denotation. Nevertheless, incommensurable ratios were similar to numbers and could therefore be viewed as numbers; in fact,

(a) an irrational number could be handled as a symbol; e.g., $\sqrt{2}$ was a symbol such that $\left(\sqrt{2}\right)^2 = 2$,

---

[353]The notion of integration as the operational inverse of differentiation and the significance of this conception within 18th-century analysis are discussed in Fraser [2003].
[354]See Section 7.2, p. 106.

(b) it could be approached by numbers as desired,

(c) even though it could not be represented rigorously by means of arithmetic, it at least had a geometrical meaning.[355]

In the same way as irrational numbers were not true numbers, transcendental quantities were not functions in the strict sense of the term and
differed from elementary functions, which were the only genuine object of
analysis. Nevertheless, transcendental quantities had properties that allowed
one to view them as entities similar to functions; in fact,

(α) transcendental quantities $f(x)dx$ arose from the integration of an elementary function $f(x)$ and could be manipulated formally under the
condition $d[\int f(x)dx] = f(x)dx$ so as to transform $f(x)dx$ into other
integrals or into infinite expressions,

(β) even when the values of these transcendental quantities were not already given and tabulated, they could be computed in an approximate
way by resorting to series or other techniques,

(γ) integrals could be thought of as geometrical (or physical) quantities:[356]
so their existence at least had a geometrical visualization or physical
basis.

---

[355]See d'Alembert [1773, 5:217–218].
[356]See Euler [1768–1770, 2:220].

26 Series and differential equations

In this chapter, I wish to investigate some important results concerning the series solution to differential equations which were derived in the second part of the 18th century. These results were of great importance for the subsequent development of mathematics since, while seeking the series solution to differential equations, mathematicians came up against many series that were not the expansions of elementary functions. Looking back with the advantage of hindsight, we might state that they gave a *de facto* treatment of many new transcendental functions. However, this is only an *a posteriori* interpretation.

We saw that mathematicians were aware that certain expansions did not derive from elementary functions; however, they did not modify their approach and, in particular, did not abandon the use of the formal methodology, which had its foundations in elementary functions. They continued to avoid considering series as autonomous objects (as Lagrange still stated in 1797, "an expression in series can always be regarded as the development of a finite expression" (Lagrange [1797, 93])). Series solutions were only viewed as an instrument for dealing with certain quantities expressed in the form of differential equations. They had an inferior status to solutions in a closed formed (which were elementary functions or nonelementary functions given in integral form). Thus, while it is sometimes possible to find the term "function" associated with quantities given by integral formulas, it is never associated with quantities given by series.

Mathematicians resorted to series solutions when they were unable to integrate a function $f(x)$ or a differential equation $F(x, y, dy/dx, \ldots) = 0$ in other ways. In this case, they applied the usual procedures that enabled them to expand elementary functions into integrals and differential equations.[357] Operating formally, one could determine the series solution $\sum a_n x^n$ that was a useful instrument in the investigation of the quantities expressed by integrals $\int f(x)dx$ or differential equations $F(x, y, dy/dx, \ldots) = 0$.

I emphasize that *this did not mean that the solution to a given differential equation was defined by the series*.[358] In the same manner as $\sum_{n=0}^{\infty} \frac{x^n}{n!}$ represented the quantity e^x and allowed one to investigate it but did not define it, a power series $\sum a_n x^n$ represented a quantity of the types $\int f(x)dx$ or $F(x, y, dy/dx, \ldots) = 0$ and allowed one to investigate it but did not define it.

When they were not reducible to closed expressions, series solutions to differential equations played two roles.

---

[357]Usual procedures could be applied since integrals and differential equations were expressions of the type $f(x)dx$ or $F(x, y, dy/dx, \ldots) = 0$, where f and F were elementary functions.

[358]The solution is defined by the differential equation, but this did not mean that it was a function in the proper sense of the term (see Chapter 23, p. 251).

First, series were instruments that provided the approximate values of a
quantity expressed by a differential equation. It was commonplace
that a solution by series was not an exact solution. According to Lagrange [1776, 301], the method of series was a method "for integrating
by approximation the differential equations whose finite integral was
impossible or, at the very least, extremely difficult." And, in his [1780,
522–523], Euler regarded the representation of a quantity by a series
as an approximate representation. Series did not provide the exact
solution and did not express a quantity exactly.

Second, series could be instruments for expressing a link between different
analytical expressions (in this role, convergence was not of importance
and even totally divergent series[359] could be used).

An example of the first role is the Eulerian treatment of the differential
equation

$$x^2(a + bx^n)\frac{d^2y}{dx^2} + x(c + ex^n)\frac{dy}{dx} + (f + gx^n)y = 0. \qquad (169)$$

In *Institutiones calculi integralis* [1768–1770, 2:177–185], he tried a solution
in the form

$$y = \sum_{j=0}^{\infty} A_j x^{\lambda+jn}, \qquad (170)$$

where $A_0 \neq 0$. By replacing (170) into (169), he found

$$\beta_0 A_0 + \sum_{j=1}^{\infty} (\alpha_{j-1} A_{j-1} + \beta_j A_j)x^{\lambda+jn} = 0,$$

where

$$
\begin{aligned}
\beta_0 &= \lambda(\lambda - 1)a + \lambda c + f, \\
\alpha_0 &= \lambda(\lambda - 1)b + \lambda e + g, \\
\alpha_j &= jn(jn + 2\lambda - 1)b + jne + \alpha_0 \ \ (\text{for } j > 0), \\
\beta_j &= jn(jn + 2\lambda - 1)a + jnc \ \ (\text{for } j > 0).
\end{aligned}
$$

Hence,

$$\lambda(\lambda - 1)a + \lambda c + f = 0 \text{ and } \beta_j A_j = \alpha_{j-1} A_{j-1} \ (\text{for } j > 0).$$

If one chooses A_0 arbitrarily, the previous relations allow one to determine
λ by solving the equation $\lambda(\lambda - 1)a + \lambda c + f = 0$ and A_j (for $j > 0$) by
recurrence.

---

[359]See p. 191.

Therefore, if $a \neq 0$, we can determine two values λ_1 and λ_2 of λ, and so, we have two series of the form (169), which furnish complete solutions to the differential equation. However, if $\lambda(\lambda - 1)a + \lambda c + f = 0$ has only one root or when the difference between the two values of λ is divisible by n,[360] the general integral cannot be expressed as the sum of two series of the form (169). In this case Euler found that the general integral had the form

$$\log x \sum_{j=0}^{\infty} C_j x^{\lambda_1 + jn} + \sum_{j=0}^{\infty} B_j x^{\lambda_2 + jn} + \sum_{j=0}^{\infty} D_j x^{\lambda_1 + jn},$$

where $\lambda_2 \leq \lambda_1$ and the coefficients A_j, B_j, C_j depend on two arbitrary constants.

An example of the second role of series is found in *Specimen transformationis singularis serierum* [1794a], where Euler showed that the solution of

$$x(1 - x)\frac{d^2 y}{dx^2} + [\gamma - (\alpha + \beta + 1)x]\frac{dy}{dx} - \alpha\beta y = 0 \qquad (171)$$

was the hypergeometric series

$$1 + \frac{\alpha\beta}{1 \cdot \gamma}x + \frac{\alpha(\alpha + 1)\beta(\beta + 1)}{1 \cdot 2 \cdot \gamma(\gamma + 1)}x^2 + \frac{\alpha(\alpha + 1)(\alpha + 2)\beta(\beta + 1)(\beta + 2)}{1 \cdot 2 \cdot 3 \cdot \gamma(\gamma + 1)(\gamma + 2)}x^3 + \ldots .$$

By using this fact and the relation between the hypergeometric series and certain appropriate expansions of

$$\int_0^\pi (1 + \alpha^2 - 2\alpha \cos \phi)^n \cos p\phi \, d\phi$$

and

$$\int_0^\pi (1 + \alpha^2 - 2\alpha \cos \phi)^{-n-1} \cos p\phi \, d\phi,$$

Euler proved the following equality between integrals

$$\binom{n + p}{p}(1 - \alpha^2)^{-n} \int_0^\pi (1 + \alpha^2 - 2\alpha \cos \phi)^n \cos p\phi \, d\phi \qquad (172)$$

$$= \binom{p - n - 1}{p}(1 - \alpha^2)^{n+1} \int_0^\pi (1 + \alpha^2 - 2\alpha \cos \phi)^{-n-1} \cos p\phi \, d\phi.$$

$$* \quad * \quad *$$

---

[360] In this case the coefficients β_j become equal to infinity for one of the two roots.

I would now like to examine some of the more interesting examples of series connected in some way with integral or differential equations.

In his *Theoremata de oscillationibus corporum filo flexili connexorum et catenae verticaliter suspensae,* Daniel Bernoulli enunciated some theorems on the oscillations of heavy chains. In one of these[361] he considered a uniform heavy flexible chain AC of length l, fixed at the upper end A and free at the lower end C. He supposed that when the chain was slightly disturbed from its position of equilibrium in a vertical plane, it underwent small oscillations. Denoted the new position of the chain by AMF and FM by x, Bernoulli took the maximum distance of the endpoint F from the vertical line to be equal to 1. He asserted that, in this case, the oscillation y of the point M from the vertical line is given by the series

$$ 1 - \frac{x}{f} + \frac{x^2}{4f^2} - \frac{x^3}{4 \cdot 9 f^3} + \frac{x^4}{4 \cdot 9 \cdot 16 f^4} - \frac{x^5}{4 \cdot 9 \cdot 16 \cdot 25 f^5} + \dots, \qquad (173) $$

where f is the solution of the equation[362]

$$ 1 - \frac{l}{f} + \frac{l^2}{4f^2} - \frac{l^3}{4 \cdot 9 f^3} + \frac{l^4}{4 \cdot 9 \cdot 16 f^4} - \frac{l^5}{4 \cdot 9 \cdot 16 \cdot 25 f^5} + \dots = 0 \qquad (174) $$

Bernoulli also stated that f is approximately equal to $0.691l$ and that f can assume infinitely other values [namely, Equation (174) has an infinite number of zeros]. He gave a proof of this theorem later in [1734-35].

In *De oscillationibus minimis funis libere suspensi* [1781a], Euler dealt with the same problem. He considered the forces acting on an element of the chain of length dx and obtained the partial differential equation, which, using modern symbols, is written as

$$ \frac{\partial^2 y}{\partial t^2} = g \frac{\partial}{\partial x}\left(x \frac{\partial y}{\partial x} \right), \qquad (175) $$

where y is the horizontal displacement of a point P of chain at the time t, x is the height of P above the bottom of the chain in its undisturbed position, and g is the gravitational constant.

He assumed the oscillation y was essentially sinusoidal with angular frequency $\omega = \sqrt{\frac{g}{f}}$ and wrote

$$ y = Av \sin\left(\xi + t\sqrt{\frac{g}{f}} \right), \qquad (176) $$

where A and ξ are constants and $v = \Phi\left(\frac{x}{f}\right)$ is an appropriate function of the only variable x. Substituting (176) into (175), he found that $\Phi\left(\frac{x}{f}\right)$ is a

---

[361] See D. Bernoulli [1732–33, 116].
[362] f is the length of the simple equivalent pendulum.

solution of the ordinary differential equation

$$\frac{d}{dx}\left(x\frac{dv}{dx}\right) + \frac{v}{f} = 0. \tag{177}$$

In this way, Euler again found Bernoulli's solution (173), which he wrote in the form

$$v = 1 - u + \frac{u^2}{4} - \frac{u^3}{4\cdot 9} + \frac{u^4}{4\cdot 9\cdot 16} - \frac{u^5}{4\cdot 9\cdot 16\cdot 25} + \cdots,$$

where $u = \frac{x}{f}$. Later Euler dealt with the series solution of Equation (177) in [1768-80, 2: Section 977] and in [1781b].

Today we recognize (173) as a Bessel function $J_0(2\sqrt{\frac{x}{n}})$ of order zero and argument $2\sqrt{\frac{x}{n}}$; however, I emphasize that for Bernoulli and Euler (173) was merely a tool for obtaining an approximate solution to a problem of physics.

The determination of f gives the solution to Equation (174) in terms of $\frac{l}{f}$. In [1781a], Euler found that the three smallest roots are $\frac{l}{f} = 1.445795$, $\frac{l}{f} = 7.6658$, $\frac{l}{f} = 18.63$ (the values of $\frac{l}{f}$ are sufficiently accurate, especially the first). In order to find the roots of $\sum_{n=0}^{\infty}(-1)^n\frac{u^n}{(n!)^2} = 0$, Euler used a technique already employed in [1734–35a]. He assumed that

$$\sum_{n=0}^{\infty}(-1)^n\frac{u^n}{(n!)^2} = \prod_{n=1}^{\infty}\left(1 - \frac{u}{a_n}\right).$$

Hence,

$$\frac{d}{du}\log\left(\sum_{n=0}^{\infty}(-1)^n\frac{u^n}{(n!)^2}\right)$$

$$= \frac{\frac{d}{du}\left(\sum_{n=0}^{\infty}(-1)^n\frac{u^n}{(n!)^2}\right)}{\sum_{n=0}^{\infty}(-1)^n\frac{u^n}{(n!)^2}}$$

$$= \log\prod_{n=1}^{\infty}\left(1 - \frac{u}{a_n}\right)$$

$$= -\sum_{n=1}^{\infty}\frac{1}{a_n - u} = -\sum_{n=1}^{\infty}\sum_{m=0}^{\infty}\frac{u^m}{(a_n)^{m+1}}$$

$$= -\sum_{m=0}^{\infty}u^m\left(\sum_{n=1}^{\infty}\frac{1}{(a_n)^{m+1}}\right) = -\sum_{m=0}^{\infty}\sigma_{m+1}u^m,$$

where $\sigma_{m+1} = \sum_{n=1}^{\infty}\frac{1}{(a_n)^{m+1}}$. Therefore,

$$\frac{d}{du}\left(\sum_{n=0}^{\infty}(-1)^n\frac{u^n}{(n!)^2}\right) = -\left(\sum_{n=0}^{\infty}(-1)^n\frac{u^n}{(n!)^2}\right)\sum_{m=0}^{\infty}\sigma_{m+1}u^m.$$

By applying the method of indeterminate coefficients, he found

$$\sigma_1 = 1, \sigma_2 = \frac{1}{2}, \sigma_3 = \frac{1}{3}, \sigma_4 = \frac{11}{48}, \sigma_5 = \frac{19}{120}, \sigma_6 = \frac{473}{4320}, \ldots$$

If a_1 is the smallest of the root and $0 < a_1 < a_2 < a_3 < \ldots$, then

$$\frac{1}{(\sigma_m)^{1/m}} < a_1 < \frac{\sigma_m}{\sigma_{m+1}}.$$

At this point, Euler computed the first values of $\frac{1}{(\sigma_m)^{1/m}}$ and $\frac{\sigma_m}{\sigma_{m+1}}$ and derived $a_1 = 1,445795$. Using similar reasoning, he found a_2 and a_3 (see Euler [1781a, 317–323]).

Another problem, which was significant to our purpose, was that of the vibrations of a stretched membrane. Euler investigated it in *De motu vibratorio tympanorum* [1764] and derived an equation equivalent to

$$\frac{1}{c^2}\frac{\partial^2 z}{\partial t^2} = \frac{\partial^2 z}{\partial r^2} + \frac{1}{r}\frac{\partial z}{\partial r} + \frac{1}{r^2}\frac{\partial^2 z}{\partial \varphi^2}, \tag{178}$$

where z is the transverse displacement at time t at the point whose polar coordinates are (r, φ) and c is an appropriate constant. He assumed that the solutions had the form

$$z = u(r)\sin(\omega t + A)\sin(\kappa\varphi + B),$$

where ω, A, κ, B are constants. By replacing $u(r)\sin(\omega t + A)\sin(\beta\varphi + B)$ in (178), he derived the equation

$$\frac{d^2 u}{dr^2} + \frac{1}{r}\frac{du}{dr} + (\alpha^2 - \frac{\kappa^2}{r^2}) = 0, \tag{179}$$

where $\alpha = \frac{\omega^2}{c^2}$. Euler assumed the existence of a power series solution of this equation and obtained

$$u(r) = \sum_{n=0}^{\infty} \frac{(-1)^n}{n!(\kappa+1)_n}\left(\frac{\alpha r}{2}\right)^{\kappa+2n} \tag{180}$$

(see Euler [1764, 344–359]).[363]

Series were crucial for the development of planetary mechanics, a very important subject in 18th-century mathematics.[364] For instance, in his *Recherches sur l'attraction des sphéroïdes homogènes* [1785], Legendre investigated the attraction of ellipsoids and, in particular, showed that if the

---

[363]Today Equation (179) is called Bessel's equation, and the solution $u(r)$ is the Bessel function $J_\beta(\alpha r)$. I again emphasize that Euler considered (180) as a tool for obtaining an approximate solution to a problem of physics.

[364]On this topic, see Todhunter [1873] and Wilson [1980] and [1985].

attraction of a solid of revolution S is known for every external point that
is on the prolongation of the axis, then it is known for every external point
P. In the proof, Legendre employed power series as an intermediate step.
Using more modern symbols and the potential function V, his proof can be
summarized as follows.

Let (r, θ, φ) and (ρ, Θ, Ψ) be the spherical polar coordinates of the attracted particle[365] and of the element of the attracting body S, respectively.
The potential function is

$$V = \iiint \frac{\rho^2}{r} \left(1 - 2\frac{\rho}{r}\cos\gamma + \frac{\rho^2}{r^2} \right)^{-1/2} \sin\Theta\, d\Theta\, d\omega\, d\rho,$$

where $\cos\gamma = \cos\theta\cos\Theta + \sin\theta\sin\Theta\sin\omega$ and ω is the difference of longitude
of the attracted point and the attracting element. If one sets

$$\frac{1}{(1 - 2x\alpha - \alpha^2)^{1/2}} = \sum_{n=0}^{\infty} P_n(x)\alpha^n,$$

namely $P_n(x)$ denotes the coefficient of α^n in the expansion of $(1 - 2x\alpha - \alpha^2)^{-1/2}$
in ascending powers of α, namely one can write

$$V = \iiint \frac{\rho^2}{r} \left(1 + P_1(\cos\gamma)\frac{\rho}{r} + P_2(\cos\theta)\frac{\rho^2}{r^2} + P_3(\cos\theta)\frac{\rho^3}{r^3} + \dots \right)$$
$$\cdot \sin\Theta\, d\Theta\, d\omega\, d\rho.$$

The coefficients $P_n(x)$ are today named Legendre's polynomials.[366]

If one now supposes that the body is symmetrical with respect its equator
and integrates with respect to ρ between the limits $-s$ and s, where $s(\Theta)$
is the radius vector of the solid corresponding to a colatitude Θ, then one
obtains

$$V = 2 \iint \frac{s^2}{r} \left(\frac{1}{3} + P_2(\cos\theta)\frac{s^2}{r^2} + P_4(\cos\theta)\frac{s^4}{r^4} + \dots \right) \sin\Theta\, d\Theta\, d\omega.$$

Since $\cos\gamma = \cos\theta\cos\Theta + \sin\theta\sin\Theta\sin\omega$, P_n is a function of θ, Θ, and ω.
Legendre shows that if one integrates with respect to ω from 0 to 2π, P_n is a
function of θ and Θ. It is of the form $2\pi f_n(\cos\theta) f_n(\cos\Theta)$ for an appropriate
function f. By using such properties, one obtains a series whose first term
depends only on r and where the nth term is

$$\frac{4\pi}{(2n+3)r^{2n+1}} f_{2n}(\cos\theta) \int_0^{\pi/2} f_{2n}(\cos\Theta) s^{2n+3} \sin\Theta\, d\Theta,$$

---

[365] One can assume $\varphi = 0$ for the particular symmetry of the problems.
[366] On Legendre's functions, see Todhunter [1873].

where $f_{2n}(\cos\theta)$ is a known function that is independent of the form of the body and is different from zero for $\theta = 0$. If the attraction is known at all points that are all on the prolongation of the axis, V is known for all such points. Therefore, the integrals

$$I_{2n} = \int_0^{\pi/2} f_{2n}(\cos\Theta) s^{2n+3} \sin\Theta d\Theta$$

are known. Consequently, the attraction is also known for every external point P.

In the same period, Laplace was working on similar questions. In particular, in *Théorie des attractions des sphéroïdes et de la figure des planètes* [1785], Laplace considered what today is known as Laplace's equation:

$$\frac{\partial^2 V}{\partial x^2} + \frac{\partial^2 V}{\partial y^2} + \frac{\partial^2 V}{\partial z^2} = 0,$$

which he expressed in the form

$$\frac{\partial}{\partial\mu}\left[(1-\mu^2)\frac{\partial V}{\partial\mu}\right] + \frac{1}{1-\mu^2}\frac{\partial^2 V}{\partial\omega^2} + r\frac{\partial^2(rV)}{\partial r^2} = 0, \tag{181}$$

where V is the potential function, r, ω, θ are spherical polar coordinates, and $\mu = \cos\theta$ [1785, 362]. Laplace sought a series expansion for V. If the point is external, he assumed that the expansion is of the type

$$V(r, \omega, \theta) = \sum_{i=0}^{\infty} \frac{U_i}{r^{i+1}}. \tag{182}$$

Laplace substituted this expression for V into in (181) and obtained

$$\frac{\partial}{\partial\mu}\left[(1-\mu^2)\frac{\partial U_i}{\partial\mu}\right] + \frac{1}{1-\mu^2}\frac{\partial^2 U_i}{\partial\omega^2} + i(i+1)U_i = 0. \tag{183}$$

He succeeded in giving an integral form to U_i by using Legendre's polynomial [1785, 362–369]. To calculate the potential of spheroids that differ only slightly from spheres, Laplace [1785, 371–373] considered the equation $r = a(1 + \alpha y)$, where α is a small constant coefficient and y is a function of ω and μ (a is the radius of a sphere). He showed that $y(\omega, \mu)$ can be developed into a series

$$y = \sum_{i=0}^{\infty} Y_0, \tag{184}$$

where the functions Y_i satisfied the equation $\frac{\partial}{\partial\mu}\left[(1-\mu^2)\frac{\partial Y_i}{\partial\mu}\right] + \frac{1}{1-\mu^2}\frac{\partial^2 Y_i}{\partial\omega^2} + i(i+1)Y_i = 0$, and that

$$Y_i = \frac{2i+1}{4\pi\alpha a^{i+3}}U_i.$$

27 Trigonometric series

The formal methodology, with regard to series, had been conceived with reference to power series. Power series are a particularly simple type of series, especially as regards the relationship between formal manipulations and convergence. In particular, an interval of convergence I_{x_0} can be associated with a power series

$$\sum a_n(x - x_0)^n.$$

The whole theory of ordinary power series was grounded upon this fact, which excluded the appearance of pathological phenomena, and on the assumption that, if $\sum a_n(x-x_0)^n$ converged to $f(x)$ near x_0, it was possible to identify $\sum a_n(x-x_0)^n$ with $f(x)$ in analytical calculations, without further regard for the interval of convergence. The situation was certainly more complicated for other types of series and, in particular, for trigonometric series.

Trigonometric series began to appear in some mathematical and physical investigation from the 1730s onwards. Astronomy was the field where the need for such series was most strongly felt; indeed, they seemed well suited to describing periodic phenomena, clearly relevant to the subject matter of astronomy.

In his paper of 1749 on irregularities of the orbits of Saturn and Jupiter [1748b], Euler investigated the following differential equations:

$$m^2 d\zeta + 2mn dx + m^2 nr d\zeta - n\frac{d^2 r}{d\zeta}$$

$$= \frac{1 + \nu}{\lambda^3} d\zeta - \frac{2nr}{\lambda^3} d\zeta + 2r \cos\frac{\omega}{\lambda} d\zeta + \frac{n(\lambda - \cos\omega)}{h(1 - g\cos\omega)^{3/2}} d\zeta$$

and

$$2m dr d\zeta + d^2 x = -n \sin\frac{\omega}{\lambda} d\zeta + \frac{\sin\omega}{h(1 - g\cos\omega)^{3/2}} d\zeta.$$

Euler observed that the integration of these two equations depended on the integral of $(1 - g\cos\omega)^{-3/2}$, but it was not possible to give a closed form for it. Thus, he dealt with the question of the integration by series of $(1 - g\cos\omega)^{-\mu}$. According to Euler, the greatest difficulty in finding the solutions to these equations was the determination of a sufficiently fast expansion of $(1 - g\cos\omega)^{-\mu}$. Indeed, he observed that the expansion of this formula, "following ordinary rules", is

$$(1 - g\cos\omega)^{-\mu} = 1 + \frac{\mu}{1} g\cos\omega + \frac{\mu(\mu + 1)}{1 \cdot 2} g^2 \cos^2\omega + \dots,$$

"but this series is not suitable for my purpose, in as much as it is not sufficiently convergent, since it contains powers of cos ω. As for the last disadvantage, one can remedy it by reducing the powers of the cosine of the angle ω, to the cosines of multiples of the angle."[367]
Since

$$
\begin{aligned}
2\cos^2\omega &= \cos 2\omega + 1, \\
4\cos^3\omega &= \cos 3\omega + 3\cos\omega, \\
8\cos^4\omega &= \cos 4\omega + 4\cos 2\omega + 3,
\end{aligned}
$$

$$\cdots$$

Euler stated that $(1 - g\cos\omega)^{-\mu}$ must have an expansion of the type

$$\sum_{i=0}^{\infty} a_i \cos i\omega.$$

Then Euler found a recursive formula for a_i $(i > 1)$ and calculated a_0 and a_1 approximately, a process that involved a long sequence of calculations (see Golland and Golland [1993, 58–64]).

In *De serierum determinatione seu nova methodus inveniendi terminos generales serierum* [1750–51b], Euler returned to his search for the expression of the general term of a series Σa_n defined by recurrence.[368] He studied several cases. The first was the series $1 + 1 + 1 + \ldots$. Euler reduced the problem to the study of the functional equation

$$y(x+1) = y(x),$$

where $y(x)$ is a function — obviously in the 18th-century sense of the term. It was solved by expanding $y(x+1)$ in Taylor series

$$y(x+1) = \sum_{n=0}^{\infty} \frac{1}{n!}\frac{d^n y}{dx^n}.$$

Since $y(x+1) = y(x)$, he obtained the differential equation of infinite degree:

$$\sum_{n=1}^{\infty} \frac{1}{n!}\frac{d^n y}{dx^n} = 0. \tag{185}$$

This enabled Euler to apply the technique for solving differential equations that he had developed in *Methodus equationes differentiales altiorum graduum integrandi ulterius promota* [1750–51a]. Indeed, by using the substitution $y = e^{zx}$, he derived the auxiliary equation $\sum_{n=1}^{\infty} \frac{1}{n!}z^n = e^z - 1 = 0$,

---

[367] See Euler [1748b, 61], translation in Golland and Golland [1993, 58].
[368] See Chapter 13.1.

which has the infinitely many roots $z = 2k\pi i$, $k \in Z$. Since $y(0) = 1$, he found that the solution to 185 was

$$y = 1 + \sum_{i=1}^{\infty} \alpha_i \sin 2i\pi x + \sum_{i=1}^{\infty} \beta_i(\cos 2i\pi x - 1)$$

(the coefficients α_i and β_i are to be determined subject to the condition $a_n = 1$, for every integer n).[369]

It should be noted that Euler, however, did not conceive of this trigonometric series as the final result. Euler, in principle, imagined that such trigonometric series could be expressed as finite functions of $\sin \pi n$ and $\cos \pi n$.

In the same way, he showed how the general term of the series

$$a + (a + g) + (a + 2g) + \ldots,$$

for a and g constants, could be determined by solving the functional equation $y(x + 1) = y(x) + g$ and found

$$a_n = A + gn + \sum_{i=1}^{\infty} \alpha_i sin2i\pi n + \sum_{i=1}^{\infty} \beta_i cos2i\pi n$$

[A, α_i, and β_i are determined subject to the condition $a_n = a + (n - 1)g$].

He then solved eight further problems, which he reduced to the following equations:

$$
\begin{aligned}
y(x + 1) &= ay(x); \\
y(x) &= ay(x - 1) + b; \\
y(x) &= ay(x - 1) + by(x - 2); \\
y(x) &= a_1y(x - 1) + a_2y(x - 2) + \ldots + a_2y(x - n); \\
y(x) &= c + a_1y(x - 1) + a_2y(x - 2) + \ldots + a_ny(x - n); \\
y(x) &= my(x) + a + bx; \\
y(x) &= y(x - 1) + F(x); \\
y(x + 1) &= xy(x);
\end{aligned}
$$

where a, b, c, g, m, a_1, a_2, $\ldots$, a_n are constants and $F(x)$ is an elementary function. In particular, he found that the solution of the equation $y(x) = y(x - 1) + F(x)$ was

$$y(x) = \int_0^x F(x)dx + 2\sum_{n=1}^{\infty} cos2\pi nx \int_0^x F(x)cos2\pi nxdx$$

$$+ 2\sum_{n=1}^{\infty} sin2\pi nx \int_0^x F(x)sin2\pi nxdx.$$

---

[369] See Euler [1750–51b, 470–480].

Trigonometric series were also used in the problem of the vibrating string.[370] This controversy is relevant to our purpose for two reasons. The first reason reason is that it showed the need to introduce quantities that were different from elementary functions in order to mathematize the study of natural phenomena; at the same time, it revealed the difficulties that 18th-century analysis (and in particular of the theory of series) had in treating nonelementary functions. The treatment of these new quantities required a change in the concept of analysis, although this was not forthcoming. The result was that analysis could not fully develop its potential.

The second, and more specific, reason for which the controversy is of importance to our purpose is that, in his *Réflexions et éclaircissemens sur les nouvelles vibrations des cordes* [1753], Daniel Bernoulli stated that all initial positions could be represented in the form

$$y = \sum_{n=1}^{\infty} a_n \sin \frac{n\pi x}{l}. \tag{186}$$

In his opinion, the trigonometric solution was general. He did not base this opinion on mathematical arguments, nor did he tackle the problem of calculating the coefficients. He only assumed that all sonorous bodies contained potentially an infinity of sounds and an infinity of corresponding ways of making their regular vibrations.[371]

Instead, in *Remarques sur les mémoires précédens de M. Bernoulli* [1753], Euler, who had already discussed the possibility of a solution of the type (186) in his *De vibratione chordarum exercitatio;*[372] rejected Bernoulli's opinion by noting that "all the curves contained in that equation, although one increases the number of the terms to infinity, have certain characteristics which distinguish them from all other curves" (see Euler [1753, 236–237]). Indeed, Euler thought that a function could be represented by a trigonometric series only if it was periodic (and that it could be represented by a sine series if it was also odd). As Grattan-Guiness and Ravetz noted,[373] periodicity was an insuperable difficulty (186).

Euler's belief derived from basic concepts of 18th-century analysis, particularly the generality of algebra[374] and the concept of function. For example, consider the equality

$$\frac{1}{2}x = \sin x - \frac{1}{2}\sin 2x + \frac{1}{3}\sin x + \ldots . \tag{187}$$

Today we think of it as an equality that is valid on a certain interval since we think that the sum of series is different from $\frac{1}{2}x$ when x varies on an

---

[370]See Chapter 23.
[371]See Bernoulli [1753, 147–172].
[372]See Euler [1749a, 50–62].
[373]See Grattan-Guiness and Ravetz [1972, 245].
[374]See Section 18.2, p. 209.

interval different from $(-\pi, \pi)$ and that the series gives rise to a function that is defined piecewise. However, equality (187) was not considered to be valid on a certain interval, but according to the principles of 18th-century analysis[375] to be valid for every value of x. As a consequence, this relation could not be thought of as a quantitative relation. A trigonometric series was understood as a formally derived relation: it might have a quantitative meaning, though this was not necessarily the case. At any event, it was not defined quantitatively, namely as the limit of the partial sums.

In his [1773, 169], Euler stated this view in a resolute way. He set

$$\cos x + \sqrt{-1}\sin x = p \quad \text{and} \quad \cos x - \sqrt{-1}\sin x = q.$$

Hence,

$$\cos nx = \frac{p^n + q^n}{2} \quad \text{and} \quad \sin nx = \frac{p^n - q^n}{2\sqrt{-1}}$$

He observed that

$$p + p^2 + p^3 + \ldots + p^n = \frac{p(1 - p^n)}{1 - p} \quad \text{and} \quad q + q^2 + q^3 + \ldots + q^n = \frac{q(1 - q^n)}{1 - q}.$$

He then summed these finite geometric progressions, replaced $\frac{p^n + q^n}{2}$ by $\cos x$, and obtained

$$\cos x + \cos 2x + \cos 3x + \ldots + \cos nx = -\frac{1}{2} + \frac{\cos nx - \cos(n+1)x}{2(1 - \cos x)}. \quad (188)$$

Similarly, he derived

$$\sin x + \sin 2x + \ldots + \sin nx = \frac{\sin x + \sin nx - \sin(n+1)x}{2(1 - \cos x)}. \quad (189)$$

Euler stated that the values of $\sin nx$ and $\cos nx$ varied between -1 and 1 when n was finite and that the last term of the sequences $\sin nx$ and $\cos nx$ when $n = \infty$ was equal to 0. Therefore,

$$\cos x + \cos 2x + \cos 3x + \ldots = -\frac{1}{2}, \quad (190)$$

$$\sin x + \sin 2x + \ldots + \sin nx = \frac{\sin x}{2(1 - \cos x)}. \quad (191)$$

He took care to show that these results corresponded to his definition of the sum. For instance, by considering

$$-\frac{1}{2} = \frac{\cos x - 1}{2(1 - \cos x)}$$

---

[375] See Chapters 18 and 19.

and expanding the last fraction, Euler obtained

$$\cos x + \cos 2x + \cos 3x + \ldots = -\frac{1}{2}.$$

Equation (190) was also derived by Lagrange in his [1759, 111–112] using similar procedures.[376] The result had been criticized by d'Alembert, who denied that $\cos x + \cos 2x + \cos 3x + \ldots = -\frac{1}{2}$. In response, Lagrange recalled the principles of power series on which trigonometric series were rooted and stated:

> I would pose the question whether every time one encounters an infinite geometric series in an algebraic formula, for example
>
> $$1 + x + x^2 + x^3 + \ldots,$$
>
> one can substitute
>
> $$\frac{1}{1-x},$$
>
> though this quantity is really equal to the sum of proposed series only when one supposes the last term x^∞ to be zero (Lagrange [1760–61, 323]).

This approach to trigonometric series prevented 18th-century mathematicians from releasing the potential of trigonometric series even though they derived many results that, in a sense, seem to anticipate Fourier's series. For instance, in his [1759], Lagrange derived the functional solution to equation $\frac{\partial^2 y}{\partial x^2} = \frac{1}{c^2}\frac{\partial^2 y}{\partial t^2}$ by a different approach. He considered n equal bodies spaced regularly along a weightless string between the fixed point and supposed that the masses satisfied the equations

$$\frac{d^2 y_k}{dt^2} = \Delta^2 y_{k-1}, \quad 1 \le k \le n,$$

with $y_0 = y_{n+1} = 0$. He solved these equations and obtained

$$
\begin{aligned}
y(x,t) = {} & \frac{2}{l}\sum_{k=1}^{\infty} Y_k \sum_{r=1}^{\infty} \sin\frac{r\pi x}{l}\sin\frac{r\pi x}{l}dx\cos\frac{r\pi ct}{l} \\
& + \frac{2}{\pi c}\sum_{k=1}^{\infty} V_k \sum_{r=1}^{\infty} \sin\frac{r\pi x}{l}dx\sin\frac{r\pi x}{l}dx\cos\frac{r\pi ct}{l},
\end{aligned}
$$

---

[376] He justified the step from (188) to (190) by stating that, in the case where n is an infinite number, the number 1 vanishes relative to n, and hence the term $\cos(n+1)x$ becomes equal to $\cos nx$ (Lagrange [1759, 111]).

where Y_k and V_k are the initial positions and velocities of the kth mass. At this point he moved from the discrete to the continuous: He replaced Y_k and V_k by $Y(x)$ and $V(x)$, and the sum

$$\sum_{k=1}^{\infty} Y(x_k) \sum_{r=1}^{\infty} \sin \frac{r\pi x}{l} dx \quad \text{and} \quad \sum_{k=1}^{\infty} V(x_k) \sum_{r=1}^{\infty} \sin \frac{r\pi x}{l} dx$$

by the integrals

$$\int_0^l Y(x) \sum_{r=1}^{\infty} \sin \frac{r\pi x}{l} dx \quad \text{and} \quad \int_0^l V(x) \sum_{r=1}^{\infty} \sin \frac{r\pi x}{l} dx.$$

Thus, he [1759, 100–101] obtained

$$
\begin{aligned}
y(x,t) &= \left(\frac{2}{l} \int_0^l Y(x) \sum_{r=1}^{\infty} \sin \frac{r\pi x}{l} dx \right) \sin \frac{r\pi x}{l} \cos \frac{r\pi ct}{l} \\
&+ \left(\frac{2}{\pi c} \int_0^l V(x) \sum_{r=1}^{\infty} \sin \frac{r\pi x}{l} dx \right) \sin \frac{r\pi x}{l} \cos \frac{r\pi ct}{l}. \quad (192)
\end{aligned}
$$

Here Lagrange seems to be very close to Fourier series. By interchanging $\int$ and $\sum$ in

$$\left(\frac{2}{l} \int_0^l Y(x) \sum_{r=1}^{\infty} \sin \frac{r\pi x}{l} dx \right) \sin \frac{r\pi x}{l},$$

one obtained the Fourier sine series. In reality, as Grattan-Guiness [1972, 16] noted, Lagrange was persuaded *a priori* of the impossibility of representing any function through trigonometric series (in his paper he also rejected Daniel Bernoulli's solution). Equation (192) "was for him only a step on the road to the Eulerian functional solution" (Grattan-Guiness and Ravetz [1972, 248]). *Lagrange used trigonometric series in a formal way according to typical 18th-century procedures and never considered them as autonomous objects, capable of defining a quantity by themselves.*

Another particularly interesting result was obtained by Clairaut in his *Sur l'orbite apparente du Soleil autour de la terre* [1754, 544–564]. He returned to the problem that Euler had examined in [1749a] and sought the expansion of the function $f(x)$ in the form

$$a_0 + 2 \sum_{k=1}^{\infty} a_k \cos kx.$$

In so doing he produced expressions for what later would be called the Fourier coefficients of the series. Indeed, he wanted to interpolate the given function $f(x)$ for $x = 2\pi/n$ and obtained the interpolation formula

$$a_0 = \frac{1}{n} \sum_{h=0}^{n-1} f\left(\frac{2h\pi}{n}\right)$$

and

$$a_k = \frac{1}{n} \sum_{h=0}^{n-1} f\left(\frac{2h\pi}{n}\right) \cos \frac{2hk\pi}{n} \quad (k > 0).$$

For $n = \infty$, Clairaut obtained

$$a_k = \frac{1}{2\pi} \int_0^{2\pi} f(x) \cos nx \, dx.$$

In 1777, in dealing with an astronomical problem, Euler showed that the coefficients of the trigonometric series

$$f(x) = a_0 + 2 \sum_{k=1}^{\infty} a_k \cos kx$$

could be obtained in a very quick way. He multiplied both sides of the last equality by $\cos mx$ and integrated the series term by term by observing that

$$\int_0^{2\pi} \cos mx \cos kx \, dx = 0$$

for $m \neq k$ (see Euler [1793]).

These results did not change the common approach to trigonometric series. The expansion of a function into a trigonometric series was always recognized as being the result of applying a formal procedure. However, while the capacity to expand a function into a power series, which was convergent on an interval, was considered to be guaranteed *a priori*, the capacity to expand a function into a trigonometric series was not to be guaranteed *a priori* by usual procedures and had to be justified even by referring to the physical meaning of the trigonometric series.[377]

---

[377]Kline observed: "[Euler] did not accept the general fact that quite arbitrary functions could be so represented [by using trigonometric series]; the existence of such a representation, where he used it, was assured by other means" [1972, 517].

28 Further developments of the formal theory of series

In this chapter, I shall illustrate some further developments of the formal theory of series. I begin by discussing the role of the binomial theorem in 18th-century analysis. I then go on to illustrate Lagrange's demonstrations of the Taylor theorem. Finally, I concentrate upon the research that led from the investigation of Leibniz's analogy to the effective rise of the calculus of operations.

A rigorous proof of the binomial theorem was the object of many attempts in the second part of the 18th century. In Chapters 4 and 8 we saw that Newton had derived the binomial theorem by interpolation and that it was accepted on the basis of the principle of infinite extension. However, when analysis developed as a deductive and self-founding system, mathematicians felt the need to prove it in a rigorous way instead of accepting it by induction or analogy. Of course, the proof had to be rigorous in the sense of the principles of 18th-century analysis; in other words, it could be grounded neither on geometric arguments nor on an arithmetical basis, but on the mere manipulation of general quantities. Moreover, a satisfying proof of the binomial theorem could not use differential methods (and, in particular, the consideration of the binomial theorem as a particular case of the Taylor theorem). Indeed, the binomial theorem was considered to belong to that part of analysis defined by Euler as the introduction to the analysis of infinities and which later became known as algebraic analysis. The introduction to the analysis of infinities occupied an intermediate position between the analysis of finites and differential and integral calculi. It investigated functions and their expansion into power series without using differential calculus; rather mathematicians employed the findings of the introduction to the analysis of infinities in the construction of the calculus. In particular, the binomial theorem was essential to the calculation of the fluxions or differentials of certain quantities.[378] Therefore the use of differential methods for proving the binomial theorem involved a *petitio principii*.

---

[378] For example, in his *De quadratura*, to prove that the fluxion of x^n is nx^{n-1}, Newton considers the increment $(x + o)^n$ of x^n and "by the Method of Infinite Series" obtains

$$x^n + nox^{n-1} + \frac{n(n-1)}{2}o^2x^{n-2} + \dots$$

The ratio between the increments o of x and $nox^{n-1} + \frac{n(n-1)}{2}o^2x^{n-2} + \dots$ of x^n is equal to $1 : \left(nx^{n-1} + \frac{n(n-1)}{2}ox^{n-2} + \dots\right)$. When the increments vanish their last ratio is $1 : nx^{n-1}$. Therefore, the fluxion of the quantity x is to the fluxion of the quantity x^n as 1 to nx^{n-1}. [1704, 336–337].

Similarly, in the Eulerian construction of the calculus, the notion of differential ratios was introduced by assuming that a function could be expanded into series and therefore by assuming the validity of the binomial theorem.

Nevertheless, during the 18th century, there were a number of proofs that used differential methods (even Euler did this). Almost all the scholars who provided such proofs admitted the vicious circle.[379] Most of these proofs (especially in the second part of the century) were mere exercises in didactic works. In some cases, they were the result of the attempts to improve some specific point of the building of analysis (see Pensivy [1987–88, 99]). Finally, in other cases, such as Euler [1755], one can hypothesize that they served to verify *a posteriori* the findings of the calculus.

Among the most interesting nondifferential proofs,[380] I would like to examine an algebraic proof by Landen. In his *Residual Analysis* [1758], Landen set

$$(1+x)^{\frac{m}{n}} = 1 + Ax + Bx^2 + Cx^3 + \dots$$

and

$$(1+y)^{\frac{m}{n}} = 1 + Ay + By^2 + Cy^3 + \dots$$

and derived

$$\frac{(1+x)^{\frac{m}{n}} - (1+y)^{\frac{m}{n}}}{x-y} = A\frac{x-y}{x-y} + B\frac{x^2-y^2}{x-y} + C\frac{x^3-y^3}{x-y} + \dots.$$

Landen set $u = 1 + x$ and $v = 1 + y$ and, using algebraic transformations, obtained

$$u^{\frac{m}{n}-1}\frac{1 + \left(\frac{v}{u}\right) + \left(\frac{v}{u}\right)^2 + \dots + \left(\frac{v}{u}\right)^{m-1}}{1 + \left(\frac{v}{u}\right)^{\frac{m}{n}} + \left(\frac{v}{u}\right)^{2\frac{m}{n}} + \dots + \left(\frac{v}{u}\right)^{(n-1)\frac{m}{n}}}$$
$$= A + B(x+y) + C(x^2 + xy + y^2) + \dots.$$

Since this equation holds for any value of x and y, he posed that $x = y$ and had

$$\frac{m}{n}u^{\frac{m}{n}-1} = A + 2B + 3Cx^2 + \dots$$

By multiplying this equation by $(1+x)$, he obtained

$$\frac{m}{n}(1+x)^{\frac{m}{n}} = (1+x)(A + 2B + 3Cx^2 + \dots).$$

Hence,

$$\frac{m}{n}(1 + Ax + Bx^2 + Cx^3 + \dots) = (1+x)(A + 2B + 3Cx^2 + \dots)$$

---

[379]On Maclaurin's proof, see Pensivy [1987–88, 102–105].
[380]I refer to Pensivy [1987–88] for a detailed examination.

Equating the coefficients, he found

$$A = \frac{m}{n}, \qquad 2B + A = \frac{m}{n}A, \quad \ldots$$

At this point Landen derived the coefficients of the sought-after expansion recursively.

A different type of proof of the binomial theorem made use of functional equations. For example, in his *Demonstratio theorematis newtoniani de binomio* [1757], Aepinus posed that

$$(x+1)^m = Ax^m + Bx^{m-1} + Cx^{m-2} + \ldots$$

and considered the second coefficient B as a function $B(m)$ of the exponent m. Then he showed that the following relation between the second coefficients $B(m+n)$, $B(m)$, $B(n)$ of the expansions of $(x+1)^{m+n}$, $(x+1)^m$, $(x+1)^n$ held:

$$B(m+n) = B(m) + B(n). \tag{193}$$

Aepinus solved Equation (193) under the condition $B(1) = 1$ and found $B(m) = m$ (in the solution he assumed that if s is an infinitesimal number, any real number r is the type $r = ns$, where n is an infinite number). Hence, it is easy to derive the binomial expansion.

In his *Demonstratio theorematis newtoniani* [1774–75], Euler attempted to prove the binomial expansion using another type of functional equation. He considered the series

$$1 + nx + \frac{n(n-1)}{2!}x^2 + \frac{n(n-1)(n-2)}{3!}x^3 + \ldots$$

and denoted it by $[n]$ (in modern terms, $1 + nx + \frac{n(n-1)}{2!}x^2 + \frac{n(n-1)(n-2)}{3!}n^3 + \ldots$ is conceived as a function of the exponent n). He did not take convergence into consideration but showed that the series $[n]$ satisfied the functional relation

$$[m+n] = [m][n].$$

The solution to this equation allowed him to obtain the binomial expansion. This method was later used by Cauchy in an entirely different context.[381]

As early as the turn of the 17th century, some mathematicians had already tried to generalize the binomial theorem and seek the expansion into power series of the mth power of a polynomial (and even an infinite polynomial), the so-called polynomial theorem. In 1697, de Moivre published

---

[381] See Chapter 33.

a paper on raising an infinite multinomial to a given power.[382] He showed that $\left(az + bz^2 + cz^3 + \ldots\right)^m$ is equal to a power series

$$Az^m + Bz^{m+1} + Cz^{m+2} + \ldots,$$

where the coefficient of the term x^{m+r} is equal to the sum of all products of the form $a^p b^q c^s \ldots$ ($p + q + r + \ldots = m$ and $p + 2q + 3r + \ldots = m + r$); each product $a^p b^q c^s \ldots$ being multiplied by $\frac{n!}{p!q!s!\ldots}$ (n is the number of factors in the product $a^p b^q c^s \ldots$).

The question was also studied by other mathematicians: Johann Bernoulli, Leibniz, Jacob Bernoulli, Colson, and Kästner. De Moivre and Leibniz also suggested the possibility of generalizing to polynomial formula in the case m is a fraction. However, it was only Carl Friedrich Hindenburg who succeeded in giving a formulation of the polynomial theorem, making it possible to apply it to the case of fractional exponents. Hinderburg's proof can be summarized as follows. Given $1 + a_1 x + a_2 x^2 + \ldots$, one sets

$$\left(1 + a_1 x + a_2 x^2 + \ldots\right)^m = (1 + z)^m$$

and expands $(1 + z)^m = 1 + mz + \frac{m(m-1)}{2} z^2 + \ldots$. Then one calculates the powers

$$z^h = \left(a_1 x + a_2 x^2 + \ldots\right)^h,$$

where h is a natural number. By rearranging, one obtains

$$\left(1 + a_1 x + a_2 x^2 + \ldots\right)^m = 1 + A_1 x + A_2 x^2 + \ldots.$$

The coefficients are given by the formula

$$A_r = \sum_{h=1}^{r} \binom{m}{h} Q_{r,n}. \tag{194}$$

$Q_{r,n}$ denotes the sum of all products of the type $a_1^{n_1} a_2^{n_2} \ldots a_k^{n_k}$, where the factors a_i belong to the set $\{a_1, \ldots, a_r\}$; each product $a_1^{n_1} a_2^{n_2} \ldots a_k^{n_k}$ is multiplied by a coefficient $\frac{n!}{n_1! n_2! \ldots n_k!}$. Since the exponent m is not found in $Q_{r,n}$ but only in the binomial coefficients $\binom{m}{h}$, formula (194) can be applied to fractional and negative exponents.[383]

$$* \quad * \quad *$$

---

[382] See de Moivre [1697].

[383] It is worthwhile noting that this proof of the polynomial theorem convinced Hinderburg that it was possible to systematize the analysis of infinities by grounding it on combinatorics. He was a clever organizer and managed to enlist the assistance of several mathematicians who agreed with the idea of the project. They founded the Combinatorial School, which developed from 1780 to 1810. The school did not include any great mathematicians (the most important were H. Bürmann, C.H. Eschenbach, C.S. Klügel, M. von Prasse, H.A. Roth, I.K. Tetens, H.A. Töpfer, C. Kramp, and Gauss's teacher

In the 1770s, Lagrange and Laplace dealt with the Taylor series. In their investigations, the Taylor series appears as an intermediate step for obtaining new results concerning finite differences and the emergent calculus of operations. We saw Laplace's proof in Chapter 21. I now examine Lagrange's demonstration of 1772.

In the *Sur une nouvelle espèce de calcul*,[384] Lagrange considered a function $f(x)$ and took for granted that

$$f(x+i) = f(x) + p_1 i + p_2 i^2 + \ldots, \tag{195}$$

where $p_i(x)$ are functions of x. From $f((x+o)+i) = f(x+(o+i))$, he derived

$$f(x+o) + \sum_{k=1}^{\infty} p_k(x+o) i^k = f(x) + \sum_{k=1}^{\infty} p_k(x)(i+o)^k.$$

Then, he set

$$f(x+o) = f(x) + \sum_{k=1}^{\infty} p_k(x) o^k$$

and

$$p_k(x+o) = p_k(x) + \sum_{n=1}^{\infty} p_{k,n}(x) o^n,$$

and obtained

$$f(x+o) + \sum_{k=1}^{\infty} \left(p_k(x) + \sum_{n=1}^{\infty} p_{k,n}(x) o^n \right) i^k = f(x) + \sum_{k=1}^{\infty} p_k(x) \sum_{n=0}^{k} \binom{k}{n} i^n o^{k-n}.$$

Hence,

$$
\begin{aligned}
p_2(x) &= \frac{1}{2} p_{1,1}(x), \\
p_3(x) &= \frac{1}{3} p_{2,1}(x), \\
p_4(x) &= \frac{1}{4} p_{3,1}(x), \\
&\cdots
\end{aligned}
$$

---

and friend J.F. Pfaff), nor did its members obtain particularly important results (compared to those of other mathematicians of the period). However, their approach "was influential in Germany at the turn of the 19th century and became the basis for the mathematical syllabus of the Prussian gymnasium in the Humboldt educational reform" (see Jahnke [1993, 265]). I refer to Jahnke [1993] and Panza [1992, 651–659] for both a general background of the school and a detailed mathematical investigation of the methods employed.

[384] See Lagrange [1772, 443–445].

According to Lagrange, this showed that all the coefficients p_i could be determined, starting from the preceding coefficient p_{i-1}, by using the same operation that made it possible to determine p_1 starting from $f(x)$. He denoted this operation by means of the symbol $'$ and the result of the operation performed on $f(x)$ by means of $f'(x)$. Consequently, the result of the operation performed on $f'(x)$ was denoted by $f''(x)$, and so on. Hence,

$$p_k(x) = \frac{1}{k!} f^{[k]}(x)$$

and

$$f(x+i) = f(x) + \sum_{k=1}^{\infty} \frac{1}{k!} f^{[k]}(x) i^k. \tag{196}$$

Here Lagrange "explicitly considered a mathematical entity that would later be referred to as an operator" (Panza [1992, 577]). Indeed, he was particularly interested in the operation O, which made it possible to calculate the coefficient p_1. He even argued that the differential calculus could be based on this operation [1772, 443]. Subsequently, Lagrange however concluded his reasoning using infinitesimals and proving that the series (196) was the same as the Taylor series. He took $i = dx$ as an infinitely small quantity and obtained

$$df = f(x+i) - f(x) = f(x+dx) - f(x). = \sum_{k=1}^{\infty} \frac{1}{k!} f^{[k]}(x) dx^k.$$

Using the principle of cancellation, he derived $du = u'(x)dx$. Hence,

$$u'(x) = \frac{du}{dx}, \quad u''(x) = (u')' = \frac{d}{dx}\left(\frac{du}{dx}\right) = \frac{d^2 u}{dx^2}, \dots$$

By replacing $\frac{d^k u}{dx^k}$ in (196), Lagrange obtained the Taylor theorem $u(x+i) = u(x) + \sum_{k=1}^{\infty} \frac{1}{k!} \frac{d^k u}{dx^k}(x) i^k$ (see [1772, 446–447]).

Lagrange developped his idea on the possibility of basing the calculus on the operation O in his *Théorie des fonctions analytiques* [1797].[385] In the first theorem of this treatise, he tried to prove equality (195):

Given a function $f(x)$, the series

$$f(x+i) = f(x) + \sum_{k=1}^{\infty} p_k(x) i^{\alpha_k} \tag{197}$$

_____

[385] For a detailed examination of Lagrange's theory of analytical functions, see Ferraro and Panza [A].

contains no fractional or negative power of i, except for partic-
ular isolated values of x, namely

$$f(x+i) = f(x) + \sum_{k=1}^{\infty} p_k(x)i^k. \qquad (198)$$

The notion that any function could be expanded into a power series had always been considered obvious by 18th-century mathematicians; Lagrange now assumed a weaker hypothesis [any function had an expansion of type (197)] and attempted to prove (198) in a general, algebraic way.

In the first part of the demonstration, Lagrange tried to show that there were no fractional powers in the expansion (197). He stated that, if the series (197) has a term of the type $\sqrt[n]{i^m}$, the only possibility is that it derives from radicals that are found in the function $f(x)$. In fact,

a. The operation $\sqrt[n]{i^m}$ does not assign only one value to i, namely $\sqrt[n]{i^m}$ is an n-valued function of i.

b. If the expansion of $f(x+i)$ contains an irrational term, then the expansion is many-valued; therefore, $f(x+i)$ is also many-valued and has the same number of values as the expansion.

c. The substitution of $x+i$ in place of x changes neither the number nor the nature of radicals, as long as x and i are indeterminate [namely, the n-valued radical $\sqrt[n]{x^m}$ becomes the n-valued radical $\sqrt[n]{(x+i)^m}$]. Consequently, if $f(x+i)$ assumes n values, then $f(x)$ assumes the same number of values as $f(x+i)$ and as the expansion $f(x)+\sum_{k=1}^{\infty} p_k(x)i^{\alpha_k}$.

d. In conclusion, if the expansion of $f(x+i)$ contains an irrational term $\sqrt[n]{i^m}$, then $f(x+i)$ and $f(x)$ have n values. However, the combination of each value of $\sqrt[n]{i^m}$ with each value of $f(x)$ gives rise to a number of values of $f(x+i)$ greater than the values of $f(x)$. This is contradictory.

To prove that there are no negative powers, Lagrange noted that if the expansion of $f(x+i)$ has a term i^{-k} with a negative exponent, then $i^{-k} = \infty$, for $i = 0$. Therefore, $f(x+i)$ is also equal to infinity for $i = 0$, namely $f(x+0) = f(x) = \infty$ when x is indeterminate. This is absurd since $f(x)$ can be equal to infinity only for particular values of x.

Lagrange knew that certain determinate values of x exist where (198) fails, but, according to the 18th-century analysis, these cases were considered exceptions that did not invalidate the rigor of the proof concerning general quantity[386] (Langrange [1797, 23]). A simple example is the expansion of

---

[386]On the treatment of exceptional values, cf. Chapter 19.3, p. 227).

the function $f(x) = \sqrt{x}$ for $x = 0$. As Lagrange showed in [1797, 26], its expansion is

$$\sqrt{x+i} = \sqrt{x} + \frac{1}{2\sqrt{x}}i - \frac{1}{8x\sqrt{x}}i^2 + \frac{1}{16x^2\sqrt{x}}i^3 - \dots, \qquad (199)$$

and the theorem does not hold for $x = 0$. In his opinion, it was the existence of such values that made it necessary to give an *a priori* demonstration of (198), even though it was verified by all known functions (Lagrange [1797, 22]).

Starting from this theorem, Lagrange [1797] constructed the theory of analytical functions with the aim of basing the calculus only on formal considerations. He first gave a reasoning similar to the one used in 1772 to show that the operation that enabled the calculation of the first coefficient p_1 starting from $f(x)$ also made it possible to calculate the other coefficients p_i starting from p_{i-1}. Thus, he could define the *derived function* (the *derivative*) of $f(x)$ as the coefficient p_1 of the expansion $\sum_{n=0}^{\infty} p_n i^n$ of $f(x)$. He denoted p_1 by $f'(x)$. He then termed the derived function of $f'(x)$ as the second derived function of $f(x)$ and denoted it by $f''(x)$ and, in general, termed the derived function of $f^{(k-1)}(x)$ as the kth derived function of $f(x)$ and denoted it by $f^{(k)}(x)$. This allowed him to write (198) in the form $f(x+i) = f(x) + \sum_{k=1}^{\infty} \frac{1}{k!} f^{(k)}(x) i^k$.

After that Lagrange calculated the derivatives of elementary functions[387] (the only functions whose existence he recognized) and introduced the notion of integration as antidifferentiation.[388]

Finally, Lagrange examined some applications of the calculus to geometry and mechanics. These applications required the consideration of the quantitative, a problem that Lagrange tackled in an explicit and innovative form, as we shall see in Chapter 30.

$$* \quad * \quad *$$

Lagrange's approach influenced several mathematicians at the turn of the 18th century. Some of them sought to provide a more convincing proof of the Taylor theorem (see Poisson[389] [1805] and Ampère [1806]). Others were stimulated to carry out research that led to an understanding of Leibniz's analogy and the rise of the calculus of operations.

In his *Du calcul des dérivations* [1800], Arbogast attempted to provide a new and more general calculus that included the differential calculus as a special case. Arbogast posed the question of finding the coefficients of the expansion of $f(a + bx + cx^2 + \dots)$ for any function f.

---

[387]He expanded elementary functions into series using algebraic methods (see Ferraro and Panza [A]) and defined the derivatives of this functions to be the first coefficient of these expansions.

[388]See Chapter 25.

[389]For Poisson, I refer to Grattan-Guiness [1990, 201–202].

He first considered a function $f(a+x)$ and sought the coefficients A_n of its power series expansion $\sum_{n=0}^{\infty} \frac{A_n x^n}{n!}$. He denoted the operation that had to be made on $f(a)$ to derive A_1 with the symbol D. By using this symbol, he wrote

$$
\begin{aligned}
A_1 &= Df(a), \\
A_2 &= DA_1 = D^2 f(a), \\
A_3 &= DA_2 = D^3 f(a), \\
&\cdots
\end{aligned}
$$

By replacing in the expansion $\sum_{n=0}^{\infty} \frac{A_n x^n}{n!}$ of $f(a+x)$, he had

$$
f(a+x) = \sum_{n=0}^{\infty} \frac{x^n}{n!} D^n f(a).
$$

Arbogast then considered increments of the form bx instead of x and obtained

$$
f(a+bx) = \sum_{n=0}^{\infty} x^n D_{c.}^n f(a),
$$

where

$$
\begin{aligned}
D_{c.}^n f(a) &= \frac{1}{n!} D^n f(a), \\
D_{.}^n f(a) &= D_{.}(D_{.}^{n-1} f(a)),
\end{aligned}
$$

and $D_{.} f(a)$ is the operation that had to be made on $f(a)$ to derive the first coefficient of the expansion of $f(a+bx)$.

Finally, Arbogast considered functions of polynomials and functions of functions and showed that the coefficients A_n of the expansion $\sum_{n=0}^{\infty} A_n x^n$ of $f(\sum_{s=0}^{\infty} a_s x^s)$ were given by

$$
A_n = D_{c.}^n f(a_0) = \sum_{h=1}^{n} \left(D_{c.}^h f(a_0) \right) \left(D_{c.}^{n-h} a_1^h \right).
$$

The introduction of the operation D allowed Arbogast to take a step forward in the calculus of operations. He explicitly spoke of the method of separating symbols of operation from quantities, which consisted of the fact that symbols of operations could be handled separately from the subjects on which they operate and treated as if they were symbols of quantities. This method, which had already been used *de facto* by Lagrange in his [1792],[390]

---

[390] See Chapter 21.

allowed him to simplify and clarify the content of Lagrange's theorem (140). Arbogast set $\delta u = \frac{du}{dx}$ and expressed the equation $\Delta u = e^{\frac{du}{dx}\xi} - 1$ in the form

$$1 + \Delta = e^{\delta \xi} \tag{200}$$

The difference between (200) and the Lagrangian

$$\Delta u = e^{\frac{du}{dx}\xi} - 1$$

is that the purely operational components are stated separately in $1 + \Delta = e^{\delta \xi}$.

Arbogast applied the calculus of operations to differential equations. Many mathematicians did the same thing, including Brisson, who, in his [1808], tackled the problem of the solution to linear partial differential equations with constant coefficients of any order and in any number of variables. He considered the equation

$$Az + B\frac{\partial z}{\partial x} + C\frac{\partial z}{\partial y} + \ldots + G\frac{\partial^2 z}{\partial x^2} + H\frac{\partial^2 z}{\partial x \partial y} + I\frac{\partial^2 z}{\partial y^2} + \ldots = 0, \tag{201}$$

where z is a function of the variables x, y, $\ldots$, and A, B, C, $\ldots$, G, H, I, $\ldots$ are constants. Brisson wrote it in the form

$$\nabla z = 0 \tag{202}$$

and used ∇ as a linear operator on the possible solution functions. He observed that the rule

$$A\frac{\partial^n \left(B\frac{\partial^m z}{\partial y^m} \right)}{\partial x^n} = AB\frac{\partial^{n+m} z}{\partial x^n \partial y^m}$$

held for repeated partial differentiations (if A and B are constant) and that the same result could be obtained by considering ∂, x, y as quantities raised to the powers n, m, $n + m$. For this reason he replaced (202) with the equation

$$\nabla' z = 0,$$

which denoted the algebraic equation that resulted from (201) after substituting repeated differentiation

$$\frac{\partial^n \left(\frac{\partial^m z}{\partial y^m} \right)}{\partial x^n}$$

by multiplication:

$$\frac{\partial^n z}{\partial x^n} \cdot \frac{\partial^m z}{\partial y^m}.$$

He factored $\nabla' z = 0$ and obtained

$$\nabla' z = \delta_0' \delta_1' \dots \delta_{n-1}' z.$$

At this point he wrote $\nabla z = \delta_0 \delta_1 \dots \delta_{n-1} z = 0$, where the δ_i are obtained from the δ_i' by inverting the previous substitution, namely replacing multiplication by iteration. Brisson stated that the solution of the given equation can be found by solving $\delta_i z = 0$ and showed how one could solve $\delta_i z = 0$ when δ_i is linear.

A further contribution to the rise of the calculus of operations was made by the mathematician Jacques-Fréderic Français. In his *Mémoire rendant à démontrer la légitimité de la sèparation des échelles*, Français applied the method of the separation of symbols to solve problems concerning differential equations and finite difference. He denoted $\frac{d\varphi}{dx}$ by $\delta\varphi$ and $\varphi(x + 1)$ by $E\varphi$ and wrote the Taylor series for φ in the form

$$E\varphi(x) = \varphi(x + 1) = \varphi x + \delta\varphi(x) + \frac{1}{2}\delta^2\varphi(x) + \dots = e^{\delta\varphi(x)}$$

By separating symbols, he obtained

$$E = e^{\delta}$$

(see Français [1812–13, 249–250]).

A simple example of how the method of the separations of symbol works is the following. Français considered the differential equation

$$\frac{d\varphi}{dx} = a\varphi,$$

which he wrote as

$$(\delta - a)\,\varphi = 0.$$

By separating symbols, he had $\delta - a = 0$ and $e^{\delta} = e^a$. Since $E = e^a$, he obtained

$$1 = e^{ak} E^{-k}.$$

Hence,

$$\varphi(x) = e^{ak} E^{-k} \varphi(x) = e^{ak} \varphi(x - k)$$

For $x = k$, he had

$$\varphi(k) = e^{ak} \varphi(0) = C e^{ak}.$$

Finally, by changing k into x in the last equation, he obtained the solution to the given differential equation

$$\varphi(x) = C e^{ax}$$

(see Français [1812–13, 244–276]).

Several mathematicians attempted to discover the mechanism that ruled these procedures. As early as 1787, Lorgna had observed that if y was a function and y' was the difference or differential of y, then the iteration of the operation of difference or differential, which he denoted by the symbol $y^{n'}$, obeyed the same law of combination as the symbol x^n that represented raising to the power of the quantity x (see Lorgna [1787, 413]). In *Mémoire rendant à démontrer la légitimité de la sèparation des échelles*, Français compared equations of the type

$$aF(x,y) + bF(x,y) + cF(x,y) + \ldots = 0, \qquad (203)$$

where a, b, c, ... are constants, with equations of the types

$$\partial^n F(x,y) + a\partial^{n-1} F(x,y) + b\partial^{n-2} F(x,y) + \ldots = 0 \qquad (204)$$

and

$$\partial^n F(x,y) + a\Delta\partial^{n-1} F(x,y) + b\Delta^2\partial^{n-2} F(x,y) + \ldots = 0 \qquad (205)$$

He stated that the symbols of operations behaved in the same manner as the constants a, b, c, ... Consequently, since one could derive

$$(a + b + c + \ldots)F(x,y) = 0$$

from (203), one could, in the same way, derive

$$(\partial^n + a\partial^{n-1} + b\partial^{n-2} + \ldots)F(x,y) = 0$$

and

$$\left(\partial^n + a\Delta\partial^{n-1} + b\Delta^2\partial^{n-2} + \ldots\right) F(x,y) = 0$$

from (204) and (205) (see Français [1812–13, 245–246]).

These observations paved the way for Servois, who gave the first satisfactory explanation of the Leibniz's analogy in two papers, entitled *Essai sur un nouveau mode d'exposition des principes du calcul différentiel* [1814–15a] and *Réflexion sur les divers systèmes d'exposition des principes du calcul différentiel* [1814–15b], in which, following Lagrange, he attempted[391] to give an algebraic basis to the calculus and reject the use of infinitesimals.[392]

---

[391]Lagrange's concept had been attacked by Wronski. The Polish philosopher and mathematician asserted that all mathematics was rooted in the "absolute law", according to which $F(x) = \sum_{n=0}^{\infty} A_n \Omega_n(x)$, where $\Omega_n(x)$ were arbitrary generating function. Servois's paper is also a response to Wronski, who never actually attempted to prove his claims.

[392]In [1814–15b, 148] Servois stated: "[The infinitesimals] neither have nor can have theory; in practice it is a dangerous instrument in the hands of beginners anticipating, for my part, the judgement of posterity, I would predict that this method will be accused one day, and rightly, of having retarded the progress of the mathematical sciences" (translation in Grattan-Guiness [1990, 137]).

In [1814–15a, 98], Servois introduced the notion of distributivity and commutativity. He defined a function f as *distributive*[393] if

$$f(x + y + z + \ldots) = f(x) + f(y) + \ldots,$$

and the functions h and g as *commutative*[394] if

$$f(g(x)) = g(f(x)).$$

Servois proved several results on distributive and commutative functions (even though he used the term "function" in all cases, he made a clear distinction between functions of functions and usual functions). In particular, he stated that *if*

$$F(z) = f(z) + g(z) + \ldots,$$

where f and g are distributive and pairwise commutative, then F^n is distributive for any integer n and one can find F^n applying the ordinary laws of algebra to

$$(f(z) + g(z) + \ldots.)^n$$

(see Servois [1814–15a]).

Servois went on to examine the operator $\Delta f(x) = f(x + \Delta x) - f(x)$ and the differential operator, which he defined as

$$dz = \Delta z - \frac{1}{2}\Delta^2 z + \frac{1}{3}\Delta^3 z - \ldots.$$

According to Servois, *the reason why the symbols of operations Δ and d could be handled as if they were the symbols of quantities was that they were distributive and commutative with each other and with constant factors.* In his opinion, this provided a simple explanation for the analogy between the iteration and powers.

Servois's work was the basis of later developments in the calculus of operations, which underwent remarkable development during the 19th century, mainly in Great Britain.[395] It is worth emphasizing that while the algebraic aspects were clear, Servois (and all those who followed him) considered the infinite series of operators and ignored problems of convergence. They considered the distributivity to be valid for infinite series and assumed that a rule that was valid for power series also remained valid for infinite series of operations. For this reason the calculus of operations can be considered a legacy (and perhaps an extreme result) of the 18th-century theory of series.

---

[393] "The functions which ... are such that the function of the sum of any number of quantities is equal to the sum of the corresponding functions of each of these quantities, will be called distributive" (Servois [1814a, 98]).

[394] "The functions which ... are such that they give identical results, whatever be the order in which one applies them to the subject, will be called commutative between each other" (Servois [1814a, 98]).

[395] On this topic and its link to the rise of abstract algebra, I refer to Goldstine [1977] and Koppelman [1971].

29 Attempts to introduce new transcendental functions

In the previous chapters, I dwelled on the aspects of analysis that remained unaltered throughout the 18th century and that, according to Fraser's expression, "constitute evidence of a shared conception significantly different from the modern one" (see Fraser [1989, 318]). However, there was a remarkable growth in mathematical knowledge during the century that gave rise to an accumulation of results, problems, and techniques. In this chapter and the one that follows, I shall deal with two aspects of this evolutionary process: the introduction of some new basic functions around the year 1800 and the use of the inequality technique in order to determine error estimates. I shall argue that such developments remained within the overall structure of 18th-century analysis: New findings accumulated and were added to older ones without challenging or changing key concepts.

I have already noted that the 18th-century notion of a function did not exclude the possibility of introducing new transcendental functions that had the same status as elementary functions, provided that they were considered as known objects.[396] Euler mentioned the question in the *De plurimis quantitatibus transcendentibus quas nullo modo per formulas integrales exprimere licet* [1780], a short note published in 1784. In this paper, Euler suggested the consideration of elliptic integrals as new basic functions. Indeed, he stated that logarithm and arctangent arose from the formulas

$$\int \frac{1}{x} dx$$

and

$$\int \frac{1}{1+x^2} dx;$$

however, they were considered similar to algebraic quantities since they could be treated as easily as algebraic quantities. In the same way, quantities concerning the rectification of conics (which were included in certain integral formulas of the type

$$\int \sqrt{\frac{f + gx^2}{b + kx^2}} dx \Big)$$

had been analyzed to the extent that, if a problem was reduced to these quantities, it could then be regarded as completely solved (Euler [1780, 522]).

Euler went further and stated that, according to a widely held opinion, all transcendental quantities could be represented (geometrically) by the

---

[396]See Chapter 18.

quadrature of certain curves or could be reduced (analytically) to integral formulas $\int f(x)dx$. He thought that this opinion was to be revised. Indeed, he pointed out the existence of transcendental quantities that could not be expressed by integral formulas although their values could be determined by approximation, an example being the "curve" expressed by the "equation"

$$y = \sum_{n=0}^{\infty} x^{\frac{n(n+1)}{2}}$$

for $|x| < 1$ (see Euler [1780, 523–525]).

    De plurimis quantitatibus transcendentibus contains some elements that enriched 18th-century analysis and shows that mathematicians were open to add on new elements to the structure of analysis, provided — and this is of fundamental importance — the foundations of that structure were not undermined. In effect, the paper retained the main points of the 18th-century approach:

(a) At any moment, the number of basic functions that were effectively accepted was fixed;

(b) it was problematic to increase this number because every new function had to be an entirely known entity;

(c) integral formulas on their own did not generate new functions;

(d) there is a difference in status between the expressions of quantities using elementary functions and using integral functions;

(e) the representation of a quantity by a series was regarded as approximate and as having a status inferior even to that of integral representation, though Euler's paper suggested that it was the only possible analytical representation in certain cases.

    De plurimis quantitatibus transcendentibus had no practical consequences on Euler's work. However, from the last decades of the 18th century to the beginning of the new century, various mathematicians thought that the acquired knowledge made it possible to consider certain quantities as new functions. For instance, Lacroix stated that all expressions involving integrals of the types

$$\int \frac{dx}{(x^2 + a)\sqrt{a + bx^2 + cx^4}}, \quad \int \frac{dx}{\sqrt{a + bx^2 + cx^4}}, \quad \int \frac{x^2 dx}{\sqrt{a + bx^2 + cx^4}}$$

were to be regarded as integrated in the same way as an integration that gave rise to expressions composed of logarithms and inverse trigonometric functions was regarded as complete. The integrals $\int \frac{dx}{(x^2+a)\sqrt{a+bx^2+cx^4}}$,

$\int \frac{dx}{\sqrt{a+bx^2+cx^4}}$, and $\int \frac{x^2\,dx}{\sqrt{a+bx^2+cx^4}}$ were therefore new transcendental functions to be introduced in the calculus (see Lacroix [1797–1800, 2:59]).

Legendre was the scholar who introduced a more accurate analysis of elliptic integrals. In his *Exercises de calcul intégral* (the three volumes of the treatise were published in 1811, 1816, and 1817,[397] at roughly the same time as Gauss's paper on the hypergeometric function,[398] which I will deal with below), he explained that the scope of his research was to compare the integrals of the type $\int \frac{P\,dx}{R}$ (where P is a rational function and $R = \sqrt{a + bx + cx^2 + dx^3 + ex^4}$) with one another, to classify them into different species, to reduce each species to the simplest form, to evaluate them by fastest and easiest approximations, and to develop an algorithm concerning these integrals in order to contribute to the extension of the domain of analysis (Legendre [1811, 1:3–4]). In other words, he aimed to reduce all the integrals to certain basic integrals that satisfied the conditions[399] (C1) and (C2).

Legendre also investigated the gamma function and other closely connected functions. In the first and second volumes of his *Exercises de calcul intégral*, many pages were devoted to them although his approach remained traditional. For instance, the gamma function was defined as

$$\Gamma(a) = \int_0^1 \left(\log \frac{1}{x}\right)^{a-1}$$

with $0 < a < \infty$ (Legendre [1811–17, 2: 4]). He easily derived $\Gamma(a + 1) = a\Gamma(a)$ and other relations concerning Γ. Shortly afterwards, he stated that, when a was positive, the function $\Gamma(a)$ could be viewed as the area under the curve $y = \left(\log \frac{1}{x}\right)^{a-1}$ between $x = 0$ and $x = 1$ [1811–17, 2:60]. However, this representation was geometrical, in a sense (*en quelque sorte géométrique*), and gave no idea of the function when a was negative. In such cases, the meaning of the function was derived by considering the formula

$$\Gamma(a + 1) = a\Gamma(a)$$

(which contained a general property of the function) to be valid beyond its original interval $0 < a < \infty$.

In [1811–17, 1:295], Legendre considered the quantity[400] M interpolating the discrete sequence

$$1 + \frac{1}{2} + \frac{1}{3} + \frac{1}{4} + \ldots + \frac{1}{x}$$

---

[397]He had already published an essay *Mémoire sur les transcendantes elliptiques* (see Legendre [1794]) on this topic in 1793 or 1794 (the booklet is dated "the second year of the republic").

[398]See Gauss [1812b].

[399]See p. 208.

[400]It is the function $\Psi(x) + \gamma$, where γ is Euler's constant.

and regarded M as a "continuous function"[401] of x since it could be expressed in the form

$$M = \gamma + \log x + \frac{1}{2x} + \sum_{n=1}^{\infty} (-1)^n \frac{B_{2n}}{2nx^{2n}}. \qquad (206)$$

In his opinion, (206) furnished the approximate values of M for $x > 1$.[402] In order to find a relation between M and $\Gamma(x)$, Legendre used the formula

$$\log \Gamma(x) = \left(x - \frac{1}{2}\right)\log x - x + \frac{1}{2}\log 2\pi + \sum_{n=1}^{\infty} (-1)^{n+1} \frac{B_{2n}}{2n(2n-1)x^{2n-1}}. \qquad (207)$$

He said that (207) could be employed to calculate the value of the function gamma for $x > 1$ (Legendre [1811–17, 1:294]). By differentiating (207), he obtained

$$\frac{d(\log \Gamma(x))}{dx} = \log x - \frac{1}{2x} + \sum_{n=1}^{\infty} (-1)^n \frac{B_{2n}}{2nx^{2n}}. \qquad (208)$$

According to Legendre, equations (208) and (206) provided the value of M and $\frac{d(\log \Gamma(x))}{dx}$ for $x > 1$. However, by comparing them he obtained

$$\frac{d(\log \Gamma(x))}{dx} = -\frac{1}{x} + M - \gamma;$$

"an equation that must take place whatever the value of x, since M can be regarded as a continuous function of x" (Legendre [1811–17, 1:297]).

It is clear that Legendre was referring to the generality of algebra,[403] and thus, the equation

$$\frac{d(\log \Gamma(x))}{dx} = -\frac{1}{x} + M - \gamma$$

was conceived to be valid in general, although numerically it only held for certain values of x (and in an approximate way). The definition lacked $\Gamma(a)$ for negative numbers a, but the generality of algebra also provided the values in this case.

Furthermore, Legendre gave no theoretical importance to the difference between divergent and convergent series. The only condition he imposed

---

[401] See Section 18.3.

[402] Legendre knew the divergent and asymptotic nature of the series (206), (207), and (208). He termed them semiconvergent because they first decrease (converge, in the language of his time) and then increase (diverge) [1811-17, 1: 267]. He also knew that they could be used in computing the values of gamma and digamma functions. He did not clarify why he assumed 1 as the bound of the interval of approximation.

[403] See Section 18.2, p. 209.

was that series provided approximate values of the expressed quantities for certain values of variables, independently of their divergence or convergence.

In his [1984, 103] and [1986, 16], Gray observed that it was not true that Legendre never considered the inverse functions of elliptic integrals[404] but that he did not think of them as functions of complex variables and, for this reason, did not realize the importance of such an inversion. This is true: In effect, Legendre followed the principle of the generality of algebra and did not consider functions of complex variables as entities enjoying properties which differed from those of real variables (e.g., double periodicity, which does not hold for functions of real variables, was beyond the boundaries of Legendre's mathematical world).

Legendre's treatment of new functions was an integral part of many studies about transcendental quantities. These developed, enriched, and attempted to consolidate the 18th-century theory, and Legendre's work probably represented the utmost advance of this theory (this is even true of his *Traité des fonctions elliptiques*, which was published in 1825 and 1826). Those mathematicians who followed the traditional line of investigation were not aware of the need to rethink 18th-century methods. They were convinced that additions had to be made to the traditional theory of functions and that the set of basic functions had to be enlarged; nevertheless, they thought that the core of analysis (the formal methodology connected to elementary functions) could and needed to remain unaltered. In effect, they investigated new transcendental quantities $\int f(x)dx$ formally by manipulating and recombining the basic components of the analytical expressions $f(x)$ with the help of the theory of integration.

It should also be noted that mathematicians often resorted to extra-analytical arguments and, in particular, geometrical interpretations when dealing with transcendental quantities (see the notion of discontinuous functions and the integral in Chapters 23 and 25, Legendre's definition of the gamma function in this chapter), but this contradicted the declared independence of analysis from geometry, one of the cornerstones of 18th-century mathematics. Around 1800, a satisfactory *analytical* theory, which organized the acquired knowledge unitarily, was lacking.

---

[404] In the 18th century, the consideration of the inverse of a given function $y = f(x)$ was not a problem, even though the analytical expression $x = g(y)$ could not be made explicit. In this case mathematicians operated on the analytical expression $f(x)$ and, of course, if a table of values of x and y was given, it was understood to express both the relation $y = f(x)$ and the relation $x = g(y)$.

30 D'Alembert and Lagrange and the inequality technique

I conclude the third part of the present work by discussing another interesting feature of the evolutionary process of 18th-century analysis, namely the use of the inequality technique to estimate errors in approximation.

Prior to the late 1760s, many methods of approximations were known. They mainly consisted in deriving infinite analytical expressions or in giving a recursive process for estimating a certain quantities. According to Grabiner [1981, 56]: "Most mathematicians preferred to write down infinite expressions, which, since they appeared in equations, seemed to give the solution exactly, rather than to write inequalities." In general, little attention was paid to computing general bounds on the errors made in approximations (see Grabiner [1981, 57]).

In 1768, d'Alembert published an innovative paper, *Réflexions sur les suites et sur les racines imaginaires*; its novel feature was the fact that the problem of approximation was associated with the determination of an explicit error estimate. The paper was composed of two parts. The first part dealt with the approximate calculation of the values of the function

$$(1+x)^m$$

by means of the series

$$\sum_{n=0}^{\infty} \binom{m}{n} x^n.$$

The second is devoted to questions concerning imaginary numbers (I shall not treat it).

D'Alembert started by observing that the series $\sum_{n=0}^{\infty} \binom{m}{n} x^n$ had to be decreasing[405] in order to compute the values of $(1+x)^m$. Consequently, the ratio between the nth and $(n+1)$th terms of the series had to be less than 1 (in absolute value) and then

$$|x| < \frac{n}{n-m-1} \text{ if } n > m+1.$$

Setting $n = \infty$, he had that the last terms of series, at least, are decreasing (convergent) if $|x| < 1$. On the contrary, the last terms are increasing (divergent) when $|x| > 1$ (d'Alembert [1768, 173]).

---

[405] D'Alembert used the term "convergent."

Then d'Alembert determined the bounds of errors. He observed that if

$$|x| < 1,$$
$$\nu > 1 + m,$$
$$S = \sum_{n=\nu-1}^{\infty} \binom{m}{n} x^n,$$
$$A = \left| \binom{m}{\nu-1} \right| x^{\nu-1},$$

one had

$$S < \sum_{i=0}^{\infty} A|x|^i = \frac{A}{1-|x|} \tag{209}$$

and

$$S > \sum_{i=0}^{\infty} A \left(\frac{\nu-1-m}{\nu} \right)^i |x|^i = \frac{A}{1-|x|\frac{\nu-1-m}{\nu}}.$$

He stated that the sum of the series from A on lay between the bounds

$$\frac{A}{1-|x|}$$

and

$$\frac{nA}{\nu - (\nu-1-m)|x|},$$

which gave "a practicable enough way" of summing the series by approximation, and that the error was less than

$$\frac{A(m+1)|x|}{(1-|x|)(\nu - (\nu-1)|x| + m|x|)},$$

if one assumed S to be equal to a value between these bounds (d'Alembert [1768, 177–178]).

D'Alembert then discussed the improvement of convergence of the series and, finally, criticized the use of divergent series in certain demonstrations (see d'Alembert [1768, 181–183]).

Some remarks are appropriate. First, d'Alembert did not depart from the basic tenets of the 18th-century conception: a series was not an autonomous object but the result of a transformation of a given closed analytical expression. Indeed, d'Alembert did not determine the sum of

$$\sum_{n=0}^{\infty} \binom{m}{n} x^n :$$

He assumed the development of the function $(1+x)^m$ is $\sum_{n=0}^{\infty} \binom{m}{n} x^n$ (it is to be imagined that, according to d'Alembert, this relation was derived by usual formal methods).

Second, it is true that d'Alembert used the technique of inequalities, but this technique is a tool for numerical evaluation of a function. In no case did he use this technique to prove the existence of a limit.[406] *D'Alembert's did not know the ratio test*, if by this term we intend a convergence criterion by which we establish if the series has a finite sum. For him the condition

$$\frac{a_{n+1}}{a_n} < 1$$

served to establish where the series approximated its known sum, and (209) is not used to prove the existence of the sum but was only a procedure to determine the bounds of errors.

*   *   *

The use of the techniques of inequalities was not connected to d'Alembert's view on convergence.[407] Indeed, Lagrange argued with d'Alembert over this point but followed him in determining explicit error estimates by resorting to inequalities. In 1798, Lagrange wrote a remarkable treatise on numerical solution to equations, *Traitè de la résolution des équations numériques de tous les degrés*, where he "presented the study of algebraic approximations and the corresponding inequality technique as a coherent subject" (Grabiner [1981, 64]).[408]

Even in the theory of analytical functions, inequalities were important for estimating errors. Similarly to d'Alembert, Lagrange did not use inequalities in order to establish if the series has a finite sum: He assumed that the sum exists and sought to approximate it. Further estimates of bounds on errors and the connected technique of inequalities were regarded as important in applications of analysis. In effect, Lagrange's use of inequalities was entirely inserted in the fabric of the 18th-century analysis and totally subordinate to formal methodology. The following excerpt from *Leçon sur le calcul des fonctions* [1806] illustrated his approach adequately:

> Every function $f(x+i)$ can be expanded, as has been seen,
> as a series
>
> $$f(x) + if'(x) + \frac{i^2}{2} f''(x) + \frac{i^3}{2 \cdot 3} f'''(x) + \ldots,$$

---

[406]See also Grabiner [1981, 63].

[407]It is also worthwhile noting that even though d'Alembert argued that only convergent series could be employed, he shared all the other aspects of the 18th-century concept of series and did not offer a new methodology capable of avoiding formal methods.

[408]I refer to Grabiner for an account of Lagrange's *Traitè*.

which naturally goes to infinity, unless the derivative functions of $f(x)$ become zero, which happens when $f(x)$ is a polynomial function of x.

When this development is only used to generate derived functions, it does not matter whether the series goes to infinity or not; this is also the case when the series is only considered as a simple analytical transformation of the function. However, if it is wished to use it to obtain the value of the function in particular cases, as in giving an expression of a simpler form because of the quantity i which is removed from the function, then, only taking account of a certain, large or small, number of terms, it is important to have a means of evaluating the remainder of the series which is neglected, or at least to find limits to the error that is made in neglecting this remainder.

The determination of these limits is of great importance, especially in the applying the theory of functions to the analysis of curves and mechanics, in order to give this application the rigor of ancient geometry. (Lagrange [1806, 85–86])

According to Lagrange, the determination of bounds on errors enabled the transition from the formal manipulation of series to the quantitative interpretations of derived results, which were required by the applied sciences. This is certainly an important novelty in 18th-century analysis. Hitherto this transition had always been very intuitive and was based upon vague considerations. When mathematicians considered the quantitative aspect of a formally derived equality $f(x) = \sum_{k=0}^{n} a_k x^k$, they limited themselves to verifying whether the series $\sum_{k=0}^{\infty} a_k x^k$ converged to a function $f(x)$ for the desired values of x and assuming the approximate equality $f(x) = \sum_{k=0}^{n} a_k x^k$, for large enough n and small enough x. This approximation allowed numerical computations for "small enough" x, although there was no consideration of bounds on errors.

Without changing the basis upon which this conception was grounded (*a posteriori* determination of the limits of validity of formally derived results), Lagrange attempted to make the transition from the formal to the quantitative precise so that the applications of formal results to geometry and mechanics could be similarly precise. The heart of Lagrange's attempt was the remainder theorem.[409]

Lagrange gave two different proofs of this theorem.[410] They are both of interest for the demonstrative technique and for further aspects of the

---

[409] On the role played by the remainder theorem in the Lagrangian theory of analytical functions as an intermediary between the formal and the quantitative, see Panza [1992, 691-841] and Ferraro and Panza [A].

[410] See Lagrange [1797, 78–83] and [1806, 86–89].

notion of continuity. I shall now briefly mention the proof contained[411] in the *Leçons* [1806], which was based on the following lemma:

> *Given a function $f(x)$, if $f(0) = 0$ and $0 < f'(x) < \infty$ [$-\infty <$ $f'(x) < 0$] for $0 < x < d$ then $0 < f(x) < \infty$ [$-\infty < f(x) < 0$] for $0 < x \leq d$. If $f(0) = 0$ and $0 < f'(x) < \infty$ [$-\infty < f'(x) < 0$] for $-d < x < 0$ then $-\infty < f(x) < 0$ [$0 < f(x) < \infty$] for $-d \leq x < 0$.*

Lagrange began the proof by stating that if $f'(x)$ is finite, one can write

$$f(x + i) = f(x) + i[f'(x) + V(x, i)]$$

with $V(x, 0) = 0$. According to him, since $V(x, 0) = 0$, if $i(> 0)$ increases, then $|V(x, i)|$ also increases, and this occurs at least up to a certain value i' of i, after which $|V(x, i)|$ might begin to decrease.[412] Moreover, if the increase of i is small, the increase of $|V(x, i)|$ is small as well.[413] Therefore, if $D > 0$ is a given quantity, there exists an i' such that

$$|V(x, i)| < D, \quad i < i'.$$

Consequently, if $i < i'$, then

$$i[f'(x) - D] < f(x + i) - f(x) < i[f'(x) + D]. \tag{210}$$

Lagrange replaces the indeterminate x by $x + i$, $x + 2i$, ..., $x + (n - 1)i$, where i is taken to be sufficiently small. He thus has

$$i[f'(x + ki) - D] < f(x + (k + 1)i) - f(x + ki) < i[f'(x + ki) + D] \tag{211}$$

for $k = 0, 1, 2, \ldots, n - 1$.

Of course, this can be done if (210) holds for $i < i'$ and i' is independent of x. In this case, one can fix i such that $ni < i'$ (for a given n) and for this fixed i the inequalities (211) hold, for every k. Here Lagrange implicitly assumes a condition of uniformity in the behavior of functions, which resembles uniform continuity.[414]

---

[411]On the proof contained in the *Théorie*, see Ferraro and Panza [A].

[412]Lagrange imagines that, given a function $h(x)$, if $h(x) \neq \infty$ when i varies between certain values T_1 and T_2, then there exists a partition of $[T_1, T_2]$ such that $h(x)$ is monotone over any interval $[t_i, t_{i+1}]$ of the partition.

[413]Here Lagrange uses the principle of continuity (LC).

[414]Lagrange's assumption of uniformity cannot simply be confused with modern uniform continuity, which presupposes a different concept of analysis. Instead, it seems to depend on the fact that x is an indeterminate quantity. I also note the similarity between Lagrange's condition of uniformity and Cauchy's assumption of uniform behavior of continuous functions (see p. 350).

By summing (211) term by term, Lagrange derives

$$i \sum_{k=0}^{n-1} f'(x+ki) - niD < f(x+ni) - f(x) < i \sum_{k=0}^{n-1} f'(x+ki) + niD$$

Let

$$D = \frac{\left| \sum_{k=0}^{n-1} f'(x+ki) \right|}{n};$$

then one has

$$0 < f(x+ni) - f(x) < 2i \sum_{k=0}^{n-1} f'(x+ki)$$

or

$$2i \sum_{k=0}^{n-1} f'(x+ki) < f(x+ni) - f(x) < 0.$$

If we denote the maximum of $|f'(x+ki)|$, for $k = 0, 1, 2, \ldots, n-1$, by P we have

$$0 < f(x+ni) - f(x) < 2inP \quad \text{or} \quad -2inP < f(x+ni) - f(x) < 0.$$

Since i can be taken as small as desired, and n as large, Lagrange assumed that in is equal to any quantity z and

$$f(x+ni) - f(x) = f(x+z) - f(x).$$

The quantity $f(x+z) - f(x)$ can represent any function of z that vanishes for $z = 0$ and

$$f'(x+in) = f'(x+z)$$

is the derivative of $f(x+z)$, with respect both to x and z. Consequently,[415]

$$0 < f(x+z) - f(x) < 2zP \quad \text{or} \quad -2zP < f(x+z) - f(x) < 0,$$

and, for $x = 0$,

$$0 < f(z) < 2zP \quad \text{or} \quad -2zP < f(z) < 0,$$

provided $f'(z)$ is finite over a certain interval.

From this lemma, Lagrange deduced the remainder theorem for which, in the *Leçons*, he gave the following formulation:

---

[415]It is clear that P must now be interpreted as the supremum of $f'(x+z)$ over an appropriate interval.

> Given a function $f(x)$, if $f^{(n)}(q)$ and $f^{(n)}(q)$ are the minimum and the maximum of the derivative $f^{(n)}(x+i)$, for $i = 0$ and $i < i'$, the following inequalities holds:

$$f(x) + if'(x) + \frac{i^2}{2}f''(x) + \ldots + \frac{i^n}{n!}f^{(n)}(p) \tag{212}$$

$$< f(x+i) < f(x) + if'(x) + \frac{i^2}{2}f''(x) + \ldots + \frac{i^n}{n!}f^{(n)}(q),$$

> provided $f^{(n)}(q)$ and $f^{(n)}(q)$ are not infinite.

The proof runs as follows. Let p and q be the values of $x + i$ such that $f'(p)$ and $f'(q)$ are the least and greatest value of $f'(x+i)$, taking x as given, and letting i vary from 0 to a given value i'. By integrating

$$f'(x+i) - f'(p) > 0 \quad \text{and} \quad f'(q) - f'(x+i) > 0$$

with respect to the variable i (between 0 and i), one obtains

$$f(x+i) - f(x) - if'(p) \quad \text{and} \quad if'(q) - f'(x+i) + f(x).$$

If $f'(x+i)$ never becomes infinite for $i < i'$ [which surely occurs when $f'(p)$ and $f'(q)$ are not infinite], an application of the above lemma, with positive i, made it possible to obtain

$$f(x+i) > f(x) + if'(p) \quad \text{and} \quad f(x+i) < f(x) + if'(q).$$

We now repeat the reasoning by assuming that p and q are the values where the second derivative assumes the maximum and minimum. We have

$$f''(x+i) - f''(p) > 0 \quad \text{and} \quad f''(q) - f''(x+i) < 0,$$

provided that $f''(x+i)$ is not infinite, which certainly occurs when $f''(p)$ and $f''(q)$ are not infinite. By applying the lemma twice, we have

$$f(x) + if'(x) + \frac{i^2}{2}f''(p) < f(x+i) < f(x) + if'(x) + \frac{i^2}{2}f''(q).$$

The reasoning can be repeated again, and it is possible to derive (212).[416]

---

[416]In the *Théorie*, Lagrange supposed that any function takes all the intermediate values between two of its values. By applying this form of the principle of continuity, Lagrange obtained

$$R_n(x,i) = \frac{i^{n+1}}{(n+1)!}f^{(n+1)}(x+\lambda i),$$

where $0 < \lambda < 1$.

Part IV

The decline of the formal theory of series

At the beginning of the 19th century the theory of series was refounded upon new bases, which mainly consisted of the attempt to avoid formal manipulation and to give an exclusively quantitative interpretation to the equality

$$f(x) = \sum_{k=0}^{\infty} f_k(x).$$

The new approach to series appeared for the first time in a paper by Gauss published in 1813. Fourier had already looked at trigonometric series from a viewpoint that differed from Euler's and Lagrange's in his work on the propagation of heat submitted to the Institut de France in 1807 (published after a complicated series of events in 1822).[417] Finally, Cauchy gave the first systematic presentation of a theory based on an exclusively quantitative approach in his famous treatise of 1821.

A very important reason for the rejection of the formal concept of series was that, at the beginning of the 19th century, this concept no longer contributed to the growth of analysis. We saw that one of the reasons for the success of the formal conception had been its capacity to produce novel and interesting results. However, this capacity was exhausted by the end of the century. In this period, the circumscribed domain of elementary functions and their power series was not sufficient for the needs of astronomy, probability, physics, etc. These sciences required the mathematical investigations of new quantities and the introduction of new functions. The formal concept of series prevented scholars from dealing with geometrical and physical quantities adequately and, in particular, prevented scholars from using series as autonomous objects that could be used to introduce and represent new quantities.

During the period from 1770 to 1820, studies of new quantities (gamma function, etc.) represented an important part of advanced mathematical research. Many results were achieved, mainly concerning certain integrals, although these results appeared as marginal additions to the analysis of the elementary functions and their power series. The potential of certain results was not appreciated. Mathematicians often succeeded only thanks to geometrical and physical considerations, but this was in contradiction to the

---

[417]It should be pointed out that Bolzano also contributed to this new approach in his *Der binomische Lehrsatz* [1816], but Bolzano's papers were little-known and had a minimal influence on later developments.

declared independence of analysis from geometry, one of the cornerstones of Eulerian and Lagrangian mathematics.

It should also be observed that various paradoxes concerning the sum of series were known, but they were not crucial in the rejection of formal methods. Mathematicians who followed formal methods considered them as local difficulties. They attempted to provide an explanation for them. However, even when an explanation was not found, such difficulties were not viewed as counterexamples, capable of invalidating the series theory, but only as exceptions.[418] Only when mathematicians adopted a new approach were certain problematic sums conceived as manifestations of the contradictions of the formal theory. Here, I point out two paradoxes that emerged at the turn of the 18th century and their attempts at explanation.

The first regards the series $1-1+1-\dots$. In 1797, J.F. Callet[419] asserted that the sum[420] of the series $1-1+1-\dots$ was equal to $\frac{2}{3}$. He argued that, for $x=1$, the expansion

$$\frac{1+x}{1+x+x^2} = 1 - x^2 + x^3 - x^5 + x^6 + \dots$$

yielded

$$\frac{2}{3} = 1 - 1 + 1 - 1 + \dots$$

In response, Bossut and Lagrange [1801–02, 1–11] noted that the series $1 - x^2 + x^3 - x^5 + x^6 + \dots$ should be interpreted as

$$1 + 0x - x^2 + x^3 + 0x^4 - x^5 + x^6 + \dots$$

and then

$$\frac{2}{3} = 1 + 0 - 1 + 1 + 0 - 1 + 1 + \dots$$

The second paradox concerns the expansion[421]

$$(2\cos x)^m = \sum_{k=0}^{\infty} \binom{m}{k} \cos(m - 2k)x,$$

which had been studied by Euler and Lagrange. In his [1811], Poisson showed that one contradiction derived from the substitution $m = \frac{1}{3}$ and $x = \pi$. Indeed, from $(2\cos x)^m$ it was possible to obtain the three complex values

$$2^{1/3}\frac{1+\sqrt{-3}}{2}, \quad -2^{1/3}, \quad 2^{1/3}\frac{1-\sqrt{-3}}{2};$$

---

[418]See Chapter 19.3, p. 227.
[419]See Bossut and Lagrange [1801–02, 1–11].
[420]On this series, see Chapter 9.
[421]On the history of this problem, see Jahnke [1987].

whereas the series gave the value $\frac{2^{1/3}}{2}$. According to Poisson, the problem arose from the fact that, when m is a rational number, the series

$$\sum_{k=0}^{\infty} \binom{m}{k} \cos(m - 2k)x$$

only represented the real part of the expansion of $(2 \cos x)^m$. However, the correct analytical sum of the series was determined only in the 1820s (see Jahnke [1993, 275]).

31 Fourier and Fourier series

Fourier began his mathematical work on the theory of heat in the early 1800s. In December 1807, he submitted a memoir, entitled *Sur la propagation de la chaleur dans le solides* [1807], to the Institut de France. Lagrange, Laplace, Monge, and Lacroix were chosen as referees. Fourier's approach was too different from Lagrange's and Laplace's, especially with regard to trigonometric series, and, thus, although Fourier made some attempts to clarify his views, the memoir remained a manuscript (it was published by Grattan-Guiness and Ravetz in *Joseph Fourier 1768–1830*[422]).

In the following years, the *Institut* announced a competition on the propagation of heat in solid bodies. Fourier submitted a new version of his memoir containing additional work. He won the prize, but the commission (Lagrange, Laplace, Malus, Haüy, and Legendre) stated that the work "leaves something to be desired on the score of generality and even rigor." (see Fuorier [*Œuvres*, 1: viii]). Thus, this version of the memoir was not published on this occasion either. Fourier wrote another version that was published by the *Académie des Sciences* in 1822 with the title *Thèorie analytique de la chaleur.*[423]

Fourier's treatise is of great interest in the history of mathematics and physics and the relationships between these sciences. Given the scope of the present book, I concentrated on Fourier's use of trigonometric series in his mathematical investigation of heat diffusion.

In *Sur la propagation de la chaleur*, Fourier considered the steady-state diffusion in a lamina and obtained the diffusion equation

$$\frac{\partial^2 z}{\partial x^2} + \frac{\partial^2 z}{\partial y^2} = 0 \qquad (x \geq 0, \quad -1 \leq y \leq 1).$$

He solved this equation by separating variables and superposing simple states. He obtained

$$z = \sum_{k=1}^{\infty} a_k e^{-n_k x} \cos n_k y$$

By considering the boundary condition $z = 0$ (when $y = \pm 1$) and $z = 1$ (for $x = 0$) and replacing y by $\frac{2u}{\pi}$, he derived the trigonometric series

$$1 = a_1 \cos u + a_2 \cos 3u + a_3 \cos 5u + \ldots,$$

where $-\frac{1}{2}\pi < u < \frac{1}{2}\pi$ (see Fourier [1807, 134–144]).

---

[422]See Grattan-Guiness and Ravetz [1972].

[423]For more details about the story of the publication of Fourier's treatise, see Grattan-Guiness and Ravetz [1972].

Fourier found the constants a_k by differentiating

$$1 = \sum_{k=1}^{\infty} a_k \cos(2k-1)u$$

term by term infinitely many times:

$$
\begin{aligned}
1 &= a_1 \cos u + a_2 \cos 3u + a_3 \cos 5u + \dots, \\
0 &= a_1 \sin u + 3a_2 \sin 3u + 5a_3 \sin 5u + \dots, \\
0 &= a_1 \cos u + 3^2 a_2 \cos 3u + 5^2 a_3 \cos 5u + \dots, \\
0 &= a_1 \sin u + 3^3 a_2 \sin 3u + 5^3 a_3 \sin 5u + \dots, \\
0 &= a_1 \cos u + 3^4 a_2 \cos 3u + 5^4 a_3 \cos 5u + \dots, \\
&\quad \dots.
\end{aligned}
$$

He set $u = 0$ in all derived equations and obtained an infinite number of equations in the unknowns a_k :

$$
\begin{aligned}
1 &= a_1 + a_2 + a_3 + a_4 + a_5 + \dots, \\
0 &= a_1 + 3^2 a_2 + 5^2 a_3 + 7^2 a_4 + 9^2 a_5 + \dots, \\
0 &= a_1 + 3^4 a_2 + 5^4 a_3 + 7^4 a_4 + 9^4 a_5 + \dots, \\
0 &= a_1 + 3^6 a_2 + 5^6 a_3 + 7^6 a_4 + 9^6 a_5 + \dots, \\
0 &= a_1 + 3^8 a_2 + 5^8 a_3 + 7^8 a_4 + 9^8 a_5 + \dots, \\
0 &= a_1 + 3^{10} a_2 + 5^{10} a_3 + 7^{10} a_4 + 9^{10} a_5 + \dots, \\
0 &= a_1 + 3^{12} a_2 + 5^{12} a_3 + 7^{12} a_4 + 9^{12} a_5 + \dots, \\
&\quad \dots.
\end{aligned}
$$

To solve this system, Fourier considered the first seven equations in the first seven unknowns and found that

$$a_1 = \frac{3 \cdot 3 \cdot 5 \cdot 5 \cdot 7 \cdot 7 \cdot 9 \cdot 9 \cdot 11 \cdot 11 \cdot 13 \cdot 13}{2 \cdot 4 \cdot 4 \cdot 6 \cdot 6 \cdot 8 \cdot 8 \cdot 10 \cdot 10 \cdot 12 \cdot 12 \cdot 14}.$$

At this point Fourier stated that if one considered more equations, one would have found an expression of a_1 similar to the previous one. In the case of eight equations, he found

$$a_1 = \frac{3 \cdot 3 \cdot 5 \cdot 5 \cdot 7 \cdot 7 \cdot 9 \cdot 9 \cdot 11 \cdot 11 \cdot 13 \cdot 13 \cdot 15 \cdot 15}{2 \cdot 4 \cdot 4 \cdot 6 \cdot 6 \cdot 8 \cdot 8 \cdot 10 \cdot 10 \cdot 12 \cdot 12 \cdot 14 \cdot 14 \cdot 16}.$$

According to Fourier, if one considered all the infinite equations, then one had

$$a_1 = \frac{3 \cdot 3 \cdot 5 \cdot 5 \cdot 7 \cdot 7 \dots}{2 \cdot 4 \cdot 4 \cdot 6 \cdot 6 \cdot 8 \dots} = \frac{4}{\pi}.$$

In the same way, he found the other coefficients:

$$
\begin{aligned}
a_2 &= -\frac{1 \cdot 1 \cdot 5 \cdot 5 \cdot 7 \cdot 7 \cdot \ldots}{2 \cdot 4 \cdot 2 \cdot 8 \cdot 4 \cdot 10 \cdot \ldots} = -\frac{1}{3}\frac{4}{\pi}, \\
a_3 &= \frac{1 \cdot 1 \cdot 3 \cdot 3 \cdot 7 \cdot 7 \cdot \ldots}{4 \cdot 6 \cdot 2 \cdot 8 \cdot 2 \cdot 12 \cdot \ldots} = \frac{1}{5}\frac{4}{\pi}, \\
a_4 &= -\frac{1 \cdot 1 \cdot 5 \cdot 5 \cdot 9 \cdot 9 \cdot \ldots}{6 \cdot 8 \cdot 4 \cdot 10 \cdot 2 \cdot 12 \cdot \ldots} = -\frac{1}{7}\frac{4}{\pi}, \\
a_5 &= \frac{1 \cdot 1 \cdot 5 \cdot 5 \cdot 7 \cdot 7 \cdot 11 \cdot 11 \cdot \ldots}{8 \cdot 10 \cdot 6 \cdot 12 \cdot 4 \cdot 14 \cdot 2 \cdot 16 \ldots} = \frac{1}{9}\frac{4}{\pi}, \\
&\quad \ldots
\end{aligned}
$$

(see Fourier [1807, 147–156]).

This procedure is rather close to typical 18th-century procedures. However, Fourier changed the interpretation of the relation

$$
f(x) = \frac{1}{2}a_0 + \sum_{n=1}^{\infty}(a_n \cos nx + b_n \sin nx), \tag{213}
$$

and this is the crucial point. He viewed this relation as a purely quantitative relation. Thus, he [1807, 158] made clear that the equality

$$
\sum_{k=0}^{\infty}(-1)^k \frac{1}{2k-1}\cos(2k-1)u = \frac{\pi}{4}
$$

did not hold when the variable u does not lie between $-\frac{\pi}{2}$ and $\frac{\pi}{2}$. Indeed, the function

$$
\sum_{k=0}^{\infty}(-1)^k \frac{1}{2k+1}\cos(2k+1)u
$$

is equal to $\frac{\pi}{4}$, over $(-\frac{\pi}{2}, +\frac{\pi}{2})$; it is 0 for $u = \pm\frac{\pi}{2}$: it is equal to $-\frac{\pi}{4}$, over $(\frac{\pi}{2}, \frac{3\pi}{2})$. To make this clear, he also provided a geometrical interpretation of the equation

$$
y = \sum_{k=0}^{\infty}(-1)^{k+1}\frac{1}{2k-1}\cos(2k-1)u,
$$

the curve having this equation being viewed as the limiting curve of the curves[424]

$$
y = \sum_{k=0}^{n}(-1)^{k+1}\frac{1}{2k-1}\cos(2k-1)u, n = 1, 2, 3, \ldots.
$$

----

[424]See Grattan-Guiness and Ravetz [1972, 169–171] and Grattan-Guiness [1990, 594–601].

Then he found the sum of $\sum_{k=0}^{\infty}(-1)^{k+1}\frac{1}{2k-1}\cos(2k-1)u$ directly by showing that the nth sum (n even) of the series is

$$\frac{1}{2}\int_0^u \frac{\sin 2nx}{\cos x}\,dx$$

and that it tended to $\frac{\pi}{4}$ as n went to infinity.

Fourier applied a similar procedure to

$$\frac{1}{2}x = \sum_{k=1}^{\infty}(-1)^{k+1}\frac{1}{k}\sin kx$$

and

$$\log(2\cos\frac{1}{2}x) = \sum_{k=1}^{\infty}(-1)^{k+1}\frac{1}{k}\cos kx$$

For example, as concerns $\frac{1}{2}x = \sum_{k=0}^{\infty}(-1)^k\frac{1}{k}\sin kx$, he considered the nth sum

$$y = \sum_{k=1}^{m}(-1)^{k+1}\frac{1}{k}\sin kx$$

and differentiated it in order to obtain

$$\frac{dy}{dx} = \sum_{k=1}^{m}(-1)^{k+1}\cos kx.$$

By multiplying the last equation by $2\sin x$ and by performing the appropriate manipulations, he obtained

$$\frac{dy}{dx} = \frac{1}{2} - \frac{\cos(m+\frac{1}{2})x}{2\cos\frac{1}{2}x}$$

By integrating by parts,

$$y = \frac{1}{2}x - \int\frac{\cos(m+\frac{1}{2})x}{2\cos\frac{1}{2}x} = C + \frac{1}{2}x + \frac{1}{2}\frac{1}{m+\frac{1}{2}}\frac{\sin(m+\frac{1}{2})x}{\cos\frac{1}{2}x} + \dots$$

and

$$y = \frac{1}{2}x + C, \quad \text{for} \quad m = \infty.$$

Since $y(0) = 0$, he obtained $y = \frac{1}{2}x$.

At this point[425] Fourier stated that the equation $y = \sum_{k=1}^{\infty}(-1)^{k+1}\frac{1}{k}\sin kx$ could be represented as a sequence of vertical and oblique straight lines[426] and

---

[425] See Grattan-Guiness and Ravetz [1972, 165].
[426] Fourier's diagram includes vertical lines ba', cb', dc', ed', fe, ... (see Fig. 21).

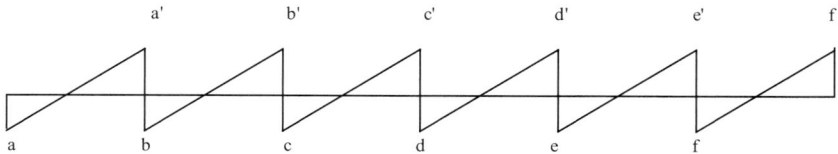

Fig. 21

made it clear that trigonometric series offered an analytical representation of Eulerian discontinuous functions.

Then Fourier investigated the behavior of the integral

$$\frac{1}{2}\int_0^u \frac{\sin 2nx}{\cos x}dx$$

to clarify that the sum

$$\sum_{k=0}^{\infty}(-1)^k \frac{1}{2k+1}\cos(2k+1)u$$

is $\frac{\pi}{4}$ only for certain values of the variable (see Fourier [1807, 159–173]).

Fourier regarded the convergence of series as lying at the heart of the question. He thought it was easy to derive trigonometric series by different procedures:

> but the essential point is to distinguish the limits within which the value of the variable is to be taken. For instance the equation $\frac{1}{2}x = \sin x - \frac{1}{2}\sin 2x - \frac{1}{3}\sin 3x + \ldots$, given by Euler, holds as long as the values of x lie between 0 and π or between 0 and $-\pi$. For all other values of x the right-hand side has a determined value very different from $\frac{1}{2}x$ ....
>
> It is by means of these observations that the contradictory consequences found by combining different series of sine and cosine are explained. (Fourier [1807, 169])

On this question, Fourier's view differed enormously from Lagrange's. Fourier also tried to explain his approach in a short note submitted to the Institut de France, which is still unpublished.[427] The discussion must have

---

[427]For a description, see Grattan-Guiness and Ravetz [1972, 169–173].

been continued privately; today Fourier's response to an objection of Lagrange is kept. Lagrange had observed that if one differentiated[428]

$$\frac{\pi - x}{2} = \sin x + \frac{1}{2}\sin 2x + \frac{1}{3}\sin 3x + \dots, \tag{214}$$

then one obtained

$$-\frac{1}{2} = \cos x + \cos 2x + \cos 3x + ..$$

Hence,

$$-\frac{x}{2} + C = \sin x + \frac{1}{2}\sin 2x + \frac{1}{3}\sin 3x + \dots$$

By setting $x = 0$, one had $C = 0$ and

$$-\frac{x}{2} = \sin x + \frac{1}{2}\sin 2x + \frac{1}{3}\sin 3x + \dots,$$

which contradicted formula (214). It is probable (we can only reconstruct Lagrange's observations through Fourier's words) that Lagrange wanted to show a contradiction in Fourier's approach by proving that if we assume that trigonometric series represent functions, then one series may represent two different functions. According to Fourier, Equation (214) did not hold for $x = 0$, and the function

$$\frac{\pi - x}{2}$$

is represented by the series

$$\sin x + \frac{1}{2}\sin 2x + \frac{1}{3}\sin 3x + \dots$$

only for a certain interval of values of x.

In the following pages of his treatise, Fourier generalized the above results and showed that an odd arbitrary function $f(x)$ can be expanded into a sine series

$$f(x) = \sum_{k=1}^{\infty} a_k \sin kx.$$

He assumed that $f(x)$ could be expanded into the series

$$f(x) = \sum_{k=0}^{\infty} \frac{f^{(2k+1)}(0)}{(2k+1)!} x^{2k+1}$$

---

[428]This equation follows from the convergent series $\frac{1}{2}x = \sin x - \frac{1}{2}\sin 2x - \frac{1}{3}\sin 3x + \dots$ by replacing x by $\pi - x$.

and wrote

$$
\begin{aligned}
f'(0) &= a_1 + 2a_2 + 3a_3 + 4a_4 + 5a_5 + \dots, \\
f^{(3)}(0) &= a_1 + 2^3 a_2 + 3^3 a_3 + 4^3 a_4 + 5^3 a_5 + \dots, \\
f^{(5)}(0) &= a_1 + 2^5 a_2 + 3^5 a_3 + 4^5 a_4 + 5^5 a_5 + \dots, \\
f^{(7)}(0) &= a_1 + 2^7 a_2 + 3^7 a_3 + 4^7 a_4 + 5^7 a_5 + \dots, \\
&\dots\dots
\end{aligned}
$$

After a long sequence of formal and rather complex manipulations, Fourier obtained

$$
\begin{aligned}
\frac{1}{2}\pi f(x) = {} & \left(f(\pi) - \frac{1}{1^2} f^{(2)}(\pi) - \frac{1}{1^4} f^{(4)}(\pi) - \dots \right) \sin x \\
& - \left(f(\pi) - \frac{1}{2^2} f^{(2)}(\pi) - \frac{1}{2^4} f^{(4)}(\pi) - \dots \right) \sin 2x \\
& \dots\dots
\end{aligned}
$$

Then he set

$$
\frac{s_n(x)}{n} = f(x) - \frac{1}{n^2} f^{(2)}(x) - \frac{1}{n^4} f^{(4)}(x) - \dots
$$

and showed that s_n satisfied the differential equation

$$
s_n(x) + \frac{1}{n^2} \frac{d^2 s_n(x)}{dx^2} = f(x).
$$

Hence,

$$
\frac{s_n(\pi)}{n} = (-1)^{n+1} \int_0^\pi f(x) \sin kudu
$$

and

$$
\frac{1}{2}\pi f(x) = \sum_{k=1}^\infty \left(\int_0^\pi f(x) \sin kudu \right) \sin kx.
$$

Fourier felt it necessary to explain the meaning of the integral

$$
\int_0^\pi f(x) \sin kudu.
$$

Indeed, the function $f(x)$ can be a discontinuous function and the notion of integration as antidifferentiation cannot applied to these functions. Fourier stated the integral $\int_0^\pi f(x) \sin kudu$ was the area under the function

$$
f(x) \sin ku
$$

and over the segment $[0, \pi]$. He did not provide an analytical notion of integration that also held for discontinuous function and seemed to be satisfied with geometrical interpretation, an approach that cannot have been shared by those who regarded analysis as separate from geometry.

Only at this point did Fourier offer an effective method for finding the coefficients of the sine series, namely the now-standard method based upon the orthogonality of sine terms of the series (see Fourier [1807, 216–217]).[429] By this method he also obtained the general cosine series (see Fourier [1807, 223–224]) and the coefficients of the full series

$$f(x) = \frac{1}{2}a_0 + \sum_{n=1}^{\infty}(a_n \cos nx + b_n \sin nx)$$

for an arbitrary function (see Fourier [1807, 260–262]). It is interesting to note that while studying the annulus, he applied the method, in particular, to the function

$$f(x) = \begin{cases} 1, & 0 \le x \le \pi \\ 0, & \pi \le x \le 2\pi \end{cases}$$

Fourier's treatise contained many other interesting results, some of which concerned those that were later referred to as Bessel's functions. I do not illustrate these results, since the above description is sufficient to clarify the novelty of his approach. Although Fourier employed formal manipulations and geometrical arguments to support his theses, he used series based on the idea that the relation between series and its sum was only a quantitative relation. In so doing he succeeded in giving an analytical form to certain discontinuous functions and in enlarging the bounds of analysis. Fourier's treatise opened up a series of new problems that his followers were to pursue vigorously.

---

[429]This method was already known to Euler and had been published posthumously in 1798 (see p. 282).

32 Gauss and the hypergeometric series

The concept of analysis as a theory of functions based upon a formal methodology was predominant in Germany during the second half of the 18th century and the first decades of the 19th century.[430] Gauss was brought up
within this cultural context. He also had a firsthand knowledge of the works
of Euler and Lagrange: Many abstracts of their writings, which dated back to
the years in which Gauss studied at Brunswick Collegium Carolinum (1792–
1795), were found in his Nachlass (see Gauss [WW, 64]). One insight into
views during the years of his cultural development is provided by an early
work, the *De Integratione formulae differentialis* $(1 + n \cos \psi)^\nu d\psi$ (Gauss
[WO]), whose final draft dates back to the years at Göttingen University
(see Gauss [WW, 8:64]).

In this paper, Gauss stated that integration of a function $f(x)$ was often
obtained by expanding $f(x)$ into finite or infinite series, but while finite
series provided a genuine expression of the integral, infinite series furnished
a less perfect relation. It is evident that Gauss still thought that infinite
series had a status different from that of closed expressions. Nonetheless,
in his opinion, infinite series provided approximations of the sought-after
relations; they were therefore sufficient for common uses (*ad vulgares usus*)
and, in a certain sense, were equivalent to equations (see Gauss [WO, 35]).

In *De Integratione*, Gauss reformulated Chapter VI, vol. I of Euler's
Institutionum calculi integralis (Euler [1768–70, 1:159–182]). He sought to
express the coefficients of the expansions of

$$(1 + n \cos \psi)^\nu \quad \text{and} \quad (1 + n \cos \psi)^\nu d\psi$$

in a more concise and elegant form as well as provide a rigorous proof (*rigida
demonstratione*) of them. According to Gauss, this rigorous proof was required by the risk of making mistakes, due to the complexity of these coefficients, and by the dignity of science (see Gauss [WO, 35–36]). In this article,
Gauss followed the style of the Combinatorial School,[431] and the rigor that
he claimed was entirely based upon the use of formal methods.

A different approach lies at the basis of *Disquisitiones generales circa
seriem infinitam* $1 + \frac{\alpha\beta}{1\cdot\gamma}x + \frac{\alpha(\alpha+1)\beta(\beta+1)}{1\cdot2\cdot\gamma(\gamma+1)}xx + \frac{\alpha(\alpha+1)(\alpha+2)\beta(\beta+1)(\beta+2)}{1\cdot2\cdot3\cdot\gamma(\gamma+1)(\gamma+2)}x^3 + etc.$
[1812b] and *Determinatio seriei nostrae per aequationem differentialem sec-
ondi ordinis* [WA]. In the first paper, Gauss listed some frequently used
functions that could be obtained from the hypergeometric series

$$1 + \frac{\alpha\beta}{1\cdot\gamma}x + \frac{\alpha(\alpha+1)\beta(\beta+1)}{1\cdot2\cdot\gamma(\gamma+1)}xx + \frac{\alpha(\alpha+1)(\alpha+2)\beta(\beta+1)(\beta+2)}{1\cdot2\cdot3\cdot\gamma(\gamma+1)(\gamma+2)}x^3 + \cdots$$

$$(215)$$

---

[430]On this topic, see Jahnke [1993].
[431]See Chapter 28.

by giving particular values to the parameters α, β, γ[432] and then explained the goal of his research in the following way:

> the previous functions are algebraic, logarithmic and circular transcendent. In no case, however, do we undertake our *general* inquiry because *of these*, but rather to promote the theory of higher transcendental functions, our series containing a very large number of them. (Gauss [1812b, 128], italics in original)

Similar expressions are found in his *Anzeige der Disquisitiones generales circa seriem infinitam* [1812a, 197], where Gauss announced his [1812b]. To achieve his aim, Gauss began by trying to define hypergeometric functions appropriately. In effect, he gave two different definitions (the first is found in [1812b], the second in [WA]); both are of great interest since they *per se* implied the rejection of the principle of the generality of algebra[433] and a radical change of the notions of functions, series, integrals, and complex numbers.

In *Disquisitiones generales*, Gauss denoted series (215) by $F(\alpha, \beta, \gamma, x)$ and sought the conditions under which (215) could actually be considered as a function of x. He first observed that γ cannot be either 0 or a negative integer (to avoid infinitely large terms) and that, when α and β are either negative integers or zero, the series exhibits a rational function. Then Gauss [1812b, 126] determined the convergence of the series by the ratio test. Indeed, the ratio of the coefficients of x^m and x^{m+1} is

$$\frac{(m+1)(\gamma+m)}{(\alpha+m)(\beta+m)} = \left(1 + \frac{\gamma+1}{m} + \frac{\gamma}{m^2}\right) : \left(1 + \frac{\alpha+\beta}{m} + \frac{\alpha\beta}{m^2}\right), \qquad (216)$$

and (216) becomes ever closer to unity as m increases. Therefore, if x is a real number such that $|x| < 1$, the series is decreasing to zero,[434] at least

---

[432]Gauss provided 23 examples of elementary functions that are particular cases of (215). Among them:

$$(t+u)^n = t^n F(-n, \gamma, \gamma, -u/t),$$
$$\log(1+t) = tF(1, 1, 2, -t),$$
$$e^t = F(1, k, 1, t/k),$$
$$\sin t = tF(k, k', \frac{3}{2}, \frac{-t^2}{4kk'}),$$
$$\cos t = F(k, k', \frac{1}{2}, \frac{-t^2}{4kk'}),$$

where γ is any number, k and k' denote infinitely large numbers (he intended $k \to \infty$ and $k' \to \infty$), and the variable t assumes appropriate values (Gauss [1812b, 127]).

[433]See Section 18.2, p. 209.

[434]Here Gauss used the expression "convergent". In [WC, 400], Gauss defined a series $\sum a_n$ to be convergent if the sequence a_n approaches zero. As stated in footnote no. 2, I use the term "convergent" in Cauchy's sense.

after a certain term, and has a finite sum. The same occurs if x has the form $a + b\sqrt{-1}$, provided $|x|^2 = a^2 + b^2 < 1$. Instead, if x is a real or complex number such that $|x| > 1$, then the series is increasing (if not initially, at least after a certain term); consequently, one cannot refer to the sum of the series. After postponing the more complicated case $x = 1$ to Section 3 of [1812b],[435] Gauss stated:

> It is clear that since our function is defined to be the sum of the series, the inquiry as to its nature is restricted to the case in which the series actually converges. Therefore it is inappropriate to ask about the value of the series for x greater than unity. (Gauss [1812b, 126])

Even though Gauss only briefly mentioned that the function $F(\alpha, \beta, \gamma, x)$ was defined as the sum of the series (215) and that he restricted himself to considering the values of x for which the series (215) was convergent,[436] this statement implied a radical shift in the 18th-century concept of function and series. Gauss rejected the restricted conditions a relation had to satisfy in order to be considered a function; rather than regarding the hypergeometric series as the expansion of a generating function, he viewed it as a function in its own right. Thus, the Eulerian notion of the sum was discarded and was replaced by the idea that the sum of a series was the limit of partial sums.[437] The role of convergence also changed: from being an *a posteriori* condition for the application of formally derived results, it became the preliminary condition for using a series (for this reason Gauss used convergence criteria in a modern sense). Moreover, unlike the few 18th-century mathematicians who criticized the use of divergent series, Gauss actually developed an adequate methodology for making this rejection concrete.

The second definition of hypergeometric functions also introduced new features into 18th-century analysis. In *Determinatio seriei nostrae,* Gauss observed that

$$\frac{dF(\alpha, \beta, \gamma, x)}{dx} = \frac{\alpha\beta}{\gamma} F(\alpha + 1, \beta + 1, \gamma + 1, x)$$

and

$$\frac{d^2 F(\alpha, \beta, \gamma, x)}{dx^2} = \frac{\alpha\beta(\alpha + 1)(\beta + 1)}{\gamma(\gamma + 1)} F(\alpha + 2, \beta + 2, \gamma + 2, x).$$

---

[435] See p.333.

[436] Gauss underlined this point in a letter to Bessel on November 21, 1811, where he considered the definition of a function by means of a divergent series to be utterly despicable (see Gauss [WW, 10:362–363]).

[437] See also Gauss [1812b, 143].

Since

$$\gamma(\gamma + 1)F(\alpha, \beta, \gamma, x) - (\gamma + 1)(\gamma - (\alpha + \beta + 1)x)F(\alpha + 1, \beta + 1, \gamma + 1, x)$$
$$- (\alpha + 1)(\beta + 1)x(1 - x)F(\alpha + 2, \beta + 2, \gamma + 2, x) = 0,$$

he set $P = F(\alpha, \beta, \gamma, x)$ and derived

$$\alpha\beta P - (\gamma - (\alpha + \beta + 1)x)\frac{dP}{dx} - x(1 - x)\frac{d^2 P}{dx^2} = 0. \qquad (217)$$

Gauss stated that this differential equation could be considered as another definition of (215). However, $P = F(\alpha, \beta, \gamma, x)$ was not the general integral but a particular solution; therefore, it was appropriate to add the conditions $P(0) = 1$, $\frac{dP}{dx} = \frac{\alpha\beta}{\gamma}$, and $\frac{d^2 P}{dx^2} = \frac{\alpha\beta(\alpha+1)(\beta+1)}{\gamma(\gamma+1)}$ to this definition (Gauss [WA, 207]).

According to Gauss, the solution P to (217) was a more general function than $F(\alpha, \beta, \gamma, x)$, which was defined for $|x| < 1$. Indeed, P is a many-valued function defined in all of the plane except for 0 and 1.[438] Gauss described this by stating that if one moved from 0 to x in a continuous way but did not arrive at 1 (where the coefficient $x - x^2$ was null), then P was a perfectly determinate quantity. However, one could achieve the real values of x greater than 1 only if one followed a path that went around the point 1 by passing through imaginary numbers. Since this could be performed in a continuous way according to different paths, then P assumed many different values.

Euler already knew that the hypergeometric series was the solution to differential equation (217). In [1794a], Euler considered the hypergeometric series without taking convergence into account and showed that it satisfied (217). He then used the hypergeometric differential equation to prove[439] (172). Euler did not use (217) to introduce a new type of functions (as we saw, a relation had to satisfy certain conditions in order to be considered a function). Instead, Gauss thought that a relation by itself was a function: Since (217) provided a relation between x and P, he could consider it as a definition of hypergeometric functions.

It is clear that, since Gauss defined hypergeometric functions by (217), his notion of integration differed from those of Euler and Lagrange.[440] A brief reference to his interpretation of the integral can be found in his letter to Bessel dated December 18, 1811. Even though Gauss was discussing

---

[438] Gauss did not consider the case $x = \infty$.

[439] See p. 269.

[440] The notion of integration as antidifferentiation presupposed that the function (which was the solution to the differential equation) was known before it was recognized as the solution to the equation. Apart from the considerations in Chapter 25, this notion did not allow an integral or a differential equation per se to be used as the definition of a new function. Gauss's concept of integral removed this difficulty.

complex integration and the letter was mainly devoted to the problem of many-valued functions, the idea of the integral that emerges is a general one. Gauss wrote:

> What should one think of for $\int \varphi x dx = a + ib$? Evidently, if one wants to proceed by clear concepts, one must assume that x passes by infinitely small increments (each of the form $\alpha + \beta i$) from the value for which the integral is equal 0 to $a + bi$; and then that all the products $\varphi x dx$ are summed. In this way, the previous written expression is meaningful .... The transition from one point to another occurs without touching a point where $\varphi x = \infty$. I therefore require that one avoids the points where the original concept of the integral $\int \varphi x dx$ (the basis of the whole) manifestly loses its own clarity and easily arouses contradictions. Apart from this, it is evident that if a function is introduced by means of $\int \varphi x dx$, the value corresponding to the same x can be more than one. (Gauss [WW, 10:366])

Gauss expressed the ancient Leibnizian definition of the integral in an abstract form[441] and assumed that the relation

$$x \rightarrow \int_a^x \varphi(x)dx$$

led to a new function provided $\varphi(x) \neq \infty$ along the path of integration. Similarly, Equation (217) provided a relationship between certain quantities x and P except for the points 0 and 1, where the coefficients were infinite and therefore led to a new function.

Both definitions of hypergeometric functions highlight other novelties in Gauss's conception concerning the principle of the generality of algebra. For instance, in *Disquisitiones generales*, Gauss explicitly verified that hypergeometric functions were also defined for complex values of variables. In this way, Gauss revealed his awareness of the difference between complex and real variables and rejected the automatic extension of formulas valid for real quantities to imaginary quantities.[442] This was possible because his approach to complex numbers differed from that of 18th-century mathematicians. To use Gauss's words, hitherto imaginary numbers were conceived of

---

[441]In a manuscript entitled *Bestimmung der convergenz der reihen, in welche die periodischen functionen einer veränderlichen grösse entwickelt werden* [WC, 407–419], Gauss stated that the integral $\int f(t)dt$ could be defined in two different senses. In a first sense, the integral could be considered to be a function of t such that $f(t)dt$ resulted from its differentiation. In the second sense, a definite integral from t_0 to $T = t_n$ was defined to be the limit of the sums $\sum_{i=1}^{n}(t_i - t_{i-1})f(t_{i-1})$ as $(t_i - t_{i-1}) \rightarrow 0$. This paper was written after 1831 and might have been influenced by Cauchy. In 1811, Gauss seems to be close to the idea of integral as the sum of infinitely many infinitesimals.

[442]See Legendre's definition of gamma function, Chapter 29.

as "excrescences" (*Überbein*) of real quantities (see Gauss [WW, 10:366]). In [1831], he stated:

> The imaginary quantities — which are contrasted with the real ones, and which were formerly, and are still occasionally (although improperly) called *impossible* — are still tolerated rather than fully accepted, and therefore appear more like a game with symbols, in itself empty of content, to which one unconditionally denies a thinkable substrate — without, however, wishing to scorn the rewards which this game with symbols achieved for our understanding of the relationships of the real quantities. (Gauss [1831, 175])

Gauss thought that complex numbers were to be regarded as just as possible as real quantities and that they had the same legitimacy as real numbers.[443] He justified his view by stating:

> One does not discuss practical aims, since, in my opinion, analysis is an autonomous science; by leaving out imaginary magnitudes, it would lose its beauty and its truths, which are valid in general, and would be strongly bounded. (Gauss [WW, 10:366])

According to Gauss, complex numbers have no practical scope. The fact that he used them in scrutinizing hypergeometric functions may appear to contradict the above observation that remarkable stimuli for the introduction of new functions derived from mathematical applications. However, while practical incentives for pure mathematics are one matter, pure mathematics is another question. Gauss required analysis to be general and beautiful, where generality and beauty seem to indicate the capability of developing a deductive theory that derived from a few clear principles, had an internal coherence and harmony,[444] and was linked to the real by means of the intuitiveness of its principles. Complex numbers allowed a higher degree of generality and beauty; for this reason, they were to be retained in analysis — independently of the fact that they were used in applications or not — provided they could be established on a sufficiently solid basis (see p. 32).

*   *   *

The introduction of the factorial and digamma functions in Section 3 of *Disquisitiones generales* and Gauss's criticisms of their previous "definitions"

---

[443]See Gauss [1799, 6], [1831, 171], and [WW, 10:366].

[444]This idea of analysis was widely shared at the turn of the 18th century, in particular in the cultural environment of Germany (see Jahnke [1993]).

offers further remarkable points of interest for understanding his approach. In [1812b, 144–145], Gauss set

$$\Pi(k, z) = \frac{1 \cdot 2 \cdot 3 \cdot \ldots \cdot k}{(z+1)(z+2)(z+3)\ldots(z+k)} k^z, \tag{218}$$

where k is a positive integer, and proved that the limit of (218) for $k \to \infty$ exists. Indeed, given a number $h > |z|$, he obtained

$$\log \Pi(h+1, z) - \log \Pi(h, z) = \sum_{j=1}^{\infty} \frac{z(1 + (-1)^{j+1} z^j)}{j+1} \sum_{i=1}^{n} \frac{1}{(h+1)^{j+1}}$$

Since this series is finite as n goes *ad infinitum*, $\lim_{k \to \infty} \Pi(k, z)$ is finite[445] and exhibits a determinate function of z, provided z differs from a negative integer number [1812b, 145]. At this point Gauss defined the factorial function[446] by setting

$$\Pi(z) = \lim_{k \to \infty} \Pi(k, z)$$

or

$$\Pi(z) = \prod_{k=1}^{\infty} \frac{(k+1)^{z+1}}{k^z(z+k)}. \tag{219}$$

Gauss then noted that, since

$$\Pi(k, z+1) = \Pi(k, z) \frac{1+z}{1 + \frac{1+z}{k}},$$

one had

$$\Pi(z+1) = \Pi(z+1)\Pi(z).$$

Hence, $\Pi(z+n) = (z+1)(z+2)(z+n)\Pi(z)$ and

$$\Pi(z) = 1 \cdot 2 \cdot \ldots \cdot z$$

(see Gauss [1812b, 145–146]). He stated:

> [T]his property of our function has badly been passed off as its definition, since it is limited to integer values by its own nature and, beside our function, is common to an infinity of others (e.g., $\Pi(z) \cos 2\pi z, (z) \cos^{2n} \pi z, \ldots$).[447] (Gauss [1812b, 146])

---

[445] Gauss used the symbol $\Pi(\infty, z)$.

[446] See footnote no. 223.

[447] Similarly, Gauss proved that the limit of $\log k - \sum_{j=1}^{k} \frac{1}{z+j}$ existed and defined the digamma function as $\Psi(z) = \lim_{k \to \infty} \left(\log k - \sum_{j=1}^{k} \frac{1}{z+j} \right)$ (see [1812b, 153–154]). In this chapter, by "digamma function", I mean the logarithmic derivative of the factorial function (see footnote no. 347). Gauss denoted this function by the sign $\Psi(z)$ but gave no name to it.

Gauss's criticism was directed toward the definition of the factorial function as a quantity that derived from the process of interpolation of the discrete sequence[448] $n!$. Gauss explicitly referred to Eulerian inexplicable functions in [1812a, 200–201] and in his correspondence.[449] He dealt diplomatically with Euler's work. However, he was more explicit in his criticisms of Kramp, who followed Euler's approach in his *Analyse des réfractions astronomiques et terrestres* [1799]. For instance, in [WW, 10:362], Gauss stated that if one did not wish to expose oneself to endless paralogism, paradoxes, and contradictions, as Kramp had done, then $1 \cdot 2 \cdot \ldots \cdot z$ could not be used as the definition of $\Pi(z)$.

It is worthwhile noting that, in the second part of the 18th century, the objection that the function interpolating $n!$ was not unique was already known. Indeed, in his *De serierum determinatione seu nova methodus inveniendi terminos generales serierum*, Euler already observed that if a sequence a_n were interpolated by a function $f(n)$, it may also be interpolated by

$$y = f(n) + g(n)\sin \pi n$$

(Euler [1750–51b, 465–467]), where $g(n)$ is any function. In *De curva hypergeometrica hac aequatione expressa* $y = 1 \cdot 2 \cdot 3 \cdot \ldots \cdot x$, Euler explicitly stated that numberless curves interpolated the points $(n, n!)$, for $n = 0, 1, 2, \ldots$, but the nature of the equation (*ratio equationis*) $y(n) = n!$ required

$$y(x + 1) = (x + 1)y(x)$$

or

$$y(x - 1) = \frac{y(x)}{x},$$

and therefore the curve had to satisfy this condition. In effect, Euler rethought the problem of Wallis's interpolation in a form that can be expressed as follows:[450]

(INT) *If a sequence a_n is given in a recurring form*

$$a_n = F(a_{n-1}, a_{n-2}, \ldots, a_{n-r}),$$

for $n > r$, find the function $y(x)$ such that

$$y(x) = F(y(x - 1), y(x - 2), \ldots, y(x - r)).$$

---

[448]See Chapters 1, 4, and 13.
[449]See the letters to Olbers on October 17, 1811, and to Bessel on November 11, 1811, in Gauss [WW, 10: 361–363].
[450]See Euler [1750–51b].

It might seem that Gauss's criticism cannot be applied to this way of approaching the question. However, there is another sense in which this criticism can be interpreted.

According to Euler, the interpolating function derived from "the nature" of the sequence $n!$. He made this assertion because a sequence a_n was not viewed during the 18th century as an arbitrary relation between the set of natural numbers and the set of real numbers but as the result of some fixed analytical operations upon natural numbers;[451] in other words, the numbers a_n were regarded as being generated by an analytical formula $f(n) = a_n$. Consequently, "finding the interpolating function $f(n)$ of a_n" meant "finding the function $f(n)$ that generates a_n." For instance, in his *De termino generali serierum hypergeometricarum*, Euler derived

$$\Delta(n) = \frac{1 \cdot 2 \cdot \ldots \cdot i}{(n+1)(n+2)\ldots(n+i)}(i+\alpha)^n, \qquad (220)$$

where i is an infinite number and α is an arbitrary number (Euler [1789, 141–146]).

Formula (220) is *de facto* the relation Gauss used to define the factorial function; however, for Euler, it was not a definition of the quantity interpolating $1 \cdot 2 \cdot \ldots \cdot n$, but only a property of this quantity.

To give a firmer foundation to this procedure, in his [1768], [1789], and [1819], Euler illustrated the problem of interpolation by referring to curved lines. In *De termino generali* [1789], for instance, he stated that the points $(n, n!)$ could be considered as points belonging to a curved line:

> Clearly, there is no doubt that determinate ordinates correspond to abscissas expressed by means of fractions or surd [irrational] numbers as well. (Euler [1789, 140])

However, he noted that the equation $y = 1 \cdot 2 \cdot \ldots \cdot n$ could not be used to determine these ordinates but rather that it was necessary to have an appropriate expression of them. Apart from any considerations about the possibility of a geometrical construction of the curve, it should be observed that an explicitly geometrical foundation was not satisfactory for mathematicians such as Euler who wanted to separate analysis from geometry.

Gauss viewed the matter differently. Functions were our own creations and their definition was a question of suitability. No function $f(x)$ was the extension of a given sequence $a_n = f(n)$ as a result of its own nature (see Gauss [WW, 10:363]). Thus, the definition of the factorial function implied a choice between the numberless functions satisfying the condition $f(n) = 1 \cdot 2 \cdot 3 \cdot \ldots \cdot n$. This choice originated from our requirement that the factorial function satisfied certain constraints such as $f(n+1) = (n+1)f(n)$,

---

[451] See Section 13.1.

although this did not mean that the function derived from the "nature" of the given sequence.

Gauss did not consider a sequence to be generated by an analytical expression. He explicitly discussed the notion of a sequence in *Grundbegriffe der Lehre von den Reihen*, where he observed that, in a very general sense, the totality of any number of arbitrary quantities might be referred to as a sequence (Gauss [WC, 390]). According to Gauss, this definition was of little use and he therefore gave a less general definition:

(D1) A sequence is the totality of quantities which are determined by a *law* that established the first, the second, ... (Gauss [WC, 390]).

Gauss compared (D1) with the common definition:

(D2) A sequence is the totality of the values of a function of a variable, when one makes this variable equal to 1, 2, 3, ... (Gauss [WC, 390]).

In this case Gauss used the term "function" to refer to 18th-century functions. He argued that (D1) was more general than (D2), unless the term "function" was understood in a broader sense than previously was the case. Gauss noted that, according to the common definition, the sequence of prime numbers 1, 2, 3, 5, 7, 11, ... and each sequence deriving from it (such as 1, $\frac{1}{4}$, $\frac{1}{9}$, $\frac{1}{25}$, $\frac{1}{49}$, $\frac{1}{121}$, ...) were not to be considered as sequences since they were not obtained from analytical expressions. He thought that this was not acceptable and therefore that definition (D1) was a better one. To differentiate the traditional and more restricted sense of the term from his wider meaning, Gauss proposed the name "analytical sequences" to those sequences whose general term was an analytical function[452] of the index (see Gauss [WC, 390–391]).

$$* \quad * \quad *$$

In the 18th century, infinitely large and infinitesimal quantities were thought to be entities that could be manipulated in the same manner as numbers except for the principle of cancellation. Instead, according to Gauss,

> a lot of circumspection was required in treating infinite quantities. In our opinion, these are to be allowed in analytical reasoning in so far as they can be reduced to the theory of limits. (Gauss [1812b, 159])[453]

---

[452]On the meaning of term "analytical function", see footnote no. 340.

[453]More explicitly, in a letter to Schumacher, Gauss stated: "I protest in the first place against the use of infinite magnitude as something completed, which is never allowed in mathematics. The infinite is but a *façon de parler* in that one actually speaks of limits to which certain relations come as close as one desires, while others are allowed to increase without bound" (Gauss [WW, 8:216] translated in Ewald [1996, 303]).

In *Disquisitiones generales*, these words annotated a paradoxical situation concerning the integral $\int_0^1 \frac{1-x^\lambda}{1-x}dx$. By the substitution $x = y^k$, Gauss obtained

$$\int_0^1 \frac{1-x^\lambda}{1-x}dx = \int_0^1 \frac{ky^{k-1} - ky^{\lambda k+k-1}}{1-y^\lambda}dy.$$

Hence, by changing x into y, he had

$$\int_0^1 \frac{1-x^\lambda}{1-x}dx = \int_0^1 \frac{kx^{k-1} - kx^{\lambda k+k-1}}{1-x^\lambda}dx. \tag{221}$$

Gauss observed that equality (221) is legitimate only if $\int_0^1 \frac{1-x^\lambda}{1-x}dx$ is a finite quantity (namely, if $\lambda + 1 > 0$). Indeed, the difference between the two integrals is infinitely large for $\lambda + 1 \leq 0$ and, therefore, Equation (221) is paradoxical if one considers it without constraints.

Gauss's notion of limit is similar to the modern notion in many aspects but also presents crucial differences. To make this clear, let us consider the following proposition.

> **Proposition (A).** *Given a sequence M_k, $k = 0, 1, 2, \ldots$ such that the ratios $\frac{M_{k+1}}{M_k}$ are of the type*
>
> $$\frac{M_{k+1}}{M_k} = \frac{P_\lambda(m+k)}{p_\lambda(m+k)}, \qquad k = 0, 1, 2, ..,$$
>
> *where*
>
> *m is an positive integer,*
>
> $P_\lambda(t) = t^\lambda + A_1 t^{\lambda-1} + A_2 t^{\lambda-2} + \ldots + A_\lambda,$
>
> $p_\lambda(t) = t^\lambda + a_1 t^{\lambda-1} + a_2 t^{\lambda-2} + \ldots + a_\lambda,$
>
> *then the series $\sum M_k$ converges if $A_1 - a_1 < -1$ (see Gauss [1812b, 139–143]).*

According to Gauss, this proposition enables one to investigate convergence of the hypergeometric series at $x = 1$ "with all rigor, for the benefit of those who are inclined to rigorous methods of the ancient geometers" [1812b, 139]. In proving Proposition (A), Gauss [1812b, 139–143] employed inequalities in a clear and precise way. In the first step of the proof, Gauss demonstrated that one has $|M_{k+1}| > |M_k|$ or $|M_{k+1}| < |M_k|$ for large enough k. He used the following unproved assumption, which contains an implicit definition of $\lim_{t\to\infty} f(t) = \infty$.

> **Lemma 1.** *Given a number a, if $F(t) \to \infty$ as $t \to \infty$, it is possible to determine a number L such that $F(t) > a$ for $t > L$.*

Indeed, Gauss observed that if $P_\lambda(t)$ and $p_\lambda(t)$ are not identical, then one of the differences $A_k - a_k$, $k = 1, 2, 3, \ldots$, is not equal to 0. If one supposes $A_k - a_k = 0$ for $k = 1, 2, 3, \ldots, r - 1$ and $A_r - a_r > 0$, then one can find a bound (*limes*) L such that

$$P_\lambda(t) > p_\lambda(t) > 0$$

for $t > L$. It is sufficient to consider L equal to the greatest real root of the equation

$$p_\lambda(t)\,(P_\lambda(t) - p_\lambda(t)) = 0.$$

If this root does not exist, then $p_\lambda(t)\,(P_\lambda(t) - p_\lambda(t))$ and $p_\lambda(t)$ are positive for every value of t. Hence,

$$\frac{M_{k+1}}{M_k} > 1,$$

M_k always assumes the same sign, and $|M_{k+1}| > |M_k|$. Similarly, if $A_k - a_k = 0$ for $k = 1, 2, 3, \ldots, r - 1$ and $A_r - a_r < 0$, then $|M_{k+1}| < |M_k|$.

In the second step of the proof, Gauss showed that

$$|M_k| \to \infty \qquad \text{for } A_1 - a_1 > 0$$

and

$$|M_k| \to 0 \qquad \text{for } A_1 - a_1 < 0$$

(see [1812b, 140]). In particular, he used the implicit lemma:

> **Lemma 2.** *Given the sequences $\alpha_k \to \infty$ and $\beta_k > 0$ and the number $\Theta > 0$, if $\beta_k > \Theta$, for large enough k, then $\alpha_k \beta_k \to \infty$; if $0 < \alpha_k \beta_k < \Theta$, for large enough k, then $\beta_k \to 0$.*

In third step, Gauss demonstrated that if $A_1 - a_1 = 0$, then $|M_k|$ goes to a nonzero finite limit, and he implicitly used the following unproved assumption:

> **Lemma 3.** *Given an increasing sequence λ_k, if this sequence has an upper bound, then there exists a real number Λ that is the limit of λ_k for $k \to \infty$.*

He indeed supposed, for instance, that the first difference $A_i - a_i$ different from zero is $A_2 - a_2$ and that $A_2 - a_2$ is greater than 0. He considered a positive integer h such that $h + a_2 - A_2 > 0$ and posed

$$M_k \left(\frac{m + k}{m + k - 1} \right)^h = N_k, \qquad k = 0, 1, 2, \ldots.$$

$$
\begin{aligned}
Q &= (t^2 - 1)^h P_\lambda(t) = t^{\lambda + 2h} + A_1 t^{\lambda + 2h - 1} + (A_2 - h)t^{\lambda + 2h - 2} + \ldots \\
q &= t^{2h} p_\lambda(t) = t^{\lambda + 2h} + a_1 t^{\lambda + 2h - 1} + a_2 t^{\lambda + 2h - 2} + \ldots
\end{aligned}
$$

By applying the first step (it is $A_2 - a_2 - h < 0$), he had that the series $|N_k|$ becomes decreasing. Since $|M_k|$ increases ($A_2 - a_2 > 0$) and $|M_k| < |N_k|$, he concluded that $|M_k|$ is convergent to a finite limit.

In the successive steps, Gauss completed the proof by using the comparison test.

This proof is exemplary of Gauss's conception for two reasons. First, he rejected infinite quantities as something complete and replaced them with the inequality technique. As we saw previously, the inequality technique was not new, but d'Alembert and Lagrange considered it only in relation to approximations. Instead, Gauss used them to prove the existence of a limit, and this represented a crucial breakthrough. This does not mean that the importance of 18th-century studies on inequalities should be underestimated. They were part of Gauss's background, and a certain knowledge of them might have helped Gauss in developing his approach by providing an important technical tool. Nevertheless, Gauss's approach differed conceptually from d'Alembert and Lagrange's. It is impossible to reduce the former to the latter.

Second, Gauss's idea of the continuum differed from the modern notion. Consider Lemma 3. Gauss did not provide the proof of the existence of the number Λ (the supremum of the series). A theorem concerning the existence of the supremum is found in his *Grundbegriffe der lehre von den Reihen*, but it is based upon reasoning that does not provide a solution to the question. Indeed, Gauss considered a series a_n such that $a_n < \lambda$, for a given number λ. He defined this number to be the upper bound of a sequence (in Gauss's terms: *limes supra seriem*, or *obere Grenze* or *une limite en plus*). He stated that if λ is an upper bound, then any number greater than λ is also an upper bound. Moreover, given a number κ less than a fixed upper bound λ, if κ was not an upper bound and was allowed to assume the value λ by moving through all the intermediate quantities in a continuous way, it necessarily reached the least upper bound L (the supremum of the series). Gauss noted that no term of the series could be greater than L, but, when n was sufficiently large, there are terms that exceed any number $\xi < L$ (Gauss [WC, 391]).

Gauss characterized the supremum adeptly; nevertheless, an adequate construction of real numbers was lacking. (Note that the requirement of a proof of Lemma 3 is precisely the starting point of Dedekind's theory of real numbers in his *Stetigkeit und irrationale Zahlen* [1872].) Gauss did not feel the need to construct the set of real numbers, as his concept of the continuum was still linked to a revised notion of continuous quantity, which was a legacy of the 18th century.

*   *   *

In the previous chapters, it was observed that from about the 1740s, mathematicians based analysis upon the notion of a general (continuous) quantity, which was merely the result of a process of abstraction practized upon geometrical quantities (which, in turn, were abstracted from physical quantities). There was a close link between the abstract notion of general quantity and concrete and specific geometrical quantities. Thus, even if analysis was conceived to be independent of geometry, continuous quantities were substantially intuited as segments of a straight line or a piece of a curved line (in Kantian terms, they derived from empirical intuition).

Instead, Gauss's analysis no longer dealt with a notion of quantity drawn from geometrical sources but was based on the intuitively given numerical continuum:[454] continuous quantity was reduced to the mere intuition of a sequence of numbers, which flows from one value to another assuming all intermediate values, without jumps. In Gauss's opinion, the numerical continuum was the basic and intuitive object of analysis and was connected with any sort of *a priori* intuition (namely, with a form of knowledge that does not appeal to any particular experience). His conception was based upon modified Kantian notions.[455] Kant believed that space and time were *a priori* intuitions from which the certainty of geometry and arithmetic derived. It is well known that Gauss did not accept the whole of Kant's theory of mathematical knowledge and, in particular, the idea that space is an *a priori* intuition (he wrote to Bessel on January 27, 1829: "We cannot establish geometry entirely *a priori*" [WW, 8:200]). However, he agreed with some crucial points of Kantian philosophy: the importance of intuition, the existence of an *a priori* intuition that guarantees the certainty of mathematical knowledge, the fact that mathematical intuition concerns abstract objects (universal objects) but can only be disclosed by empirical objects (universal *in concreto*).

Thus, in his review of J.C. Schwab's *Commentatio in primum elementorum Euclidis librum,* Gauss stressed the importance of the intuition:

> A great part of the text [of Schwab] turns on the contention against Kant that the certainty of geometry is not based on intuition but on definitions and on the *principium identitatis* and the *principium contradictionis.* Kant certainly did not wish to deny that use is constantly made in geometry of these logical aids to the presentation and linking of truths: but anybody who is acquainted with the essence of geometry knows that they are able to accomplish nothing by themselves, and they put forth only sterile blossoms unless the fertilizing living intuition of the

---

[454]Here, I use the term "intuition" in the sense that Gauss gave to it in *Zur Metaphysik der Mathematik*: "We can have an idea of quantity in two ways, either by immediate intuition (an immediate idea), or by comparison with other quantities given by immediate intuition (mediate idea)" (Gauss [WD, 57] translated in Ewald [1996, 294]).

[455]Gauss wrote little about this question, although he displayed knowledge of Kant's mathematical concept, which he criticized only partially.

object itself prevails everywhere. (Gauss [1816, 172] translated
in Ewald [1996, 299–300])

Intuition as an immediate representation of an object seems to be what
actually allows one to produce and understand mathematics. According to
Gauss, this intuition is, at least partially, *a priori*. On this point he viewed
the matter differently from Kant. On September 4, 1830, he wrote:

> It is my deepest conviction that the theory of space has a
> completely different position in our *a priori* knowledge than does
> the pure theory of quantity. Our knowledge of the former utterly
> lacks the complete conviction of necessity (and also of absolute
> truth) that belongs to the latter, for number is merely the prod-
> uct of our mind. (Gauss [WW, 8:201] translated in Ewald [1996,
> 299–302])

It is probable that, following Kant,[456] Gauss thought that the numerical
continuum originated from a more basic intuition: the *a priori* intuition of
time. In *Zur Metaphysik der Mathematik*, Gauss stated that quantities are
time, number,[457] and geometrical quantities (Gauss [WD, 57]). He thought
that a quantity could be represented by numbers by measuring it with regard to a fixed quantity, which was considered to be the unit. Gauss did
not clarify what he meant by the measure of an incommensurable quantity
and irrational numbers (his *Zur Metaphysik der Mathematik* stops after a
few pages). However, he seems to have considered irrational numbers as
determinations of a continuum that can be approximated by rationals and
individuated by means of an endless number of steps. Thus, time is crucial
for avoiding geometrical quantity and for giving the impression of analysis as
a mere creation of our mind. Nevertheless, in mathematical practice, time
can be easily replaced by its numerical representation, and Gauss directly
operated upon a continuum viewed as an unbroken flow of numbers.

By describing the continuum as an unbroken flow of numbers, I do not
offer a definition but merely a circumlocution to suggest Gauss's idea of the
continuum: Gauss's continuum cannot be really defined but only intuited.
It is worthwhile noting that this notion definitively enables a continuous
quantity to be separated from a geometrical substratum; continuity, which
earlier was referred to as an empirical intuition (by means of geometry), is
now reduced to the realm of numbers, even though by means of the *a priori*
intuition of time.

This does not mean that the numerical continuum is reduced to the
discrete. There is no construction of the continuum: It is immediately given

---

[456]Kant had provided a theoretical basis for the numerical continuum by means of the
notion of time as a pure intuition.

[457]Here number means whole number or, at most, rational number.

(with its ordinate structure and the property of completeness). Numbers do not generate the continuum; rather, the numerical continuum is given and one can cut it and determine any single, specific number (see Bussotti [2000]).

Apart from Kant's epistemological influence, Gauss's concept of the continuum is similar to the notion of the continuum that was widely held in the 17th and 18th centuries.[458] For this reason, the main consequences of the 17th and 18th centuries concept of the continuum also held for Gauss's numerical continuum, which, in principle, embodies the flow of time. For instance, an interval was always closed, pointwise definitions of function were impossible, and functions were considered on closed intervals.

At this juncture, a remark on Gauss's concept of analysis is appropriate. Gauss thought that analysis regarded arithmetic relations between quantities. He intended arithmetic relations in a broad sense, in contrast to geometrical relations. The latter concerned geometrical quantities and considered them with respect to location; the former considered quantities "only insofar as they are quantities". In Gauss's opinion, analysis was grounded upon some quantities of which we have an immediate —intuitive— idea. Natural numbers $1, 2, 3, \ldots$ and continuous quantity (the continuum) seem to be the only immediate quantities. However, Gauss employed other simple intuitive principles and notions (to be part of, greater than, relationship, to be right or left, ...). Basic principles and notions of analysis were not provided by axioms in the modern sense (arbitrary propositions) but were provided by intuition.[459] More precisely, with regard to analysis, they were provided by *a priori* intuition. This is a sophisticated, Kantian way of saying that mathematics expressed reality, where reality is taken as the mental interpretation of sensorial perception.[460]

According to Gauss, other quantities must be reduced to immediate quantities:

> The duty of the mathematician is accordingly either actually to represent the sought-for quantity (geometric representation or construction), or to indicate the way and manner in which, from the idea of an immediately given quantity, one can achieve the idea of the sought quantity (arithmetical representation). (Gauss [WD, 57–58])

Thus, one can introduce new quantities, but these are not arbitrary creations. They are connected with original intuition and must have a model (a sensible representation that allows one to understand them intuitively). An example is the introduction of complex numbers.

---

[458]See Section 7.2, p. 102.

[459]Analysis has a different status from geometry (see [WW, 8:201]).

[460]See also Gray [1992, 229].

In his [1831], Gauss assumed that certain objects A, B, C, D, ..., were ordered into a sequence and that this was unbounded on both sides. He stated that if the relation (or the transition) of A to B counted as $+1$, then the relation of B to A had to be represented by -1. So every real integer represented the relation of an arbitrary member chosen as the origin and a determinate member of the sequence. Then he supposed the objects were of such a sort that they could not be ordered in a sequence but only in sequences of sequences. In this case, for the measurement of the transition from one term of the system to another one needed besides the previous units $+1$ and -1 two others, also inverse to each other, $+i$ and $-i$ (Gauss [1831, 176]). Complex numbers were viewed as abstract relations among abstract objects. However, these relationships could "be brought to intuition only by a representation in space" [1831, 176–177] and, for this reason, Gauss constructed the model of complex number in the plane today associated with this name:

> Here, therefore, an intuitive meaning of $\sqrt{-1}$ is completely established, and one needs no other to admit these quantities in the domain of arithmetic (Gauss [1831, 177] translated in Ewald [1996, 313]).

In such a way new quantities were not arbitrary creations but were connected with original intuition and had a model (a concrete representation). In his [1799, 6], Gauss distinguished between imaginary quantities, which were to be accepted, and impossible quantities, which were to be rejected. He compared these impossible quantities with a right-angle equilateral triangle[461] and defined them as those quantities that satisfied conditions that could not be satisfied even by imaginary quantities. He realized that his argument about impossible quantities could be employed for imaginary quantities and proposed to vindicate them elsewhere. This vindication was published in [1831] and [1832] and consisted of providing the concrete model mentioned above. The existence of a model seemed to guarantee that, unlike the right-angle equilateral triangle, certain objects were possible.

$$* \quad * \quad *$$

Gauss considered functions as defined over a segment of the real line or over a region of the whole complex plane (and, if necessary, over the whole real line or the whole complex plane) except for a certain (finite or infinite) number of isolated singularities. However, as the continuum was not reduced to a set of numbers, Gauss's functions were not conceived of as relations between elements of two sets, namely, they were not pointwise-defined

---

[461] It is clear that Gauss did not consider the possibility of a geometry where such triangles exist.

relations. In Gauss's writings, a function[462] $y = f(z)$ was a relation between the variable quantity z and the variable quantity y; both variables were intended as continuous flows; of course, if a value of the variable quantity z was determinate, then there was (at least) a value of the corresponding quantity y.

This concept is particularly evident in reference to the continuity of functions. In *Disquisitiones generales*, Gauss stated:

> Continuity of the function Πz breaks, every time that the value is infinitely large, i.e. for negative integer value of z. (Gauss [1812b, 147])

The flow that the variable y describes can be broken for certain (exceptional) values of z; thus, $y = (z)$ is discontinuous when $z = -1, -2, -3, \dots$. A point can break continuity though continuity is not a property of a point: Any reasoning implying continuity did not concern a single point, but a part of the continuum. What is important for Gauss is that the independent variable z could move from a point A of the domain of definition to another B (including the endpoints) in a continuous way and that the corresponding variation of the dependent variable y is continuous as well.[463]

One could argue that this idea is unclear, in particular the relationship between the single number and the flow of numbers. This is true, and I think that this is probably linked to some problematic derivations of *Disquisitiones generales*. For instance, Gauss established that

$$F(\alpha, \beta, \gamma, 1) = \frac{(\gamma - \alpha)_k (\gamma - \beta)_k}{(\gamma)_k (\gamma - \alpha - \beta)_k} F(\alpha, \beta, \gamma + k, 1), \tag{222}$$

where $(x)_n = x(x+1)(x+2)(x+n-1)$. Applying (222) repeatedly, he had

$$F(\alpha, \beta, \gamma, 1) = \frac{\Pi(k, \gamma - 1) \cdot \Pi(k, \gamma - \alpha - \beta - 1)}{\Pi(k, \gamma - \alpha - 1) \cdot \Pi(k, \gamma - \beta - 1)} F(\alpha, \beta, \gamma + k, 1).$$

As $F(\alpha, \beta, \gamma + k, 1)$ went to 1 as k tended to infinity, he derived

$$F(\alpha, \beta, \gamma, 1) = \frac{\Pi(\gamma - 1) \cdot \Pi(\gamma - \alpha - \beta - 1)}{\Pi(\gamma - \alpha - 1) \cdot \Pi(\gamma - \beta - 1)}.$$

---

[462]In his arithmetical works, Gauss introduced some arithmetical functions, although he avoided using the term "function" with reference to them. For instance, in [1801, 30–31], he designated the number of positive numbers which are relatively prime to the given number A and smaller than it by the "prefix" or the "symbol" ϕ, but did not give the name function to ϕA (see also [1801, 43]). In [1801] he spoke of functions if the variables included in certain expressions could be thought of as continuous. In the present paper, I always refer to functions of a continuous variable.

[463]See also Ferraro [2007b, Section 14].

Gauss stated that it was evident that $F(\alpha, \beta, \gamma + k, 1)$ went to 1 as $k \to \infty$ [1812b, 147]. Since

$$F(\alpha, \beta, \gamma + k, 1) = \sum_{i=1}^{\infty} \frac{(\alpha)_k (\beta)_k}{n!(\gamma + k)_k},$$

he interchanged the operation of limit; however, he did not justify this interchange, as if it was immediate and obvious.

Similarly, on p. 151 of his *Disquisitiones generales*, Gauss considered the equality

$$\int_0^{\nu} y^{\lambda-1}(1 - \frac{y}{\nu})^{\nu} dx = \frac{\Pi(\nu, \lambda)}{\lambda}.$$

For $\nu \to \infty$ he obtained

$$\int_0^{\infty} y^{\lambda-1}(1 - e^{-y}) dx = \Pi(\lambda - 1)$$

(see Gauss [1812b, 151]).

Gauss performed some operations that are legitimate in these specific cases, but their legitimacy was not discussed. He behaved as if certain theorems on the interchange of limits, integrals, etc. were evident. Of course, in contrast to 18th-century mathematicians, Gauss rejected the infinite extension of finite rules.[464] It is likely that Gauss viewed the matter in a manner that resembled Cauchy's approach;[465] however, we lack the evidence for a more precise hypothesis.[466] In any case, Gauss's concept of the continuum made it impossible to distinguish among continuity and uniform continuity, convergence and uniform convergence, which are necessary for a better formulation of the theorems to which I referred above.

Although Gauss had a fairly general concept of a function, he was interested in "special" functions that had "particular" characteristics and that played a special role in mathematics, such as $\Pi(z)$. This special role is due to the utility and simplicity of certain functions. In [1812b, 146–147], Gauss noted that there were other functions similar to $\Pi(z)$, such as the function of the variables a, b, c that Kramp denoted by $a^{b|c}$ and that was equal to

$$\frac{c^b \Pi(\frac{a}{c} + b - 1)}{\Pi(\frac{a}{c} - 1)};$$

---

[464]See p. 117.
[465]See Chapter 33.
[466]Gauss possessed sophisticated knowledge of complex analysis —see, e.g., [WW, 10:368]— though he did not expound it in a systematic way. It is therefore difficult to state which theorems he was aware of.

however, $\Pi(z)$ was preferable since a function of one variable was easier than a function of three variables. In the letter to Bessel on December 18, 1811, Gauss asserted that the function

$$\int_0^x \frac{e^x - 1}{x} dx$$

was simpler and could give more incisive results than

$$\int_0^x \frac{1}{\log x} dx$$

because the former was a single-valued function (see Gauss [WW, 10:368]).

Gauss seems to think that a useful function is linked to any form of analytical representation: in his work, a function is given by a series, by the limit of a sequence [for example, $\Pi(z)$ and $\Psi(z)$], by an integral,[467] by a differential equation or by the inversion of another function.[468] However, Gauss is aware that a function can have different representations that are valid in different domains. Thus, he notes that the digamma function $\Psi(\lambda) = \lim_{k \to \infty} \left(\log k - \sum_{j=1}^k \frac{1}{\lambda + j} \right)$ has the integral representation

$$\Psi(\lambda) = \int_0^1 \left(-\frac{1}{\log z} - \frac{z^\lambda}{1 - z} \right) dz \qquad (223)$$

and that (223) is valid if $\lambda + 1 > 0$, whereas the function $\Psi(\lambda)$ is defined for every λ but $-1, -2, -3, \ldots$.

In *Determinatio seriei nostrae*, Gauss explicitly discussed a seeming paradox connected with different representations of $F(\alpha, \beta, \gamma, x)$.[469] Setting

$$\gamma = \alpha + \beta + \frac{1}{2} \quad \text{and} \quad x = 4y - 4y^2,$$

Gauss transformed Equation (217) into

$$4\alpha\beta P - \left(\alpha + \beta + \frac{1}{2} - (2\alpha + 2\beta + 1) \right) \frac{dP}{dy} - (y - y^2) \frac{d^2 P}{dy^2} = 0.$$

This equation has the particular solution $F(2\alpha, 2\beta, \alpha + \beta + \frac{1}{2}, y)$. Hence,

$$F(\alpha, \beta, \alpha + \beta + \frac{1}{2}, 4y - 4y^2) = F(2\alpha, 2\beta, \alpha + \beta + \frac{1}{2}, y).$$

---

[467] He stated the best definition of the factorial function is $\Pi(m) = \int_{-\infty}^{+\infty} e^{(m+1)x} e^{e^x} dx$ (see Gauss [WW, 3:230]).

[468] It is the case of the sinus lemniscaticus, which was defined as the inverse function of lemniscatic integral $\int_0^x \frac{1}{\sqrt{1-t^4}} dt$.

[469] On this paradox, see Gray [1986, 13–14].

By changing y by $1 - y$, he had

$$F(2\alpha, 2\beta, \alpha + \beta + \frac{1}{2}, y) = F(2\alpha, 2\beta, \alpha + \beta + \frac{1}{2}, 1 - y),$$

which is certainly false[470] (Gauss [WA, 225–226]). Gauss explained that one had to distinguish between two meanings of the symbol $F(\alpha, \beta, \gamma, x)$, namely between $F(\alpha, \beta, \gamma, x)$ as the sum of infinite series (215) and $F(\alpha, \beta, \gamma, x)$ as the solution to differential equation (217). If one considers $F(\alpha, \beta, \gamma, x)$ as the sum of a series, it has a meaning only in the circle of convergence with radius 1 and center $(0, 0)$ and $F(\alpha, \beta, \alpha + \beta + \frac{1}{2}, 4y - 4y^2)$ only in the circle of convergence with radius

$$\frac{1}{2} \sqrt{\frac{1}{2}}$$

and center

$$\left(\frac{1}{2} - \frac{1}{2} \sqrt{\frac{1}{2}}, \ 0 \right)$$

(Gauss [WA, 226–227]).

∗   ∗   ∗

In *Disquisitiones generales*, Gauss tried to offer a reformulation of 18th-century results (and, especially, Euler's results) about gamma and digamma functions, accordingly his new methodology. I refer to Ferraro [2007b, Section 13] for an account. Here I limit myself to pointing out that Gauss did not always meet with success in rederiving 18th-century results. Indeed, in [1812b, 152 and 154], he discussed the Euler–Maclaurin expansions of $\log \Pi(z)$ and $\Psi(z)$:

$$\log \Pi(z) \ = \ \left(z + \frac{1}{2} \right) \log z - z + \frac{1}{2} \log 2\pi \qquad (224)$$

$$+ \sum_{n=1}^{\infty} (-1)^{n+1} \frac{B_{2n}}{2n(2n-1)z^{2n-1}}$$

$$\Psi(z) = \log z + \frac{1}{2z} + \sum_{n=1}^{\infty} (-1)^n \frac{B_{2n}}{2nz^{2n-1}}, \qquad (225)$$

where B_n are the Bernoulli numbers. He stated that the first terms of a series of this sort decrease in absolute value promptly enough for large values of z:

---

[470] See also the paradox connected to formula (221) on p. (333).

This allowed one to compute the approximate values of the functions with relative ease and with a sufficient degree of precision. Nevertheless, for any value z, such series were certainly increasing, if they continued sufficiently. "But one cannot deny that the theory of these divergent series has been hidden up to now by certain difficulties, of which I, perhaps, shall discuss more fully in other circumstances" (Gauss [1812b, 152]). Gauss seems to have planned a theory of series that also included divergent series but did not give any further consideration to this question.[471]

In his *Disquisitiones generales,* Gauss made another reference to divergent series. Indeed, he had treated the expansion of hypergeometric series into continued fractions and had proved that

$$F(\alpha, 1, \gamma, x) = \frac{1}{1-} \frac{a_1 x}{1-} \frac{a_2 x}{1-} \frac{a_3 x}{1-} \frac{a_4 x}{1-} \cdots, \qquad (226)$$

where

$$a_{2n+1} = \frac{(\alpha + n)(\gamma + n - 1)}{(\gamma + 2n - 1)(\gamma + 2n)}, \quad a_{2n} = \frac{n(\gamma + n - 1 - \alpha)}{(\gamma + 2n - 2)(\gamma + 2n - 1)}, \quad n = 0, 1, 2, \ldots$$

(see Gauss [1812b, 134–138]). By applying (226), for infinitely large γ, he derived

$$\begin{aligned}
F(\frac{m}{n}, 1, \gamma, -\gamma n t) &= 1 - mx + m(m + n)x^2 - m(m + n)(m + 2n)x^3 + \ldots \\
&= \frac{1}{1+} \frac{nx}{1+} \frac{nx}{1+} \frac{(m + n)x}{1+} \frac{2nx}{1+} \frac{(m + 2n)x}{1+} \frac{3nx}{1+} \cdots
\end{aligned}$$

The series

$$1 - mx + m(m + n)x^2 - m(m + n)(m + 2n)x^3 + \ldots \qquad (227)$$

did not satisfy the condition of convergence $|x| = |-\gamma n t| < 1$; however, Gauss mentioned it without annotations (see [1812b, 138]).

In Part III, I noted that both the series of type (224) and those of type (227) had been investigated in the 18th century because of their usefulness and that such a use of divergent series had represented a remarkable success of formal methodology, which helped to allay d'Alembert's criticisms. In Gauss's eyes, these successes had become insufficient (when he wrote *Disquisitiones generales* [1812b], they were already dated; rather, formal methods were an obstacle to the introduction of new functions in analysis

---

[471]Some attempts at summing divergent series can be found in Gauss's manuscripts; however, they are merely youthful attempts following the 18th-century style (see, for instance, *Neue methode die summe der divergierenden reihe* $1 - 1 + 2 - 6 + 24 - etc. [=]0, 5963$ *zu finden,* in [WC, 382–385]). This material is only an example of the approach Gauss used in his youth.

and to the development of mathematics). Even if Gauss was aware of the historical importance of formal methodology and of the applications of divergent series in certain fields of the exact sciences, he marginalized those findings of the old theory that were difficult to adapt to new concepts in *Disquisitiones generales.*

33   Cauchy's rejection of the 18th-century theory of series

This final chapter is an analysis of how Cauchy introduced innovations to the 18th-century theory of series; it is therefore not a comprehensive study of all different aspects of his series theory. I limit myself to examining Cauchy's works in the early 1820s and, in particular, to the *Cours d'analyse a l'École Royal Polytechnique* and *Résumé des leçons a l'École Royal Polytechnique sur le calcul infinitésimale*, published in 1821 and 1823, respectively. These treatises represented the definitive abandonment of the 18th-century concept and provided a systematic exposition of the new quantitative approach, which we have already seen taking shape in the work of Fourier and Gauss.

Unlike Gauss, who almost hid the radical novelty of his approach in his *Disquisitiones generales* [1812b], Cauchy openly declared it in the introduction to the *Cours d'analyse*:

> As for methods, I have sought to give them all the rigor that one requires in geometry, so as never to have recourse to the reasons drawn from the generality of algebra. Reasons of this kind, although commonly admitted, particularly in the passage from convergent series to divergent series, and from real quantities to imaginary expressions, can, it seems to me, only sometimes be considered as inductions suitable for presenting the truth, but which are little suited to the exactitude so vaunted in the mathematical sciences. We must at the same time observe that they tend to attribute an indefinite extension to algebraic formulas, whereas in reality the larger part of these formulas exist only under certain conditions and for certain values of the quantities that they contain. Determining these conditions and these values, and fixing in a precise way the sense of the notations I use, I make any uncertainty vanish; and then the different formulas involve nothing more than relations among real quantities, relations which are always easy to verify on substituting numbers for the quantities themselves. In order to remain faithful to these principles, I admit that I was forced to accept several propositions which seem slightly hard at first sight. For example ... a divergent series has no sum. (Cauchy [1821, ii–iii])

This excerpt clearly shows that Cauchy regarded the rejection of divergent series as a consequence of the rejection of the formal methodology and of the acceptance of an entirely quantitative conception. As I have explained at some length, the use of the divergent was unavoidable within the context of 18th-century methodology.

Though Cauchy's infinitesimal analysis was profoundly innovative, it was still firmly based on the notion of variable quantity. Variable quantities were the primary objects of analysis and functions were defined as relations between variable quantities.[472] According to Cauchy,

> a variable quantity is a quantity that can be considered capable of receiving several different values successively. (Cauchy [1821, 19])

Cauchy's use of this notion shows that variable quantities were regarded as entities capable of varying with continuity between fixed limits. In Cauchy's view, variable quantities could be represented as sequences of numbers that varied without jumps, although a variable quantity did not vary upon sets of numbers. The legacy of 18th-century analysis is particularly strong in this matter. Cauchy continued to consider numbers as the ratios of the variable quantity to a fixed unity. The continuum was a primitive notion and was not constituted by numbers; on the contrary, numbers expressed specific determinations of quantity. Variable quantities varied on this primitive continuum and could potentially receive all the specific determinations one wished.[473] Laugwitz has expressed this concept as follows: "Cauchy's intervals are not sets but loci where a variable can move freely" (Laugwitz [1987, 273]).[474]

If Cauchy's notion of a variable quantity is similar to the 18th-century one, a crucial difference emerges as concerns the way quantities were dealt with. When 18th-century mathematicians examined a relation between quantities, they made the values of these quantities abstract and only considered the way that they combined with each other (see, e.g., Lagrange [1806, 1]). In other words, they operated upon analytical expressions. Instead, "Cauchy's proof and other concepts ... did not rely on the concept of an analytic expression" (Lüzten [2003a, 157]). According to Cauchy, an equality $A(x) = B(x)$ was not a formal relation that was valid when the variable quantity x was indeterminate; it was a quantitative relation that was valid only for the specific, determinate values of the variable that satisfied it.

This did not mean that analytical expressions played no role in Cauchy's analysis. As Lützen observes, Cauchy always conceived of functions expressed by analytical expressions in his mind (see Lützen [2003a, 157]). However, unlike 18th-century mathematicians, Cauchy thought that the

---

[472] "If variable quantities are so joined between themselves that, the values of one of these being given, one can derive the values of all the others, one ordinarily conceives these diverse quantities expressed by means of the one among them, which then takes the name of *independent variable*; and the other quantities expressed by means of the independent variable are those which one calls *functions* of this variable" (Cauchy [1821, 31], translation in Rüthing [1984, 74]).

[473] This concept is clearly similar to Gauss's, which is illustrated in the previous chapter.

[474] See also Breger [1992a, 251].

analytical expressions that were suitable for representing a function were constituted not only by the composition of elementary functions but also by integrals[475] and series (see below).

The above concept of the continuum is reflected in the notion of continuous functions. Laugwitz [1987, 273] stated: "[C]ontinuity (at least piecewise) of any function appearing in the calculus is a deep-rooted conviction with Cauchy" [1987, 273]. Indeed, Cauchy's functions, like Euler's and Lagrange's functions, are intrinsically continuous and, at most, have isolated points of discontinuity. While Euler and Lagrange viewed these points as exceptional points that were ignored in the formulation of the theorems, Cauchy specified the limits of continuity and investigated functions only on the interval where they were continuous. He felt the need for an explicit definition of continuous functions:

> Let $f(x)$ be a function of the variable x and suppose that for each value of x between two given limits this functions always takes a unique and finite values. If, having a value of x between these limits, one can attribute to the variable x an infinitely small increase α, the function itself increases by the difference
>
> $$f(x + \alpha) - f(x),$$
>
> which depends simultaneously on the new variable α and the value of x. This done, the function $f(x)$ will be, between the two limits assigned to the variable x, a *continuous* function of this variable if, for each value of x intermediate between these limits the numerical value of difference
>
> $$f(x + \alpha) - f(x)$$
>
> decreases indefinitely with α. In other words, *the function f(x) will remain continuous with respect to x between the given limits if, between these limits an infinitely small increase in the variable always produces an infinitely small increase in the function itself.* (Cauchy [1821, 43], Cauchy's emphasis, translation in Fauvel–Gray [1987, 566–567])

This definition merely served to give an explicit description of the property[476] (LC) that 18th-century mathematicians assumed held for any function $f(x)$ and that was used without referring to the specific intervals where the property actually held.

Cauchy also defined continuity in the neighborhood of a particular value of x:

---

[475] For this reason, Cauchy considered $\frac{2}{\pi} \int_0^\infty \frac{x^2 \, dt}{t^2 + x^2}$ as an adequate counterexample to Euler's notions of continuity (see p. 351).

[476] See Chapter 18.3, p. 212.

One says furthermore that the function $f(x)$ is, in the neighborhood of a particular value attributed to x, a continuous function of this variable, whenever it is continuous between two limits of x, however close, which contain that value of x. (Cauchy [1821, 43])

Two remarks should be made here. First, providing a definition of continuity in the neighborhood is different from defining continuity at a point. There is no room for pointwise continuity in Cauchy's concept, which still belonged to the time when the continuous was merely the unbroken. Continuity was a property connected to variables (and intervals), rather than to single points. It implied a variation, without jumps, of the variable x which corresponded to a variation, without jumps, of the variable $y = f(x)$.

Second, Cauchy's definition of continuity in the neighborhood of a point —or, to use Cauchy's words, a particular value— x_0, means that there exists an interval (a, b) containing x_0, such that if x and $x + \alpha$ vary over (a, b) and α is infinitesimal, then $f(x + \alpha) - f(x)$ is infinitesimal. Setting $z = x + \alpha$, Cauchy's definition can be formulated:

(Ca) $f(x)$ *is continuous in the neighborhood of a particular value x_0 if an interval (a, b) containing x_0 exists such that when the variables x and z vary over (a, b) and their difference $x - z$ is infinitesimal, then $f(z) - f(x)$ is infinitesimal.*

Cauchy assumed that a continuous function displayed uniform behavior, with respect to the variables x and z, in a whole interval containing x_0. There is a very strong temptation to interpret (Ca) as

(UC) $f(x)$ *is continuous in the neighborhood of a point x_0 if an interval (a, b) containing x_0 exists such that if $x \in (a, b)$, $z = x + \alpha \in (a, b)$, and $x - z$ is infinitesimal, then $f(z) - f(x)$ is infinitesimal.*

This is the uniform continuity in the interval (a, b). But this interpretation presupposes the notion of belonging, the concept of the interval (a, b) as (a set of) made of points, pointwise-defined functions, etc. Even though Cauchy's definition resembles uniform continuity, it cannot be interpreted as such unless notions are ascribed to Cauchy that he did not possess.[477]

Cauchy's assumption of uniform behavior recalls the similar assumption to be found in Lagrange's proofs of the remainder theorem.[478] It seems to be more a legacy of the past (closely connected with the notion of variable quantity) than an anticipation of the future.

Of course, when I underline the similarity between Cauchy's notion of continuity and the 18th-century one, I do not refer to Euler's definition of a

---

[477] On this question, see the interesting analysis of Laugwitz in his [1987, 273].
[478] See p. 307 and, for more details, Ferraro and Panza [A].

continuous function[479] but to property (LC). Euler's definition, which was closely connected with the generality of algebra,[480] was explicitly rejected by Cauchy.[481] In his [1844, 145], Cauchy based his rejection on the observation that a simple change of notation is often enough to transform a continuous function (in Euler's sense) into a discontinuous function (in Euler's sense). For instance, the function[482]

$$f(x) = \begin{cases} x & \text{for } x \geq 0 \\ -x & \text{for } x \leq 0 \end{cases}$$

can be represented both as

$$\frac{2}{\pi} \int_0^\infty \frac{x^2 dt}{t^2 + x^2}$$

and as

$$\sqrt{x^2}.$$

*   *   *

Cauchy's treatment of series was one of crucial novelties of his *Cours d'analyse* and *Résumé des leçons*. He gave the following definition of the sum of a series $\sum_{n=0}^\infty u_n$:

Let
$$s_n = u_0 + u_1 + u_2 + \ldots + u_{n-1} \tag{228}$$

the sum of the first n terms, where n is an arbitrary integer. If the sums s_n approach a certain limit s indefinitely for increasing values of n, the series is said to be *convergent*, and the limit in question is called the sum of the series. (Cauchy [1821, 114], translation in Lüzten [2003a, 159])

As an example, he considered the geometric series $\sum_{n=0}^\infty x^n$ and showed that it converges or diverges according to whether the absolute value of x is less than or greater than 1. Immediately afterwards, Cauchy provided

---

[479]See Section 18.3.
[480]See Section 18.2, p. 209.
[481]I would like to point out that, at the end of the 18th century, Arbogast had also discussed the meaning of the term "continuity". His point of view differed slightly from that of Euler. In particular, he had stated that the continuity of a curve can be broken in two ways:
1. If the analytical expression of the curve changes;
2. if the curve is composed of different parts and these parts are not joined to each other (see Arbogast [1791, 9–11]).
[482]See footnote no. 290.

certain necessary and sufficient conditions for convergence.[483] The first is trivial:

  1. The series is convergent if and only if the sequence of the partial sums is convergent toward a fixed limit s;

Cauchy thought that this condition could be reformulated as

  2a. the series is convergent toward s if and only if $s_{n+k} - s \approx 0$, for infinitely large n and every k,

or

  2b. the series is convergent if and only if $s_{n+k} - s_{n+h} \approx 0$, for infinitely large n and every k and h.

Moreover from 2a Cauchy deduced that

  3. the series is convergent if and only if, for every ε, one has

$$|s_{n+k} - s_n| = r_{n+k} = |u_n + u_{n+1} + u_{n+2} + \ldots + u_{n+k-1}| < \varepsilon,$$

  for enough large n and every k.[484]

---

[483] "[I]n order for series

$$u_0, \; u_1, \; u_2, \; \ldots, \; u_n, \; u_{n+1}, \; \ldots$$

to be convergent, it is necessary and it suffices that increasing values of n make the sum

$$s_n = u_0 + u_1 + u_2 + \ldots + u_{n-1}$$

converge indefinitely toward a fixed limit s; in other words, it is necessary and it suffices that, for infinitely large values of the number n, the sums

$$s_n, \; s_{n+1}, \; s_{n+2}, \; \ldots$$

differ from the limit s, and in consequence among each other, by infinitely small quantities" [1821, 115].

[484] In his own words:

"[T]he successive differences between the first sum s_n and the following ones are determined by the equations

$$s_{n+1} - s_n = u_n,$$
$$s_{n+2} - s_n = u_n + u_{n+1},$$
$$s_{n+3} - s_n = u_n + u_{n+1} + u_{n+2},$$
$$\ldots\ldots$$

Therefore, in order for the series [228] to be convergent, it is firstly necessary that the general term u_n decreases indefinitely, when n increases; but this condition is not sufficient,

Cauchy considered 1, 2a, 2b, and 3 as trivial and gave no proof of them. To modern eyes, 1 and 2a are really trivial, whereas the sufficiency of 2b and 3 is not. The difference between 2b and 3, on one side, and 2a, on the other, is that in the formulation 2a the existence of the number s satisfying the condition $s_{n+k} - s \approx 0$, for large n and every k, is assumed by the hypothesis. If such a number exists, it is trivial to state that it is the sum. Instead, in the formulations 2b and 3 the existence of this number is not assumed by the hypothesis. The condition

$$|s_{n+k} - s_n| \;=\; r_{n+k} = |u_n + u_{n+1} + u_{n+2} + \ldots + u_{n+k-1}| < \varepsilon,$$
$$\text{for every } \varepsilon \;>\; 0, \text{ for every } k, \text{ and for large } n$$

and the equivalent one in terms of infinite and infinitesimal numbers

$$s_{n+k} - s_n \approx 0, \text{ for every } k \text{ and } h, \text{ and infinitely large } n$$

do not imply that the sum s exists, but only that the series is bounded.[485] Instead, according to Cauchy, 2a and 2b are trivially equivalent. The only possible explanation is that he presumed that the condition $s_{n+k} - s_n \approx 0$ implies the existence of a value s such that $s_{n+k} - s \approx 0$. This implicit assumption is grounded on Cauchy's concept of continuum and, more precisely, on the principle that a variable quantity, by passing from a value A to a value B, must receive all intermediate values between A and B. For instance, if $s_n > 0$ and $s_{n+k} - s_n \approx 0$, for every k and infinitely large n, then the sequence s_n is bounded and a variable quantity by passing from the value $A = s_1$ to a value $B > s_n$, for all n, must become equal to the value s that separates the values greater than s_n from the other values.[486]

An emblematic theorem of Cauchy's theory of series is the following one:

> Theorem 1. When the different terms of the series [228] are functions of the same variable x, and continuous with respect to

---

and it must also be true for increasing values of n the different sums

$$u_n + u_{n+1},$$
$$u_n + u_{n+1} + u_{n+2},$$
$$\ldots,$$

that is, the sums of the quantities

$$u_n, \; u_{n+1}, \; u_{n+2}, \; \ldots$$

taken from the first, in whatever number we wish, will always end up having numerical values [absolute values] that are constantly smaller than any assignable limit. Conversely, when these various condition are satisfied, the convergence of series is assured." [1821, 115–116]

[485]The only trivial deduction that follows from $s_{n+k} - s_n \approx 0$ is that the series is not infinite.

[486]The situation is similar to the existence of maximum in Gauss's proof of the Gauss's criterion; see p. 335.

this variable in the neighborhood of a particular value for which
the series is convergent, them the sum s is also a continuous
function in the neighborhood of this particular value. (Cauchy
[1821, 120], translation in Lüzten [2003a, 169])

A theorem of such a kind was not conceivable in the 18th century for at
least two reasons.

First, in the 18th century, series theory concerned the expansion of functions into series by formal methods or, if a series was given, the problem was
to find the generating function of that series.[487] Theorem 1 presupposes that

(a) a series is an autonomous object,

(b) it is suitable for providing a function, and

(c) one is interested in knowing whether the function is continuous.

Second, in the 18th century, functions were thought to be always continuous at least over an interval, and the generality of algebra made it possible
to consider the whole function as if it were continuous. A theorem that
provided conditions for continuity over an interval was useless.

Cauchy's proof runs as follows. Let s be the sum of the series $\sum u_n$ and
let s_n be the nth sum of that series. The sum s can be written in the form
$s = s_n + r_n$. Consider the equality

$$s(x + \alpha) - s(x) = s_n(x + \alpha) - s_n(x) + r_n(x + \alpha) - r_n(x).$$

If α is an infinitesimal, then $s_n(x + \alpha) - s_n(x)$ is infinitesimal, for all n.
Moreover, if the series $\sum u_n$ is convergent in the neighborhood of x, then
$r_n(x + \alpha)$ and $r_n(x)$ are infinitesimal[488] for large n. This implies that $s(x)$
is continuous.

Cauchy's theorem and his proof are ambiguous. Indeed, one can note
that

a. Cauchy uses infinitesimal neighborhoods of x in a decisive way.[489] Infinitesimals are not thought as a mere *façon de parler*, but they are
   conceived as numbers, though a theory of infinitesimal numbers is
   lacking.[490]

b. Cauchy does not give separate definitions of the sum of function series
   and of the sum of numerical series. In the above-mentioned definition,

---

[487]See Chapter 8, p. 120.

[488]As Laugwitz [1987, 264–265] notes, Cauchy assumes that if the series $\sum u_n$ is convergent in the neighborhood of x, then it is convergent for $x + \alpha$, where α is an infinitesimal
number.

[489]On this question, see Laugwitz [1987].

[490]Cauchy tried to provide a classification of infinitesimal quantities in [1823, 250–256].
It remains rather vague and any attempt to transform it into the modern theory of infinitesimals implies the rejection of Cauchy's notion of quantity. Even Laugwitz, who had
made some attempts of such kind, was forced to admit that such attempts "should be
taken with reluctance and reserve" [1987, 273].

the terms u_n can denote both numbers and functions. Cauchy considered function series as a series of variable quantities that vary over an interval. In his view, convergence is not pointwise,[491] but a matter that concerns a variable over a whole interval.[492]

The combination of a and b leads one to think that the pointwise convergence of functions (for every real number in an interval $[a, b]$) is not the same as the convergence of functions (between the limits a and b) in Cauchy's sense, which involved convergence in infinitesimal neighborhood of a value of x. Unfortunately, Cauchy did not clarify this point: It is probable that his notion of variable quantity and number prevented him from doing so.[493]

$$*\quad*\quad*$$

Having rejected formal methodology, Cauchy could not accept the principle of infinite extension[494] and so, in his [1821], he, for the first time in the history of series theory, gave an appropriate definition of operations between series and also proved many results that had previously been considered obvious on the basis of the analogy between finite and infinite sums. For instance, in the 18th century, it seemed obvious that

$$(a + b + c + d + \text{etc.}) \cdot (A + B + C + D + \text{etc.})$$

was equal to

$$aA+$$
$$(aB + Ab)+$$
$$(aC + bB + cA)+$$
$$(aD + bC + cB + dA)+$$
$$\text{etc.}$$

independently of the meaning of "etc." in such expressions: The rule of the ordinary multiplication between two polynomials was extended to infinite series without the difference between finite and infinite series having been pointed out. Instead, Cauchy gave explicit theorems for the sum and multiplication of series:

---

[491]Of course, this also follows from the fact that Cauchy's functions are not defined pointwise.

[492]I emphasize that the condition 2a, when applied to function series, does not imply the consideration of the specific value of the independent variable. It can be interpreted as follows: The series $\sum u_n$ is convergent toward s if and only if $s_{n+k} - s \approx 0$, for infinitely large n and every k, independently of x.

[493]On Abel's interpretations of Cauchy's theorem, see Pensivy [1987–88, 20–28].

[494]See p. 117.

(S) If the series $\sum_{n=0}^{\infty} u_n$ and $\sum_{n=0}^{\infty} v_n$ are convergent and

$$\sum_{n=0}^{\infty} u_n = s \quad \text{and} \quad \sum_{n=0}^{\infty} v_n = S,$$

then the series

$$\sum_{n=0}^{\infty} (u_n + v_n)$$

is convergent and its sum is $s + S$ (Cauchy [1821, 127 and 132]).

(P) Given the series $\sum_{n=0}^{\infty} u_n$ and $\sum_{n=0}^{\infty} v_n$, let $\sum_{n=0}^{\infty} a_n$, define as follows:

$$a_0 = u_0 v_0,$$
$$a_1 = u_0 v_1 + u_1 v_0,$$
$$a_2 = u_0 v_2 + u_1 v_1 + u_2 v_0,$$
$$\ldots,$$
$$a_n = u_0 v_n + u_1 v_{n-1} + \ldots + u_{n-1} v_n + u_n v_0,$$
$$\ldots$$

If $\sum_{n=0}^{\infty} |u_n|$ and $\sum_{n=0}^{\infty} |v_n|$ are convergent and

$$\sum_{n=0}^{\infty} u_n = s \quad \text{and} \quad \sum_{n=0}^{\infty} v_n = S,$$

then the series

$$\sum_{n=0}^{\infty} a_n$$

is convergent and its sum is sS (Cauchy [1821, 127–128 and 133]).

Cauchy also provided a counterexample that shows how the theorem on the product of series was not valid if one only assumed convergence of series $\sum_{n=0}^{\infty} u_n$ and $\sum_{n=0}^{\infty} v_n$. He showed that

$$\sum_{n=1}^{\infty} (-1)^{n+1} \frac{1}{\sqrt{n}}$$

is convergent, whereas the product

$$\left(\sum_{n=1}^{\infty} (-1)^{n+1} \frac{1}{\sqrt{n}} \right) \left(\sum_{n=1}^{\infty} (-1)^{n+1} \frac{1}{\sqrt{n}} \right)$$

is divergent (Cauchy [1821, 134–135]).

Cauchy's new approach (as well as Gauss's) placed great importance on convergence tests. According to Cauchy, before summing any series, I had to examine when the series can be summed, or, in other terms, what are the conditions for their convergence. (Cauchy [1821a, v])

Some convergence tests were already used in the 18th century, but their use occasionally appears in few specific proofs to ascertain the conditions that enabled the use a series to approximate a quantity. They were not employed to guarantee the *a priori* existence of a sum (in effect, this had no sense in the formal conception, according to which summing a series meant exhibiting the generating functions). Instead, in Cauchy's view, convergence tests served to guarantee the existence of the sum in the quantitative sense.[495] In [1821], Cauchy considered several other tests (apart from the Cauchy criterion discussed above). Indeed, he proved

- the root test (the series $\sum_{n=0}^{\infty} u_n$ is convergent if $\limsup \sqrt[n]{|u_n|} < 1$ and divergent if $\limsup \sqrt[n]{|u_n|} > 1$) (Cauchy [1821, 121–123 and 129]),

- the ratio test, already used by Gauss (Cauchy [1821, 123 and 129]),

- the condensation test (if $u_{n+1} > u_n > 0$, then $\sum_{n=0}^{\infty} u_n$ and $\sum_{n=0}^{\infty} 2^n u_{2^n}$ are convergent or divergent together) (Cauchy [1821, 123–124]),

- the logarithmic test (if $u_n > 0$ and $\lim \frac{\log u_n}{\log \frac{1}{n}} < 1$, then the series is convergent, it is divergent if $\lim \frac{\log u_n}{\log \frac{1}{n}} > 1$) (Cauchy [1821, 125–127]),

- the alternating test (Cauchy [1821, 130]).

Later, in his [1827], he also published the integral test.[496]

$$* \quad * \quad *$$

The expansion of elementary functions into power series was the main problem in 18th-century series theory. This problem continued to be important for Cauchy, although it changed. In the 18th century, the problem of the expansion was a direct problem that was solved by applying the procedures P1–P4, while the problem of the sum was a converse problem that consisted of seeking the generating function.[497] The problem of the sum now became the direct problem, which consisted of seeking the limit of the partial sums, while the problem of expansion was the inverse problem, which consisted of seeking a series whose sum is the given function.

---

[495] See also p. 325.

[496] Given $u(x)$, if $u_n = u(n) > 0$, for $n = 1, 2, \ldots$, then $\sum_{n=1}^{\infty} u_n$ is convergent [divergent] if and only if $\int_1^{\infty} u(x)dx$ is convergent [divergent]. This criterion was already known as a sum estimator.

[497] See Chapter 8.

Cauchy's treatment of the expansions of elementary functions is based on various theorems concerning power series. In particular, he proved that a power series $\sum_{n=0}^{\infty} a_n x^n$ is convergent if the values of x are between $-\frac{1}{A}$ and $\frac{1}{A}$ (where

$$A = \limsup \sqrt[n]{|a_n|}$$

is the radius of convergence) and that it diverges if $x < -\frac{1}{A}$ and $x > \frac{1}{A}$ [1821, 136]. Cauchy also showed that the expansion into power series is unique [1821, 144–145]. His proof is similar to the proof provided by Euler in [1740, 417–418] and [1748a, 1:230–231].

As an example of how Cauchy dealt with the expansion of elementary functions, I shall briefly illustrate his proof of the binomial theorem (Cauchy [1821, 146–147]). Given the series

$$\sum_{n=0}^{\infty} \binom{\mu}{n} x^n,$$

Cauchy first proved that it is convergent for $|x| < 1$. Then he considered $\sum_{n=0}^{\infty} \binom{\mu}{n} x^n$ as a series of the variable μ for a fixed x ($|x| < 1$). The sum of $\sum_{n=0}^{\infty} \binom{\mu}{n} x^n$ is a continuous function for any value of μ. Cauchy denoted this function by $\phi(\mu)$ and observed that

$$\phi(\mu + \mu') = \phi(\mu)\phi(\mu'). \tag{229}$$

He had already proved in Chapter 5 of his *Cours d'analyse* [1821, 100–103] that the solution of the functional equation (229) is

$$\phi(\mu) = [\phi(1)]^\mu.$$

Since $\phi(1) = (1 + x)$, he obtained

$$\phi(\mu) = (1 + x)^\mu;$$

therefore,

$$(1 + x)^\mu = \sum_{n=0}^{\infty} \binom{\mu}{n} x^n$$

for $|x| < 1$.

Cauchy's rejection of the generality of algebra also implies his refusal to extending a formula from the real values to the complex values of variables. In contrast to 18th-century mathematicians, Cauchy considered functions of complex variables as a subject in its own right and offered appropriate demonstrations for theorems concerning functions of complex terms. For example, he provided a proof of the binomial theorem

$$(1 + x)^\mu = \sum_{n=0}^{\infty} \binom{\mu}{n} x^n$$

in the case that x is a complex variable ($|x| < 1$) and μ is a real number (Cauchy [1821, 243–247]). To prove this theorem, he considered the functional equation

$$\omega(x + y) = \omega(x)\omega(y),$$

where $\omega(x) = \phi(x) + \sqrt{-1}\zeta(x)$ is continuous complex-valued function. In [1821, 222–226] Cauchy proved that the solution of this equation is

$$\omega(x) = A^x(\cos bx + \sqrt{-1}\sin bx),$$

where A and b are real numbers and $A > 0$. He applied this result to the series

$$\sum_{n=0}^{\infty} \binom{\mu}{n} z^n(\cos n\theta + \sqrt{-1}\sin n\theta), \tag{230}$$

and showed that the sum $\omega(\mu)$ of (230) is

$$\omega(\mu) = r^\mu(\cos \mu\theta + \sqrt{-1}\sin \mu\theta) \quad \text{(for } -1 < z < 1\text{)},$$

where r and t depend on z and θ. He set $r\cos t = 1 + z\cos\theta$ and $r\sin t = z\sin\theta$ and proved that

$$r^\mu(\cos \mu\theta + \sqrt{-1}\sin \mu\theta) = \left[1 + z(\cos\theta + \sqrt{-1}\sin\theta)\right]^\mu.$$

Of consequence,

$$\sum_{n=0}^{\infty} \binom{\mu}{n} z^n(\cos n\theta + \sqrt{-1}\sin n\theta) = \left[1 + z(\cos\theta + \sqrt{-1}\sin\theta)\right]^\mu,$$

for $-1 < z < 1$.

While *Cours d'analyse* is devoted to algebraic analysis, the subject matter of *Résumé des leçons* is the differential and integral calculus. In this book, Cauchy dealt with the Taylor series. He had already investigated this series in his *Sur le développement des fonctions en séries et sur l'intégration des équations différentielles ou aux différences partielles*, where he had given a counterexample to the uniqueness of the expansion in the Taylor series, a cornerstone of Lagrange's theory of analytical functions (at least, he thought so). Indeed, Cauchy considered a function $h(x)$ such that its derivatives were equal to 0 for $x = 0$ (for instance, $e^{-\frac{1}{x}}$, $e^{-\frac{1}{x^2}}$, ...). The expansion of $h(x+i)$ for $x = 0$ is

$$h(i) = h(0) + h'(0)i + \frac{1}{2}h''(0)i^2 + \ldots = 0. \qquad (231)$$

Cauchy observed that the function $f(x) + h(x)$ has the same expansion as $f(x)$ so that a single Taylor series can represent more than one function. For this reason, "one cannot substitute series for functions indistinctly" (Cauchy [1822, 278]).

It is worth noting that Cauchy's criticism assumed that (198) is a quantitative relation, valid for any value of x (whereas Lagrange thought that it held when x is indeterminate). If we assume that (198) is a formal relation, derived by means of one of the 18th-century methods of expansion, this criticism is difficult to accept, not just for the reason that Lagrangian formulas could fail at some exceptional points, but also because Lagrange's aim was to show that it was possible to obtain the function $f'(x)$ from the function $f(x)$ by means of formal procedures. Cauchy did not prove that this was impossible.[498] For instance, as Panza noted in [1992, 724], if one considers the function e^{-1/x^2}, one can easily obtain

$$
\begin{aligned}
e^{-(x+i)^{-2}} &= e^{-[x^{-2}-2x^{-3}i+3x^{-4}i^2+\ldots]} = (e^{-x^{-2}})(e^{+2x^{-3}i})(e^{-3x^{-4}i^2})\ldots \\
&= e^{-x^{-2}} + 2e^{-x^{-2}}x^{-3}i + e^{-x^{-2}}[2x^{-6} - 3x^{-4}]i^2 + \ldots
\end{aligned}
$$

This expansion provides the derivative of the function e^{-1/x^2} when x is indeterminate, but it fails to provide a quantitative representation of the function e^{-1/x^2} in a neighborhood of $x = 0$.

In his *Résumé des leçons*, Cauchy gave his interpretation of the Taylor theorem. While this theorem had previously only concerned the derivation of the coefficients of the power series $\sum_{n=0}^{\infty} a_n(x - x_0)^n$, namely in proving that

$$a_n = \frac{f^{(n)}(x_0)}{n!},$$

Cauchy mainly focused his attention on proving that the series

$$\sum_{n=0}^{\infty} \frac{f^{(n)}(x_0)}{n!}(x - x_0)^n$$

converged to $f(x_0 + x)$ for appropriate values of x. He [1823, 214–217] showed[499] that

$$f(x + h) = \sum_{k=0}^{n-1} \frac{f^{(k)}(x)}{k!}h^k + \int_0^h \frac{(h - z)^{n-1}}{(n - 1)!}f^{(n)}(x + z)dz. \qquad (232)$$

---

[498]Consequently, Cauchy's counterexample is not actually a counterexample if it is referred to as Lagrange's theory.

[499]He used repeated integrations according to a method that dated back to Johann Bernoulli (see Chapter 3).

Since

$$\int_0^h \frac{(h-z)^{n-1}}{(n-1)!} f^{(n)}(x+z)dz = \frac{h^n}{n!} f^{(n)}(x+\theta h),$$

with $0 < \theta < 1$, he rewrote (232) in the form

$$f(x+h) = \sum_{k=0}^{n-1} \frac{f^{(k)}(x)}{k!} h^k + \frac{h^n}{n!} f^{(n)}(x+\theta h).$$

Afterwards, Cauchy [1823, 220–221] observed that the Taylor series

$$\sum_{k=0}^{\infty} \frac{f^{(k)}(x)}{k!} h^k$$

was convergent to $f(x+h)$ if

$$\lim_{n\to\infty} \int_0^h \frac{(h-z)^{n-1}}{(n-1)!} f^{(n)}(x+z)dz = \lim_{n\to\infty} \frac{h^n}{n!} f^{(n)}(x+\theta h) = 0. \qquad (233)$$

Only when this condition is satisfied could one write

$$f(x+h) = \sum_{k=0}^{\infty} \frac{f^{(k)}(x)}{k!} h^k.$$

For instance, the functions $\sin x$ and $\cos x$ satisfy (233) and, therefore, they admit the expansions

$$\sin x = \sum_{k=0}^{\infty} \frac{(-1)^k}{(2k+1)!} x^{2k+1}$$

and

$$\cos x = \sum_{k=0}^{\infty} \frac{(-1)^k}{(2k)!} x^{2k}.$$

In conclusion, I shall discuss how, in his *Résumé*, Cauchy dealt with the problem of integration of series. This method had been fundamental in the rise and development of the calculus but had been used *de facto* as a trivial consequence of the principle of infinite extension. Cauchy attempted to provide a proof of it, and this, in itself, is an innovative step. He presented the following theorem:

If x_0 and X are finite numbers, $u_n(x)$ are continuous functions over $[x_0, X]$ and the series $s(x) = \sum_{n=0}^{\infty} u_n(x)$ is convergent for all values of x over $[x_0, X]$, then

$$\int_{x_0}^{X} s(x)dx = \sum_{n=0}^{\infty} \int_{x_0}^{X} u_n(x)dx \qquad (234)$$

In the proof, Cauchy set

$$r_n(x) = s(x) - \sum_{n=0}^{\infty} u_n(x).$$

By integrating, he obtained

$$\int_{x_0}^{X} s(x)dx = \sum_{i=0}^{n} \int_{x_0}^{X} u_i(x)dx + \int_{x_0}^{X} r_n(x)dx.$$

Cauchy stated that since

$$\int_{x_0}^{X} r_n(x)dx = r_n(\xi)(X - x_0),$$

where $x_0 < \xi < X$, and the series $\sum_{n=0}^{\infty} u_n(x)$ was convergent, the sequence $r_n(\xi)$ became zero for $n = \infty$. Hence, he obtained (234) (see Cauchy [1823, 237–238]).

This theorem is subject to the same difficulties as in other of Cauchy's theorems. Here, Cauchy assumes that the way $r_n(x)$ tends toward 0, for $n \to 0$, does not depend on the choice of a particular value of the variable x, but only on the variable itself. Once again he presupposes a uniformity in the behavior of functions and function sequences.

A few years later, mathematicians considered this theorem, as well as Theorem 1 and other of Cauchy's theorems, as mistaken and began the process that was to lead toward the concept of uniform convergence and pointwise convergence. However, the investigation of the development of this notion is beyond the scope of the present book.

References

Aepinus, F.U.T. [**1757**] Demonstratio theorematis newtoniani de binomio ad potentiam indefinitam elevando, *Novi Commentarii Academiae Scientiarum Petropolitanae*, **8** (1757), pp. 169–180.

Agostini, A. [**1941**] Le serie sommate da Pietro Mengoli, *Bollettino Unione Matematica Italiana*, (2) 3 (1941), pp. 231–251.

d'Alembert, J.B. [**1747**] Recherches sur la courbe que forme une corde tenduë mise en vibration, *Mémoires des l'Académie Royale de Berlin*, 3 (1747), pp. 214–219.

—— [**1768**] *Réflexions sur les suites et sur les racines imaginaires* in Opuscles mathématiques, Paris: David Briasson, Jombert, t.V, 1768, pp. 171–215.

—— [**1773**] *Mélange de littérature, d'histoire, et de philosophie*, 2nd edition, Amsterdam: Z. Chatelain et fils, 5 vols.

Ampère, A.M. [**1806**] Recherches sur quelques point de la théorie des fonctions dérivés qui conduisent à une nouvelle démonstration de la série de Taylor et à l'expression finie des termes qu'on néglige lorsqu'on arrête cette série à travers un terme quelconque, *Journal de l'École Polytechnique*, 6 (1806), pp. 148–181.

Arbogast, L.-F.-A. [**1791**] *Mémoire sur la nature des fonctions arbitraires qui entrent dans les intégrales des équations aux dérivées partielles*, St. Petersburg: de l'imprimerie a l'Academie Imperiale des Sciences, 1791.

—— [**1800**] *Du calcul des dérivations*, Strasbourg: Levrault, 1800.

Archimedes [QA] Quadrature of parabola, in *The Works of Archimedes*, translated by T. L. Heath, Cambridge: University Press, 1897, pp. 233–252.

Aristotle, *Aristotle's Categories and De Interpretatione*, translated by L. Ackrill, Oxford: Clarendon Press, 1963.

—— *Metaphysics*, translated by W. D. Ross, in *The Basic Works of Aristotle*, Richard McKeon (ed.), New York: Random House, 1941.

—— *Aristotle's Physics*, translated by Joe Sachs. Rutgers, NY: University Press, 1994.

Barrow, I. [**1670**] *Lectiones geometricae: in quibus (praesertim) generalia curvarum linearum symptomata declarantur*, Londini: Typis Gulielmi Godbid, & prostant venales apud Johannem Dunmore, 1670.

Bayes, T. [**1763**] A letter from the late Reverend Mr. Thomas Bayes, F.R.S. to John Canton, M.A. and F.R.S, *Philosophical Transactions,* 53 (1763), pp. 269–271.

Bernoulli, Daniel [**1728**] Observationes de seriebus quae formantur ex additione vel subtractione quacunque terminorum se mutuo consequentium,

Commentarii Academiae Scientiarum Imperialis Petropolitanae, 3 (1728), published in 1732, pp. 85–100.

— [**1732–33**] Theoremata de oscillationibus corporum filo flexili connexorum et catenae verticaliter suspensae, *Commentarii Academiae Scientiarum Petropolitanae*, 6 (1732–1733), published in 1738, pp. 108–122.

— [**1734–35**] Demonstrationes Theorematum suorum de Oscillationibus corporum filo flexili connexorum et catenae verticaliter suspensae, *Commentarii Academiae Scientiarum Petropolitanae*, 7 (1734–1735), published in 1740, pp. 162–173.

— [**1753**] Réflexions et éclaircissemens sur les nouvelles vibrations des cordes exposées dans les mémoires de l'Académie de 1747 et 1748, *Mémoires des l'Académie Royale de Berlin*, 9 (1753), pp. 147–172.

— [**1771**] De summationibus serierum quarunduam incongrue veris earumque interpretatione atque usu, *Novi Commentarii Academiae Scientiarum Imperialis Petropolitanae*, 16 (1771), published in 1772, pp. 71–90 (Summarium 12–14).

Bernoulli, Jacob [1689–1704] *Positiones arithmeticae de seriebus infinitis, earumque summa finita*, J. Conradi a Mechel, Basileæ, 1689–1704. In *Jacobi Bernoulli basileensis Opera*, Genevae: Cramer et fratum Philibert, 1744, 2 vols., vol. 1: pp. 375–402 and 517–542, vol. 2: pp. 745–767, 849–867 and 955–975.

— [**P**] *Ars conjectandi. Opus posthuma. Accedit Tractatus de seriebus infinitis et Epistola Gallice scripta de ludo pilae reticularis*, Basilae: Thurnisiorum Fratrum, 1713.

Bernoulli, Johann [1694] Additamentum effectionis omnium Quadraturarum et Rectificationum Curvarum per seriem quandam generalissimam, *Acta Eruditorum*, 1694, pp. 437–441.

— [**1718**] Remarques sur ce qu'on a donné jusqu'ici de solutions de problêmes sur les isoperimétres, in *Opera omnia*, Lausannae et Genevae: Marci-Michaelis Bousquet et Sociorum, 1742, vol. 2, pp. 235–269.

Bernoulli, Nikolaus II [1738] Inquisitio in summa seriei $1 + \frac{1}{4} + \frac{1}{9} + \frac{1}{16} + \frac{1}{25} +$etc., *Commentarii Academiae Scientiarum Imperialis Petropolitanae*, 10 (1738), published in 1747, pp. 19–21.

Bolzano, B [1816] *Der binomische Lehrsatz, und als Folgerung aus ihm der polynomische, und die Reihen, die zur Berechnung der Logarithmen und Exponentialgrößen dienen, genauer als bisher erwiesen*, Prague: Enders 1816.

Bombelli, R. [1572] *L'Algebra parte maggiore dell'aritmetica divisa in tre libri*. Nuovamente posta in luce. Bologna: Giovanni Rossi 1572. Reprinted in *L'Algebra*, Opera di Rafael Bombelli da Bologna, Prima edizione integrale, Milano: Feltrinelli 1966.

Bos, H. [1974]. Differentials, higher order differentials and the derivative in the Leibnizian Calculus, *Archive for History of Exact Sciences*, 14, pp. 1–90.

— [**1993**] *Lectures in the history of mathematics*, Providence, RI: American Mathematical Society, 1993.

— [**2001**] *Redefining Geometrical Exactness: Descartes' Transformation of the Early Modern Concept of Construction*, New York: Springer, 2001.

Bossut, C and **Lagrange, J.L.** [**1801–02**] Rapport sur un mémoire présenté à la classe par le citoyen Callet, *Mémoires de l'Institut national des sciences et arts. Sciences Mathématiques et Physiques*, 3 (An IX (1801–02)): pp. 1–11.

Breger, H. [**1992a**] A restoration that failed: Paul Finsler's theory of sets. In Gillies [1992, 249–264].

— [**1992b**] Les Continu chez Leibniz. In Salanskis–Sinaceur [1992, 75–84].

Brezinski, C. [**1980**] *History of Continued Fractions and Padé Approximants*, New York: Springer–Verlag, 1980.

Brisson, B. [**1808**] Sur l'intégration des équations différentielles partielles, JEP [Journal de l'École Polytechnique, Paris.] 7, 191–261.

Burchard, J. [**1721**] Epistola ad Virum Clarissimum Brook Taylor, *Acta Eruditorum*, (1721), pp. 195–228.

Burn, R.P. [**2001**] Alphonse Antonio de Sarasa and Logarithms, *Historia Mathematica*, 28 (2001), pp. 1–17.

Bussotti, P. [**2000**] Aritmetica e aritmetizzazione. La via indicata da Gauss e Kronecker, *Epistemologia*, 23 (2000), pp. 23–50.

Cardano, G. [**1545**] *Artis Magnae, sive de regulis algebraicis liber unus*, Norimbergae: apud Iohannem Petreium, 1545. English translation by T. Richard Witmer, *The great art or the rules of algebra*, Cambridge: Cambridge University Press, 1968.

Cataldi, Pietro Antonio [1613] *Trattato del modo brevissimo di trovare la radice quadra delli numeri, et regole da approssimarsi di coninuo al vero nelle radice de' numeri non quadrati, con le cause, & inuentioni loro, et anco il modo di pigliarne la radice cuba*, Bologna: Bartolomeo Cochi, 1613

Cauchy, A.–L. [**1821**] *Cours d'analyse de l'école royale polytechnique. 1re partie. Analyse algébrique*, Paris: Debure frères, 1821. In [Œa, (2) 3].

— [**1822**] Sur le développement des fonctions en séries et sur l'intégration des équations différentielles ou aux différences partielles, *Bulletin de la Société Philomatique*, 1822, pp. 49–54. In [Œa, (2) 2: 276–282].

— [**1823**] *Résumé des leçons données à l'École royale polytechnique sur le calcul infinitésimal. Tome Premier*, Paris: Debure frères, 1823. In [Œa, (2) 4: 9–261].

— [**1827**] Sur la convergence des séries. In [Œa, (2) 7: 267–279].

— [**1844**] Mémoire sur les fonctions continues, *Comptes rendus de l'Académie des Science*, 18 (1844), pp. 116–131. In [Œa, (1), 8:145–160].

— [**Œa**] Œuvres complètes de Augustin Cauchy, ed. by Acadèmie des Sciences, Paris: Gauthiers–Villars, 1882–1974, 27 vols.

Chabert, J.–L. (Ed.) **[1999]**, *A History of Algorithms. From the Pebble to the Microchip*, Berlin: Springer–Verlag, 1999.

Clairaut, A.C. **[1754]** Sur l'orbite apparente du Soleil autour de la terre, en ayant égard aux perturbations produites par les actions de la lune et des planètes principales, *Histoire de l'Academie des Science de Paris*, 1754, pp. 521–564.

Dedekind, R. **[1872]** *Stetigkeit und irrationale Zahlen*, Braunschweig: Vieweg, 1872.

Descartes, R. **[1637]** *Discours de la méthode*, Leyden: J. Maire, 1637.

Dhombres, J. **[1988]** Un texte d'Euler sur les fonctions continues et les fonction discontinues, véritable programme d'organisation de l'analyse au XVIIIème siècle, *Cahiers du Séminaire d'Historie des Mathématiques*, 9 (1988), pp. 23–97.

—— **[1995]** L'innovation comme produit captif de la tradition. Entre Apollonius et Descartes, une théorie des courbers chez Grégoire de Saint–Vincent, in Panza-Roero [1995, 13–83].

Dutka, J. **[1982]** Wallis's product, Brouncker's continued fraction, and Leibniz's series, *Archive for History of Exact Sciences*, 26 (1982), pp. 115–126.

—— **[1984–85]** The early history of the hypergeometric function, *Archive for History of Exact Sciences*, 31 (1984–85), pp. 15–34.

—— **[1995]** On the early history of Bessel functions, *Archive for History of Exact Sciences*, 49 (1995), pp. 105–134.

Engelsman, S.B. **[1984]** *Families of Curves and Origins of Partial Differentiation*, Amsterdam: North–Holland, 1984.

Euclid [E], *The Thirteen Books of Euclid's Elements*, Translated by Thomas L. Heath, 2nd ed., Cambridge: Cambridge University Press, 1926.

Euler, L. [1730–31a] De progressionibus transcendentibus seu quarum termini generales algebraice dari nequeunt, *Commentarii Academiae Scientiarum Imperialis Petropolitanae*, 5 (1730–31), published in 1738, pp. 36–57. In [EOO, (1) 14:1–24].

—— **[1730–31b]** De summatione innumerabilium progressionum, *Commentarii Academiae Scientiarum Imperialis Petropolitanae*, 5 (1730–1731), published in 1738, pp. 91–105. In [EOO, (1) 14:25–41].

—— **[1732–33]** Methodus generalis summandi progressiones, *Commentarii Academiae Scientiarum Petropolitanae*, 6 (1732–1733), published in 1738, pp. 68–97. In [EOO, (1) 14:42–72].

—— **[1734–35a]** De summis serierum reciprocarum, *Commentarii Academiae Scientiarum Petropolitanae*, 7 (1734–1735), published in 1740, pp. 123–134. In [EOO, (1) 14:73–86].

—— **[1734–1735b]** De progressionibus harmonicis observationes, *Commentarii Academiae Scientiarum Petropolitanae*, 7 (1734–1735), published in 1740, pp. 150–156. In [EOO, (1) 14:87–100].

— [**1736a**] Methodus universalis serierum convergentium summas quam proxime inveniendi, *Commentarii Academiae Scientiarum Petropolitanae,* 8 (1736), published in 1741, pp. 3–9. In [EOO, (1) 14:101–107].

— [**1736b**] Inventio summae cuiusque seriei ex dato termino generali, *Commentarii Academiae Scientiarum Petropolitanae,* 8 (1736), published in 1741, pp. 9–22. In [EOO, (1) 14:108–123].

— [**1736c**] Methodus universalis series summandi ulterius promota, *Commentarii Academiae Scientiarum Petropolitanae,* 8 (1736), published in 1741, pp. 147–158. In [EOO, (1) 14:124–137].

— [**1737a**] De fractionibus continuis dissertatio, *Commentarii Academiae Scientiarum Petropolitanae,* 9 (1737), published in 1744, pp. 98–137. In [EOO, (1), 14:187–215].

— [**1737b**] Variae observationes circa series infinitas, *Commentarii Academiae Scientiarum Petropolitanae,* 9 (1737), published in 1744, pp. 160–188. In [EOO, (1) 14:216–244].

— [**1737c**] De variis modis circuli quadraturam numeris proxime exprimendi, *Commentarii Academiae Scientiarum Petropolitanae,* 9 (1737), published in 1744, pp. 222–236. In [EOO, (1) 14:245–259].

— [**1739a**] De productis ex infinitis factoribus ortis, *Commentarii Academiae Scientiarum Petropolitanae,* 11 (1739), published in 1750, pp. 3–31. In [EOO, (1) 14:260 – 290].

— [**1739b**] De fractionibus continuis observationes, *Commentarii Academiae Scientiarum Petropolitanae,* 11 (1739), published in 1750, pp. 32–81. In [EOO, (1) 14:291–349].

— [**1739c**] Consideratio progressionis cuiusdam ad circuli quadraturam inveniendam idoneae, *Commentarii Academiae Scientiarum Petropolitanae,* 11 (1739), published in 1750, pp. 116–127. In [EOO, (1) 14:350–365].

— [**1740**] De seriebus quibusdam considerationes, *Commentarii Academiae Scientiarum Imperialis Petropolitanae,* 12 (1740), published in 1750, pp. 53–96. In [EOO, (1) 14:407–462].

— [**1743a**] De summis serierum reciprocarum ex potestatibus numerorum naturalium ortarum dissertatio altera in qua eaedem summationes ex fonte maxime diverso derivantur, *Miscellanea Berolinensia,* 7 (1743), published in 1743, pp. 172–192. In [EOO, (1) 14:138–159].

— [**1743b**] Démonstration de la somme de cette suite $1 + \frac{1}{4} + \frac{1}{9} + \frac{1}{16} + \frac{1}{25} + \frac{1}{36} + \ldots$, *Journ. lit. d'Allemange, de Suisse et du Nord,* 2:1 (1743), published in 1743, pp. 145–127. In [EOO, (1) 14:176–186].

— [**1748a**] *Introductio in analysin infinitorum,* Lausannae: M. M. Bousquet et Soc., 1748. In [EOO, (1), vols. 8–9].

— [**1748b**] Recherches sur la question des inégalités du mouvement de Saturne de Jupiter, *Piece qui a remporte le prix de l'academie royale des sciences* (1748), published in 1751, pp. 1–123. In [EOO, (2), 25: 45–157].

— [**1749a**] De vibratione chordarum exercitatio, *Nova Acta Eruditorum* (1749), published in 1749, pp. 512–527. In Euler [EOO, (2), 10:50–62].

— [**1749b**] De la controverse entre Mrs. Leibnitz et Bernoulli sur les logarithmes des nombres negatifs et imaginaires, *Mémoires des l'Académie des Sciences de Berlin*, 5 (1749), published in 1749, pp. 139–179. In [EOO, (1), 17:195–232].

— [**1750–51a**] Methodus aequationes differentiales altiorum graduum integrandi ulterius promota, *Novi Commentarii Academiae Scientiarum Imperialis Petropolitanae*, 3 (1750–51), published in 1753, pp. 3–35. In [EOO, (1) 22:181–213].

— [**1750–51b**] De serierum determinatione seu nova methodus inveniendi terminos generales serierum, *Novi Commentarii Academiae Scientiarum Imperialis Petropolitanae*, 3 (1750–51), published in 1753, pp. 8–10 and 36–85. In [EOO, (1), 14:463–515].

— [**1750–51c**] Consideratio quarundam serierum quae singularibus proprietatibus sunt praeditae, *Novi Commentarii Academiae Scientiarum Imperialis Petropolitanae*, 3 (1750–1751), published in 1753, pp. 10–12 and 86–108. In [EOO, (1) 14:516–541].

— [**1750–51d**] De partitione numerorum, *Novi Commentarii Academiae Scientiarum Petropolitanae*, 3 (1750–1751), published in 1753, pp. 125–169. In [EOO, (1), 2:254–294].

— [**1753**] Remarques sur les mémoires précédens de M. Bernoulli, *Mémoires des l'Académie Royale des Sciences de Berlin*, 9 (1753), pp. 196–222. In [EOO, (2), 10:233–255].

— [**1754–55a**] Subsidium calculi sinuum, *Novi Commentarii Academiae Scientiarum Petropolitanae*, 5 (1760), published in 1760, pp. 164–204. In [EOO, (1) 14:542–582].

— [**1754–55b**] De seriebus divergentibus, *Novi Commentarii Academiae Scientiarum Imperialis Petropolitanae* 5 (1754–1755), published in 1760, pp. 19–23 and 205–237. In [EOO, (1) 14: 583–617].

— [**1755**] *Institutiones calculi differentialis cum eius usu in analysi finitorum ac doctrina serierum*, Berolini: 1755. In [EOO, (1), 10].

— [**1761**] Remarques sur un beau rapport entre les séries des puissances tant directes que réciproques, *Mémoires de l'Académie des Sciences de Berlin*, 17 (1761), published in 1766, pp. 83–106. In Euler [EOO, (1) 14:83–106]

— [**1764**] De motu vibratorio tympanorum, *Novi Commentarii Academiae Scientiarum Petropolitanae*, 10 (1764), published in 1766, pp. 243–260. In [EOO, (2) 10:344–359].

— [**1765**] De usu functionum discontinarum in Analysi, *Novi Commentarii Academiae Scientiarum Petropolitanae*, 11 (1765), published in 1767, pp. 3–27. In [EOO, (1), 23:74–91].

— [**1768**] De curva hypergeometrica hac aequatione expressa $y = 1 \cdot 2 \cdot 3 \cdot \ldots \cdot x$, *Novi Commentarii Academiae Scientiarum Petropolitanae* 13 (1768), published in 1769, pp. 3–66. In [EOO, (1) 28:41–98].

— [**1768–70**] *Institutionum calculi integralis*, Petropoli: Impensis Academiae Imperialis Scientiarum, 1768–1770. In [EOO, (1), vols. 11–13].

— [**1770**] *Vollständige Anleitung zur Algebra.* In [EOO, (1), vol. 1].

— [**1773**] Summatio progressionum $\sin \varphi^\lambda + \sin 2\varphi^\lambda + \sin 3\varphi^\lambda + \ldots + \sin n\varphi^\lambda$, $\cos \varphi^\lambda + \cos 2\varphi^\lambda + \cos 3\varphi^\lambda + \ldots + \cos n\varphi^\lambda$, *Novi Commentarii Academiae Scientiarum Petropolitanae, 18* (1773), published in 1774, pp. 24–36. In [EOO, (1), 15:168–190].

— [**1774–75**] Demonstratio theorematis newtoniani de evolutione potestatum binomii pro casibus quibus exponentes non sunt numeri integri, *Novi Commentarii Academiae Scientiarum Petropolitanae, 19* (1774–75), published in 1775, pp. 103–111. In [EOO, (1), 15:207–216].

— [**1780**] De plurimis quantitatibus transcendentibus quas nullo modo per formulas integrales exprimere licet, *Acta Academiae Scientiarum Petropolitanae*, 4 pt. 2 (1780), published in 1780, pp. 31–37. In [EOO, (1) 15: 522–527].

— [**1781a**] De oscillationibus minimis funis libere suspensi, *Acta Academiae Scientiarum Petropolitanae* (1781), pt. 1, pp. 157–177. In [EOO, (2) 11:307–323].

— [**1781b**] De perturbatione motus chordarum ab earum pondere oriunda, *Acta Academiae Scientiarum Petropolitanae* (1781), pt. 1, pp. 178–190. In [EOO, (2) 11:324–334].

— [**1784**] De transformatione seriei divergentis $1 - mx + m(m+n)x^2 - m(m+n)(m+2n)x^3 + m(m+n)(m+2n)(m+3n)x^4$ —etc. in fractionem continuam *Nova Acta Academiae Scientiarum Petropolitanae, 2* (1784), pp. 36–45. In Euler [EOO, (1) 16:34–46).

— [**1787**] Dilucidationes in capita postrema calculi mei differentalis de functionibus inexplicabilibus, in L. Euler, *Institutiones calculi differentialis cum eius usu in analysi finitorum ac doctrina serierum*, 2nd edition, Pavia: Galeati, 1787, pp. 797–819. In [EOO, (1) 16 pt. 1:1–33].

— [**1789**] De termino generali serierum hypergeometricarum, *Nova Acta Academiae Scientiarum Pretopolitanae*, 7 (1789), pp. 42–63. In [EOO, (1) 16 pt. 1:139–162].

— [**1793**] Disquisitio ulterior super seriebus secundum multipla cuiusdem anguli progendientibus, *Nova Acta Academiae Scientiarum Petropolitanae, 11* (1793), published in 1798, pp. 114–132. In [EOO, (1) 16 pt. 1:333–355].

— [**1794a**] Specimen transformationis singularis serierum *Nova Acta Academiae Scientiarum Pretopolitanae, 12* (1794), pp.58–70. In [EOO, (1), 16 pt. 2:41–55].

— [**1794b**] De resolutione formulae integralis $\int x^{m-1}dx(\Delta + x^n)^\lambda$ in seriem semper convergentem, ubi simul plura insignia artificia circa serierum explicantur, *Nova Acta Academiae Scientarum Imperialis Petropolitanae* 12 (1801), published in 1801, pp. 58-70. In [EOO, (1), 19:110–143].

— [**1812**] Methodus succincta summas serierum infinitarum per formulas differentiales investigandi, *Mémoires de l'académie des Sciences de St.Pétersboug*, 5 (1812), pp. 45–56. In [EOO, (1) 16 part 2:200–221].

— [**1819**] De unciis potestatum binomii earumque interpolatione; *Mémoires de l'académie des Sciences de St.Pétersboug*, 9 (1819–20), pp. 57–76. In [EOO, 162:241–260].

— [**EOO**] *Leonhardi Euleri Opera omnia*. Soc. Sci. Nat. Helveticæ, Leipzig, Berlin, Basel, 1911–...

Ewald, W.B. (ed.) [**1996**] *From Kant to Hilbert. A Source Book in the Foundation of Mathematics, Oxford Science Publications*, Oxford: Claredon Press, 1996.

Feigenbaum, L. [**1985**] Brook Taylor and the method of increments, *Archive for History of Exact Sciences*, 34 (1985), 1–140.

Ferraro, G. [**1998**] Some aspects of Euler's series theory. Inexplicable functions and the Euler–Maclaurin summation formula, *Historia Mathematica*, 25 (1998), pp. 290–317.

— [**1999**] Rigore e dimostrazione in Matematica alla metà del Settecento, *Physis*, (2) 36 (1999), pp. 137–163.

— [**2000a**] Functions, functional relations and the laws of continuity in Euler, *Historia Mathematica*, 27 (2000), pp. 107–132.

— [**2000b**] The value of an infinite sum. Some observations on the Eulerian theory of series, *Sciences et techniques en perspective*, 4 (2000), pp. 73–113.

— [**2000c**] True and fictitious quantities in Leibniz's theory of series, *Studia Leibnitiana*, 32 (2000), pp. 43–67.

— [**2002**] Convergence and formal manipulation of series from the origins of calculus to about 1730, *Annals of Science*, 59 (2002), pp. 179–199.

— [**2007a**] Convergence and formal manipulation in the theory of series from 1730 to 1815, *Historia Mathematica*, 34 (2007), pp. 62–88.

— [**2007b**] The foundational aspects of Gauss's work on the hypergeometric, factorial and digamma functions, *Archive for History of Exact Sciences*, 61 (2007), pp. 457-518.

— [**2007c**] *L'evoluzione della matematica. Alcuni momenti critici.* Napoli. Ernesto Ummarino Editore, 2007.

— [**2007d**] La teoria formale delle serie in Euler: gli anni dal 1730 al 1755 in Ferraro [2007c, 183-208].

Ferraro, G. and **Panza, M.** [**2003**] Developing into series and returning from series. A note on the foundation 18th-century analysis, *Historia Mathematica*, 30 (2003), pp. 17–46.

— [**A**] *Lagrange's theory of analytical functions and his ideal of purity of method (1797–1813)*, unpublished manuscript.

Fraser, C. [1985] J.L. Lagrange's changing approach to the foundations of the calculus of variations, *Archive for History of Exact Sciences*, 32 (1985), pp. 151–191.

— [**1987**] Joseph Louis Lagrange's algebraic vision of the calculus, *Historia Mathematica*, 14 (1987), pp. 38–53.

— [**1989**] The calculus as algebraic analysis: Some observations on mathematical analysis in the 18th century, *Archive for History of Exact Sciences*, **39** (1989), pp. 317–335.

— [**1997**] The background to and early emergence of Euler's analysis, in Otte and Panza [1997, 47–78].

— [**2003**] "Mathematics" in *The Cambridge History of Science Volume 4 The Eighteenth Century;* Cambridge: Cambridge University Press (2003).

Friedman, M.C [**1992**], *Kant and the Exact Science*, Cambridge (Mass.): Harvard University Press, 1992.

Fourier, J. [**1807**] *Sur la propagation de la chaleur dans le solides*, manuscript 1807, published by Grattan-Guiness and Ravetz, in Grattan-Guiness and Ravetz [1972].

— [**1822**] *Thèorie Analytique de la Chaleur*, Paris: F. Didot, père et fils, 1822.

— [**Œuvres**] *Œuvres de Fourier*, edited by G. Darboux, Paris: Gauthier-Villars et fils, 1888–1890.

Français, J.–F. [**1812–13**] Mémoire rendant à démontrer la légitimité de la sèparation des échelles de différentiation et d'integration des fonctions qu'elles affectent, *Annales de Mathématiques Pures et Appliquées*, 3 (1812–13), pp. 244–273.

Fuss, P.H. [**1843**] *Correspondance Mathématique et Physique de Quelque Célèbres Géomètres du XVIIIème Siècle*, St. Pétersbourg: Académie impériale des sciences, 1843.

Gauss, C.F. [**1799**] Demonstratio nova theorematis omnem functionem algebraicam rationalem integram unius variabilis in factores reales primi vel secundi gradus resolvi posse, in [WW, 3:1–39].

— [**1801**] Disquisitiones Arithmeticae, Lipsiae: Gerh. Fleischer, 1801. In [WW, 1:1–475].

— [**1812a**] Anzeige der Disquisitiones generales circa seriem infinitam Göttingische gelehrte Anzeigen, 10 February 1812. In [WW, 3:197–202].

— [**1812b**] Disquisitiones generales circa seriem infinitam $1 + \frac{\alpha\beta}{1\cdot\gamma}x +$ $\frac{\alpha(\alpha+1)\beta(\beta+1)}{1\cdot 2\cdot\gamma(\gamma+1)}xx + \frac{\alpha(\alpha+1)(\alpha+2)\beta(\beta+1)(\beta+2)}{1\cdot 2\cdot 3\cdot\gamma(\gamma+1)(\gamma+2)}x^3$+etc. Pars Prior, *Commentationes Societatis Regiae Scientiarum Gottingensis recentiores*, 2 (1813). In [WW, 3:125–162].

— [**1816**] Review of J.C. Schwab, *Commentatio in primum elementorum Euclidis librum, qua veritatem geometriae principiis ontologicis niti evincitur, omnesque propositiones, axiomatum geometricorum loco habitae, demonstrantur* (Stuttgart, 1814) and of Matthias Metternich, *Vollständige Theorie der Parallel-Linien* (Mainz, 1815). *Göttingische gelehrte Anzeigen*, 20 April 1816. In [WW, 8:170–174].

— [**1831**] Anzeige der Theoria residuorum biquadraticorum. Commentatio secunda, *Göttingische gelehrte Anzeigen*, 23 April 1831. In [WW, 2:169–178].

— [**1832**] Theoria residuorum biquadraticorum. Commentatio secunda, *Commentationes Societatis Regiae Scientiarum Gottingensis recentiores*, 7 (1832). In [WW, 2:93–148].

— [**WO**] De integratione formulae differentialis $(1 + n\cos\psi)^\nu d\psi$. In [WW, 8:35–64].

— [**WA**] Determinatio seriei nostrae per aequationem differentialem secondi ordinis. In [WW, 3: 207–229].

— [**WB**] Zur Theorie der unendlichen Reihe $F(\alpha, \beta, \gamma, x)$. In [WW, 10 pt 1: 326–359].

— [**WC**] Zur Lehre von den Reihen. In [WW, 10 pt 1: 382–428].

— [**WD**] Zur Metaphysik der Mathematik. In [WW, 12:57–61].

— [**WW**] Werke, Göttingen, Leipzig, and Berlin: Königliche Gesellschaft der Wissenchaften, 1863–1929, 12 vols.

Giusti, E. [**1991**] Le prime ricerche di Pietro Mengoli: la somma delle serie, *Proceedings of the international meeting "Geometry and Complex Variables"*, New York: Dekker, 1991, pp. 195–213.

Gilain, Ch. [**1998**] Condorcet et le calcul integral, in *Sciences a l'Époque de la Révolution Française. Recherches Historiques*, R. Rashed (ed.), Paris: Blanchard, 1988.

Gillies, D. (eds.) [**1992**] *Revolutions in Mathematics*, Oxford: Claredon Press, 1992.

Goldbach, C. [**1720**] Specimen methodi ad summas serierum, *Acta eruditorum* (1720), pp. 27–31.

— [**1727**] De transformatione serierum, *Commentarii Academiae Scientiarum Imperialis Petropolitanae*, 2 (1727), published in 1729, pp. 30–34.

— [**1728**] De terminis generalibus serierum, *Commentarii Academiae Scientiarum Imperialis Petropolitanae*, 3 (1728), published in 1732, pp. 164–173.

Goldstine, H. [**1977**] *A History of Numerical Analysis from the 16th Century Through the 19th Century*, New York: Springer-Verlag, 1977.

Golland, L.A. and **Golland, R.W.** [**1993**] Euler's troublesome series: An early example of the use of trigonometric series, *Historia Mathematica*, 20 (1993), pp. 54–67.

Grabiner, J.V. [**1981**] *The Origins of Cauchy Rigorous Calculus*, Cambridge: MIT Press, 1981.

Grandi, G. [1703] *Quadratura circuli et hyperbolae*, Pisis: ex typographia F. Bindi, 1703.

Grattan–Guinness, I. [**1990**] *Convolutions in French Mathematics, 1800–1840*, Basel: Birkhäuser Verlag,1990.

— [**1996**] Numbers, magnitudes, ratios and proportions in Euclid's Elements: How did he handle them?, *Historia Mathematica*, 23 (1996), pp. 355–375.

Grattan–Guinness, I.and **Ravetz, J.R.** [**1972**] *Joseph Fourier 1768–1830*, Cambridge: MIT Press, 1972.

Gray, J. [**1984**] A commentary on Gauss's mathematical diary, 1796–1814, with an English translation, *Expositiones Mathematicae,* 2 (1984), pp. 97–130.

— [**1986**] *Linear Differential Equations and Group Theory from Reimann to Poincaré*, Boston: Birkhäuser, 1986. (2nd edition Boston: Birkhäuser, 2000).

— [**1992**] The nineteenth-century revolution in mathematical ontology. In Gillies [1992, 226–248].

Grégoire de Saint-Vincent [**1647**] *Opus qeometricum quadraturae circuli et sectionum coni decem libris comprehensum*, Antwerp: J. and J. Meursios, 1647.

Gregory, J. [**1667**] *Vera circuli et hyperbola quadratura*, Padova: Cadorini, 1667.

— [**1668**] *Exercitationes Geometricae*, London: Godbid and Pitt, 1668.

Guicciardini, N. [**1989**] *The Development of Newtonian Calculus in Britain 1700–1800*, Cambridge: Cambridge University Press, 1989.

— [**1999**] *Reading the Principia. The Debate on Newton's Mathematical Methods for Natural Philosophy from 1687 to 1736*, Cambridge: Cambridge University Press, 1999.

Halley, E. [**1695**] A most compendius and facile method for constructing the logarithms, exemplified and demonstrated from the nature of numbers, without any regard to the hyperbola, with a speedy method for finding the number from the logarithm given, *Philosophical Transactions*, 19 (1695–97), pp. 58–67.

Hardy, G.H. [**1949**] *Divergent Series*, Oxford: Clarendon Press, 1949.

Hayashi, T. [**2002**] Leibniz's construction of *Mathesis Universalis*: A consideration of the relationship between the plan and his mathematical contributions, *Historia Scientiarum*, 12 (2002), pp. 121–141.

Horvàth, M. [**1982**] The problem of infinitesimal small quantities in the Leibnizian mathematics, *Studia Leibnitiana, Supplementa* 22 (1982), pp. 150–157.

— [**1986**] On the attempts made by Leibniz to justify his calculus, *Studia Leibnitiana*, 18 (1986), pp. 60–71.

Huygens, C. [**P**] *Descriptio automati planetarii* in *Oevres Complètes de Christiaan Huygen*s, La Haye: M. Nijhoff, 1888–1950, vol.21, pp. 579–647.

Jahnke, H.N. [**1987**] Motive und Probleme der Arithmetisierung der Mathematik in der ersten Hälfte des 19. Jahrhunderts – Cauchys Analysis in der Sicht des Mathematikers Martin Ohm, *Archive for History of Exact Sciences,* 37 (1987), pp. 101–182.

— [**1993**] Algebraic analysis in Germany, 1780–1840: Some mathematical and philosophical issues, *Historia Mathematica*, 20 (1993), pp. 265–284.

Jahnke, H.N. (ed.) [2003] *A History of Analysis.* Providence: American Mathematical Society, 2003.

Katz, V.J. [1987] The Calculus of the Trigonometric Functions, *Historia Mathematica* 14 (1987), pp. 311–324.

Klein, J. [1968] *Greek Mathematical Thought and the Origin of Algebra*, translation by E. Brann, Cambridge: MIT Press, 1968.

Kline, M. [1972] *Mathematical Thought from Ancient to Modern Times*, New York: Oxford University Press, 1972.

Knobloch, E. [1991] Leibniz and Euler: Problems and solutions concerning infinitesimal geometry and calculus, in: *Giornate di Storia della Matematica*, M.Galuzzi (ed.), Commenda di Rende: Editel, 1991, pp. 271–293.

—— **[1989]** Leibniz et son manuscrit inédité sur la quadrature des sections coniques, in *The Leibniz Renaissance*, M. Mugnai (ed.), Firenze: 1989, pp. 127–151.

—— **[1999]** Galileo and Leibniz: Different approaches to infinity, *Archive for History of Exact Sciences*, 54 (1999), pp. 87–99.

Koppelman, E. [1971] The calculus of operations and the rise of abstract algebra, *Archive for History of Exact Sciences*, 8 (1971), pp. 155–241.

Kramp, C. [1799] *Analyse des réfractions astronomiques et terrestres.* Strasbourg: Philippe Jacques Dannbach, 1799.

Lacroix, S.F. [1797–1800] *Traité du calcul différentiel et du calcul intégral*, Paris: Duprat, 1797–1800 (3 vols).

—— **[1810–19]** *Traité du calcul différentiel et du calcul intégral*, Second edition, revue et augmentee. Paris: Courcier, 1810–1819, 3 vols.

Lagrange, J.–L. [1754] *Lettera a Giulio Carlo di Fagnano.* Torino: Stamperia Reale, 1754.

—— **[1759]** Recherches sur la nature, et la propagation du son, *Miscellanea Taurinensia*, 1, classe mathématique, pp. i–x and 1–112. In [Œuvr, 1, 39–148].

—— **[1760–61]** Addition aux premières recherches sur la nature, et la propagation du son, *Miscellanea Taurinensia*, 2 (1760–61), classe mathématique, pp. i–x and 1–112. In [Œuvr, 1:319–333].

—— **[1766]** Sur l'intégration de quelques équations différentielles dont les indéterminées sont séparées, mais dont chaque membre en particulier n'est point intégrable, *Miscellanea Taurinensia*, 4 (1766). In [Œuvr 9, 5–33].

—— **[1768]** Nouvelle méthode pour résoudre le équations littérales par le moyen des séries, *Histoire de l'Académie Royale des Sciences et des Belles-Lettres de Berlin* (1768), pp. 251–326. In [Œuvr, 3: 5–76].

—— **[1769]** Sur le problème de Kepler, *Histoire de l'Académie Royale des Sciences et des Belles-Lettres de Berlin* (1769), pp. 204–233. In [Œuvr, 3:113–140].

—— **[1772]** Sur une nouvelle espèce de calcul relatif à la différentiation et à l'intégration des quantitès variable, *Nouveaux Mémoires de l'Académie*

Royale des Sciences et de Belles-Lettres de Berlin (1772), pp. 185–221. In [Œuvr, 3: 441–476].

—— [**1773**] Solutions analytiques de quelques problèmes sur les pyramides triangulaires, *Nouveaux Mémoires de l'Académie Royale des Sciences et de Belles–Lettres de Berlin* (1773), pp. 149–176. In [Œuvr, 3: 661–692].

—— [**1776**] Sur l'usage des fractions continues dans le calcul integrals, *Nouveaux Mémoires de l'Académie Royale des Sciences et de Belles–Lettres de Berlin*, 1776. In [Œuvr, 4:301–332].

—— [**1788**] *Traité de mécanique analytique*, Paris: Desaint 1788. In [Œuvr, 12].

—— [**1792**] Mémoire sur le méthode d'interpolation, *Nouveaux Mémoires de l'Académie Royale des Sciences et de Belles–Lettres de Berlin*, 40 (1792–1793), pp. 267–288. In [Œuvr, 5:663–684].

—— [**1797**] *Théorie des fonctions analytiques*, Paris: Impr. De la République, 2nd. ed. Paris: Courcier 1813. In [Œuvr, 9].

—— [**1798**] *Traitè de la résolution des équations numériques de tous les degrés*, Paris: Duprat 1798. 2nd ed., Paris: Courcier 1806. In [Œuvr, 8].

—— [**1806**] *Leçon sur le calcul des fonctions*, 2nd ed., Paris: Courcier, 1806. In [Œuvres, 10].

—— [**Œuvr**] *Œuvres de Lagrange*, M.J.-A. Serret [et G. Darboux] (ed.), Paris: Gauthier–Villars, 1867–1892, 14 vols.

Lambert, J.H. [**1770**] Observations analytiques, *Nouveaux Mémoires de l'Académie Royale des Sciences et de Belles-Lettres de Berlin* (1770), pp. 225–244.

Landen, J. [**1758**] *A Discourse Concerning the Residual Analysis: A New Branch of Algebraic Art, of Very Extensive Use Both in Pure Mathematicks and Natural Philosophy*, London: 1758.

Laplace, P.S. [**1773**] Mémoire sur l'inclinaison moyenne des orbites des comètes, sur la figure de la Terre et sur les fonctions, *Mémoires de l'Académie royale des sciences de Paris (Savants étrangers)*, 7 (1773), pp. 503–540. In [ŒC, 8:279–321].

—— [**1777**] Mémoire sur l'usage du calcul aux différences partielles dans la théorie des suites, *Mémoires de l'Académie royale des sciences de Paris* (1777). In [ŒC, 9:313–335].

—— [**1779**] Mémoire sur les suites, *Mémoires de l'Académie royale des sciences de Paris. Mémoires mathématiques et physique* (1779–1782), pp. 207–309. In [ŒC, 10:1–89].

—— [**1784**] *Théorie du mouvement et de la figure elliptique des planètes*, Paris: Ph.-D. Pierres, 1784.

—— [**1785**] Théorie des attractions des sphéroïdes et de la figure des planètes, *Mémoires de l'Academie Royal des Sciences de Paris. Mémoires mathématiques et physique (*1782–1785), pp. 207–309. In [ŒC, 10:339–419].

—— [**1799–1825**] *Traité de mècanique céleste*, Paris: Courcier 1799–1825. In [ŒC, vols. 1–5]

— [**1809**] Mémoire sur diverses points d'analyse, *Journal de l'École Polytechnique*, XV Cahier, Tome VIII; 1809. In [ŒC, 14:178–214].

— [**1812**] T*héorie analytique des probabilités*, Paris: Courcier, 1812.

— [**ŒC**] *Œuvres Complète de Laplace*, Paris: Gauthier–Villars, 1867–1892, 14 vols.

Laugwitz, D. [**1987**] Infinitely small quantities in Cauchy's textbooks, *Historia Mathematica*, 14 (1987), pp. 258–274.

Legendre, A.M. [**1785**] Recherches sur l'attraction des sphéroïdes homogènes, *Mémoires de mathématique et de physique, présentés à l'Académie Royale des Sciences par divers savans et lus dans ses assemblées, Académie des Sciences (Paris)*, 10 (1785), pp. 411–434.

— [**1794**] *Mémoire sur les transcendantes elliptiques*, Paris: DuPont and Didot. The second year of the Republic.

— [**1811–17**] *Exercices de Calcul Intégral, sur divers ordres de Transcendantes et sur les quadratures*, Paris: Courcier, 1811–1817.

Leibniz, G. W. [**1682**] De vera proportione circulis ad quadratum circumscriptum in numeris rationalibus expressa, *Acta Eruditorum* (1682), pp. 41–46. In Leibniz [GMS, 5:118–122].

— [**1684**] Meditationes de cognitione, veritate et ideis, *Acta Eruditorum* (1684) pp. 537–542. In Leibniz [GP, 4:422–426].

— [**1691**] Quadratura arithmetica communis sectionum conicarum, quae centrum habent, indeque ducta trigonometria canonica ad quantamcumque in numeris exactitudinem a tabularum necessitate liberata, cum usu speciali ad lineam rhomborum nauticam, aptatumque illi planisphaerium, *Acta Eruditorum* (1691), pp. 178–182. In Leibniz [GMS, 5:128–132].

— [**1693**] Supplementum geometriae practicae sese ad problemata transcendentia extendens, ope novae methodi generalissimae per series infinitas, *Acta Eruditorum* (1693) pp. 178–180. In Leibniz [GMS, 5:285–288].

— [**1710**] Symbolismus memorabilis calculi algebraici et infinitesimalis in comparatione potentiarum et differentiarum, et de lege homogeneorum transcendentali, *Miscellanea Berolinensia* (1710), pp. 160–165. In Leibniz [GMS, 5:377–382].

— [**1713**] Epistola ad V. Cl. Christianum Wolfium, Professorem matheseos Halensem, circa scientiam infiniti, *Acta Eruditorum Supplem.*, 5 (1713), pp. 264–270. In Leibniz [GMS, 5:382–487].

— [**D**] *Gothofridi Guillemi Leibnitii Opera Omnia, nunc primum collecta, in classes distribuita, praefationibus et indicibus exhornata, studio Ludovici Dutens*, Genevae: Fratres de Tournes, 1768 (6 vols.).

— [**GP**] *Die philosophischen Schriften von G.W. Leibniz*, C.I. Gerhardt (ed.), Berlin: Weidmann, 1875–1890.

— [**GMS**] *Leibnizes mathematische Schriften*, C.I. Gerhardt (ed.), Berlin: Asher & Comp. 1849–1850: vols. I–II; Halle: H. W. Schmidt, 1855–1863: vols. III–VII.

— [**GBLW**] Briefwechsel zwischen Leibniz und Christian Wolff, C.I. Gerhardt (ed.), Halle: 1860, reprinted Hildesheim: 1963.

— [**KQA**] De quadratura arithmetica circuli ellipseos et hyperbolae cujus corollarium est trigonometria sine tabulis, E. Knobloch (ed.), Göttingen: Vandenhoeck & Ruprecht, 1993.

Leibniz, G.W.–Bernoulli, Johann [LB] *Virorum celeberr. Got. Gul. Leibnitii et Johan. Bernoulli Commercium philosophicarum et mathematicarum*, Lousannae et Genevae: M.M.Bousquet, 1745.

Libri, G [1846] Review of P.H. Fuss, *Correspondance mathématique et physique de quelque célèbres géomètres du XVIIIème siècle, Journal des Savants*, Janvier 1846, pp. 50–62.

Lorgna, A–M, [1787] Théorie d'une novelle espèce de calcul fini et infinitésimal, *Mémoire de l'academie Royal de Turin*, 8 (1786–1787), pp. 411–430.

Lützen, J. [1993] Euler vision of a general partial–differential calculus for a generalized kind of functions, *Mathematics Magazine*, 56 (1983), pp. 299–306.

— [**2003a**] The foundation of analysis in the 19th Century, in Jahnke [2003, 155–195].

— [**2003b**] Between rigor and applications. Developments in the concept of function in mathematical analysis, *The Cambridge History of Science*, Marry Jo Nye (ed.), vol. 5, pp. 468–487, Cambridge: Cambridge University Press, 2003.

de l'Hôpital, G.F.A. [1696] *Analyse des Infiniment petits, pour l'intelligence des lignes courbes*, 2nd ed., Paris: F. Montalant, 2 ed., 1716.

Maclaurin, C. [1742] *A Treatise of Fluxions in Two Books*, Edinburgh: Ruddimans 1742, 2 vols.

Mahoney, M.S. [1994] *The Mathematical Career of Pierre de Fermat 1601–1665*, 2nd ed., Princeton: Princeton University Press, 1994.

Maierù, L. [1994] *Fra Descartes e Newton. Isaac Barrow e John Wallis*, Cosenza: Rubbettino, 1994.

— [**1995**] *Metodi degli "Antichi" e dei "Moderni" nell'algebra di John Wallis. Considerazioni circa i nessi tra loro esistenti*, in Panza–Roero [1995, 85–130].

— [**2000**] Il "continuo" nella *Arithmetica Infinitorum* di John Wallis, Atomismo e Continuo nel XVII secolo in *Atti del Convegno Internazionale "Atomisme et Continuum au XVIII Siècle", Napoli, 28–29–30 aprile 1997*, E. Festa and R. Gatto (eds.), Napoli: Vivarium, Napoli, 2000, pp. 151–181.

Malet, A. [1991] Mathematics and Mathematization in the Seventeenth Century, *Studies in History and Philosophy of Science*, 22 (1991) pp. 673–678.

— [**1993**] James Gregorie on Tangents and the "Taylor" Rule of Series Expansions, *Archive for History of Exact Sciences*, 46 (1993), pp. 97–137.

— [**2006**] Renaissance notions of number and magnitude, *Historia Mathematica,* 33 (2006), pp. 63–81.

Massa, M.R. [**1997**], Mengoli on "quasi proportions", *Historia Mathematica,* 24 (1997), pp. 257–280.

Mazet, E. [**2003**] La théorie des séries de Nicole Oresme dans sa perspective aristotélicienne. "Questions 1 et 2 sur la Géométrie d'Euclide", *Revue d'histoire des mathématiques* 9 (2003), pp. 33–80.

Mengoli, P. [**1650**] *Novae quadraturae arithmeticae, seu de additione fractionum.* Bologna: Iacobi Montii, 1650.

— [**1659**] *Geometriae speciosae elementa.* Bologna: 1659.

— [**1672**] *Circolo.* Bologna: 1672.

Menninger, K. [**1969**] *Number Words and Number Symbols: A Cultural History of Numbers,* revised German edition translated by Paul Broneer, Cambridge: MIT Press, 1969.

Mercator, N. [**1668**] *Logarithmotechnia; sive Methodus construendi logarithmos nova, accurata et facilis,* London: G. Godbid, 1668.

de Moivre, A. [**1697**] A method of raising an infinite multinomial to any given power, or extracting any given roots of the same, *Philosophical Transaction,* 9 (no. 230, 1697), pp. 619–625.

— [**1698**] A method of extracting the root of an infinite equation, *Philosophical Transaction,* 20 (no. 240, 1698), pp. 190–193.

— [**1718**] *The Doctrine of Chances or a Method of Calculating the Probabilities of Events in Play,* London 1718.

— [**1722**] De fractionibus algebraicis radicalitate immunibus ad fractionibus simpliciores reducendis, de que summandis terminis quarundum serierum aequali intervallo a se distantibus, *Philosophical Transaction,* 32 (1722), pp. 167–172.

— [**1730**] *Miscellanea analytica de seriebus et quadraturis,* London: J. Tomson and J. Watts, 1730.

— [**1738**] *The Doctrine of Chances or a Method of Calculating the Probabilities of Events in Play,* 2nd ed., London, 1738.

— [**1756**] *The Doctrine of Chances or a Method of Calculating the Probabilities of Events in Play,* 3rd ed., London, 1756.

Netz, R. [**1999a**] *The Shaping of Deduction in Greek Mathematics,* Cambridge: Cambridge University Press, 1999.

— [**1999b**] Archimedes transformed: The case of a result stating a maximum for a cubic equation, *Archive for History of Exact Sciences,* 54 (1999), pp. 1–47.

Newton, I. [**1704**] *Tractatus de quadratura curvarum,* in [OO, 1:332–386].

— [**1707**] *Arithmetica universalis, sive de compositione et de resolutione arithmetica liber,* in [OO, 1:1:234].

— [**1711**] *De analysi per æquationes numero terminorum infinitas,* in [OO, 1:257–283].

— [**CE**] *Commercium epistolicum d. Johannis Collins et aliorum de analysi promota*, London: Iussu Soc. Regoae, typis Paeersoniani, 1712.

— [**1720**] *Universal Arithmetick, or, a treatise of arithmetical composition and resolution*, translated by J. Raphson. London: Senex, Taylor, Warner and Osborn, 1720.

— [**1736**] *The Method of Fluxions and infinite series; with its Application to the geometry*, J. Colson (ed.), London: H.Woodfall and J.Nourse, 1736.

— [**OO**] *Opera quae exstant omnia commentariis illustrabat Samuel Horsley*, Londini: J. Nichols, 1779–1885, 5 vols.

— [**PN**] *Philosophia naturalis principia mathematica. The third edition (1726) with variant readings assembled and edited by Alexandre Koyré and I. Bernard Cohen, with the assistence of Anne Whitman*, Cambridge: Cambridge University Press, 1972.

— [**M**] *The Mathematical Principles of Natural Philosophy,* book two, translated by Andrew Motte, London: Benjamin Motte, 1729.

— [**C**] *The Correspondance of Isaac Newton*, H.W. Turnbull, J.W. Scott and A.R. Hall (eds.), Cambridge: Cambridge University Press, 1959–1977, 7 vols.

— [**MP**] *The Mathematical Papers of Isaac Newton*, D.T. Whiteside (ed.), Cambridge: Cambridge University Press, 1967–1981, 8 vols.

— [**QHW**] Introduction to the quadrature of the curves, translated by John Harris, D.R. Wilkins (ed.), http://www.maths.tcd.ie/pub/HistMath/ People /Newton/ Quadratura/ HarrisIQ.pdf.

Oresme, N. [**A**] *Quaestiones super geometriam Euclidis*, H.L.L. Busard (ed.), Leiden: E.J. Brill, 1961.

Otte, M. [**1989**] The ideas of Hermann Grassmann in the context of the mathematical and philosophical tradition since Leibniz, *Historia Mathematica* 16 (1989), pp. 1–35.

Otte, M. and **Panza. M.** [**1997**] (eds.) *Analysis and synthesis in Mathematics. History and Philosophy*, Dordrecht: Kluwer A.P., 1997.

Panza, M. [**1992**] La forma della quantità. Analisi algebrica e analisi superiore: il problema dell'unità della matematica nel secolo dell'illuminismo, vols. 38–39 of the *Cahiers d'Historie et de Philosophie des Sciences*, 1992.

— [**1995**] Da Wallis a Newton: una via verso il calcolo. Qudrature, serie e rappresentazioni infinite delle quantità e delle forme transcendenti, in Panza–Roero [1995, 131–219].

— [**1996**] Concept of function, between quantity and form, in the eighteenth century, in H.N. Jahnke, N. Knoche and M. Otte (eds.), *History of Mathematics and Education: Ideas and Experiences*, Göttingen: Vandenhoeck & Ruprecht, 1996, pp. 241–274.

— [**1997a**] Classical sources for the concepts of analysis and synthesis, in Otte and Panza [1997, 365–414].

— [**1997b**] Quelques distinctions à l'usage de l'historiographie des mathématiques, in J.–M. Salanskis, F. Rastier, R. Scheps (eds.), *Herméneutique: texts, sciences*, Paris: PUF, 1997, pp. 357–382.

— [**2003**] *Newton*, Paris: Les belles lettres, 2003.

Panza, M. and **Roero, S. (eds.)** [**1995**] *Geometria, flussioni e differenziali. Tradizione e innovazione nella matematica.* Napoli: La città del sole, 1995.

Pensivy, M. [**1987–88**] Jalons historique pour une épistémologie de la série du binôme, *Sciences et Techniques en Perspective*, 14 (1987–1988).

Poisson, S.D. [**1805**] Démonstration du théorème de Taylor, *Correspondance sur l'Ecole Polytechnique*, 1 (1804–1808), pp. 52–55.

— [**1811**] Note sur le développement des puissances des sinus et des cosinus en séries es sinus ou des cosinus d'arcs multiples, *Correspondance sur l'Ecole Polytechnique*, 2 (1809–1813), pp. 212–217.

Pourciau, B. [1998] The preliminary mathematical lemmas of Newton's Principia, *Archive for History of Exact Sciences*, 52 (1998), pp. 279–295.

— [**2001**] Newton and the notion of limit, *Historia Mathematica* 28 (2001), pp. 18–30.

Reiff, R. [**1889**] *Geschichte der unendlichen Reihen*, Tübingen: H. Laupp, 1889.

Rothe, H.A. [**1793**] *Formulae de serierum reversione demostratio universalis, signis localibus, combinatorio-analyticorum vicariis exhibita*, Leipzig: Sommer 1793.

Rüthing, D. [**1984**] Some definitions of the concept of function from Joh. Bernoulli to N. Bourbaki, *Mathematical Intellingencer,* 6(4) (1984), pp. 72–77.

Salanskis, J.-M. and **Sinaceur, H. (Eds.)** [**1992**] *Le Labyrinthe du Continu*, Paris: Springer–Verlag France, 1992.

Scott, J.F. [**1938**] *The Mathematical Works of John Wallis*, 2nd ed., New York: Chelsea Publishing Company, 1981.

Servois, F.J. [**1814–15a**] Essai sur un nouveau mode d'exposition des principes du calcul différentiel, *Annales des Mathématiques Pures et Appliquées*, 5 (1814), pp. 93–140.

— [**1814–15b**] Réflexion sur les divers systèmes d'exposition des principes du calcul différentiel, *Annales des mathématiques pures et appliquées*, 5 (1814), pp. 141–170.

Smith, D. [**1959**] *A Source Book in Mathematics*, New York: Dover, 1959.

Stevin, S. [**1585**] *L'Arithmétique de Simon Stevin de Bruges*, Leyden: 1585.

Stirling, J. [**1717**] *Lineae Tertii Ordinis Neutonianae*, Oxford: e Thetro Sheldoniano, 1717.

— [**1719**] Methodus differentialis Newtoniana illustrata, *Philosophical Transactions of the Royal Society*, 30 (1719), pp. 1050–1070.

— [**1730**] *Methodus differentialis sive Tractatus de Summatione et Interpolatione Serierum Infinitarum*, London: G. Bowyer and Straham, 1730.

Taylor, B. [**1715**] *Methodus incrementorium directa et inversa*, London: G. Innys, 1715.

Todhunter, I. [**1873**] *A History of the Mathematical Theories of Attraction and the Figure of the Earth from Newton to Laplace*, London : Macmillan, 1873.

Torricelli, E. [**1644**] *Opera geometrica*, Firenze: Masse et de Laudis, 1644.

Turnbull, H.W. [**GT**] *James Gregory Tercentenary Memorial Volume*, London, 1939.

Tweddle, I. [**2003**] *James Stirling's Methodus Differentialis. An Annotated Translation of Stirling's Text*. London: Springer, 2003.

Varignon, P. [**1715**] Précautions à prendre dans l'usage des suites ou series infinies résultantes, tant da la division infinie des fractions, que du développement à l'infini des puissance d'exposants négatifs entiers, *Mémoires de l'Academie Royal des Sciences. Mémoires Mathématiques et Physique* (1715), pp. 203–225.

Viète, F. [**1591**] *In artem analyticem isagoge*, Turonis: J. Mettayers, 1691. English translation in Klein [1968, 315–353].

— [**1593**] *Variorum de rebus mathematicis responsorum. Liber VIII*, Tours 1593, in *Opera mathematica: in unum volumen congesta ac recognita opera atque studio Francisci a Schooten*, Lyede: B. et A. Elzeviriorum, 1646.

Wallis, J. [**1656**] *Arithmetica Infinitorum sive Mova Methodus Inquirendi in Curvilineorum Quadraturam, aliaque difficiliora Matheseos Problemata*, Oxon: Lichfield, 1656.

— [**1657**]. *Mathesis universalis, sive Arithmeticum Opus Integrum*, in Wallis [1693–1699, vol. 1].

— [**1668**] Logarithmotechnia Nicolae Mercatoris. Concerning which we shall here deliver the account of the Judicious Dr. I. Wallis given in a Letter to the Viscount Brouncker, *Philosophical Transactions*, 1668 (3), pp. 753–768.

— [**1693**] *De Algebra Tractatus historicus et practicus*, in Wallis [1693–1699], vol. 2.

— [**1693–1699**] *Opera Mathematica*, Oxford: Lichfield, 1693–1699.

Westfall, R. [**1980**] *Never at Rest. A Biography of Isaac Newton*, Cambridge: Cambridge University Press, 1980.

Wilson, C. [**1980**] Perturbations and solar tables from Lacaille to Delambre: The rapprochement of observation and theory, Part II, *Archive for History of Exact Sciences,* 22 (1980), pp. 53–188 and 189–304.

— [**1985**] The great inequality of Jupiter and Saturn: from Kepler to Laplace, *Archive for the History of Exact Sciences*, 33 (1985), pp. 215–290.

Wollenshläger, K. **[1933]** Der Mathematische Briefweschel zwischen Johann I Bernoulli und Abraham de Moivre, *Verhandlungen der Natur-forschenden Gesellschaft in Basel,* 43 (1933), pp. 151–317.

Youschkevitch, A.P. **[1976]** The concept of function up to the middle of the 19th century, *Archive for History of Exact Sciences,* 16 (1976), pp. 37–84.

Author Index

Aepinus, F.U.T., 285
Agostini, A., 8
d'Alembert, J.B., 202, 204, 231, 252, 280, 303–309, 335, 344
Ampère, A.M., 290
Arbogast, L.-F.-A., 290–292, 351
Archimedes, 3
Aristotle, 93–94

Barrow, I., 80
Bayes, T., 183
Bernoulli, D., 90–91, 126, 127, 128, 129, 137, 138, 155, 182, 218, 227, 270, 278
Bernoulli, Jacob, 10, 31, 79–85, 121, 127, 286
Bernoulli, Johann, 2, 33, 45–51, 205, 209, 286, 360
Bernoulli, N., 216, 217, 218, 219, 221, 226
Bessel, F.W., 271, 272, 326, 336, 342
Bolzano, B., 311
Bombelli, R., 17, 18
Bos, H., 34, 208
Bossut, C., 312
Breger, H., 102, 103
Briggs, H., 22
Brisson, B., 292, 293
Brouncker, W., 17
Burchard, J., 91
Bürmann, H., 286
Burn, R.P., 7
Bussotti, P., 338

Callet, J.F., 312
Cardano, G., 107, 108
Cataldi, P. A., 17, 18
Cauchy, A.L., 1, 165, 285, 311, 327, 347–362
Cavalieri, B., 7, 10
Clairaut, A.C., 281, 282
Colson, J., 53, 286
Condorcet, M. J.-A. N., 251

Dedekind, R., 335
Descartes, R., 97, 98, 107
Dhombres, J., 6, 211

Eschenbach, C.H., 286
Euclid, 5, 81, 94, 96
Euler, L., 32, 102, 104–108, 110, 131–132, 137–140, 146, 147, 151, 153, 155–169, 171–179, 181–199, 201–202, 205–210, 212–213, 215–216, 218–222, 224–226, 228–229, 231, 236, 240, 252–253, 254–262, 263–265, 269–272, 275–282, 283–285, 297–298, 312, 319, 322, 323, 326, 330–331, 343, 351, 358

Feigenbaum, L., 45, 46, 51, 87, 88, 89, 91, 92
Ferraro, G., 15, 58, 102, 104, 106, 118, 120, 288, 290, 306, 307, 340, 343, 350
Fourier, J., 281, 311, 315–322, 347
Français, J.-F., 293–294
Fraser, C., 207, 210, 222, 227, 263, 264, 297
Friedman, M.C., 96
Fuss, P.H., 126, 127, 156, 157, 216, 217, 219, 222, 226

Gauss, C.F., 311, 323–345, 347, 357
Gilain, C., 251
Goldbach, C., 126–127, 129–130, 155–157, 222
Goldstine, H., 22, 23, 295
Golland, L.A., 228, 276
Golland, R.W., 228, 276
Grabiner, J.V., 303, 305
Grandi, G., 25, 121, 126
Grattan-Guiness, I., 93, 278, 281, 315
Gray, J., 301, 338, 342, 349
Grégoire de Saint-Vincent, 2, 3, 6
Gregory [or Gregorie], J., 2, 20–24, 87
Guicciardini, N., 25, 99, 146, 147

Halley, E., 20
Hardy, G.H., 128, 219, 225
Hindenburg, C. F., 286
Horsley, S., 76
Horvàth, M., 34
Huygens, C., 18, 25, 35

Subject Index

Sources and Studies in the
History of Mathematics and Physical Sciences

Continued from page ii

A.W. Grootendorst
Jan de Witt's *Elementa Curvarum Linearum, Liber Primus*

A. Hald
A History of Parametric Statistical Inference from Bernoulli to Fischer 1713–1935

T. Hawkins
Emergence of the Theory of Lie Groups: An Essay in the History of Mathematics 1869–1926

A. Hermann/K. von Meyenn/V.F. Wcisskopf (Eds.)
Wolfgang Pauli: Scientific Correspondence I: 1919–1929

C.C. Heyde/E. Seneta
I.J. Bienaymé: Statistical Theory Anticipated

J.P. Hogendijk
Ibn Al-Haytham's *Completion of the Conics*

J. Høyrup
Length, Widths, Surfaces: A Portrait of Old Babylonian Algebra and Its Kin

B. Hughes
Fibonacci's *De Practica Geometrie*

A. Jones (Ed.)
Pappus of Alexandria, Book 7 of the *Collection*

E. Kheirandish
The Arabic Version of Euclid's *Optics*, Volumes I and II

J. Liitzen
Joseph Liouville 1809–1882: Master of Pure and Applied Mathematics

J. Liitzen
The Prehistory of the Theory of Distributions

G.H. Moore
Zermelo's Axiom of Choice

O. Neugebauer
A History of Ancient Mathematical Astronomy

O. Neugebauer
Astronomical Cuneiform Texts

F.J. Ragep
Naṣīr al-Dīn al-Ṭūsī's *Memoir on Astronomy*
(al-Tadhkira f ī cilm al-hay'a)

B.A. Rosenfeld
A History of Non-Euclidean Geometry

G. Schubring
Conflicts Between Generalization, Rigor and Intuition: Number Concepts Underlying the Development of Analysis in 17th-19th Century France and Germany

Sources and Studies in the
History of Mathematics and Physical Sciences

Continued from the previous page

J. Sesiano
Books IV to VII of Diophantus' *Arithmetica*: In the Arabic Translation Attributed to Qusṭā ibn Lūqā

L.E. Sigler
Fibonacci's *Liber Abaci:* A Translation into Modern English of Leonardo Pisano's Book of Calculation

J.A. Stedall
The Arithmetic of Infinitesimals: John Wallis 1656

B. Stephenson
Kepler's Physical Astronomy

N.M. Swerdlow/O. Neugebauer
Mathematical Astronomy in Copernicus's De Revolutionibus

G.J. Toomer (Ed.)
Appolonius *Conics* Books V to VII: The Arabic Translation of the Lost Greek Original in the Version of the Banū Mūsā, Edited, with English Translation and Commentary by G.J. Toomer

G.J. Toomer (Ed.)
Diocles on Burning Mirrors: The Arabic Translation of the Lost Greek Original, Edited, with English Translation and Commentary by G.J. Toomer

C. Truesdell
The Tragicomical History of Thermodynamics, 1822–1854

K. von Meyenn/A. Hermann/V.F. Weisskopf (Eds.)
Wolfgang Pauli: Scientific Correspondence II: 1930–1939

K. von Meyenn (Ed.)
Wolfgang Pauli: Scientific Correspondence 111: 1940–1949

K. von Meyenn (Ed.)
Wolfgang Pauli: Scientific Correspondence IV, Part I: 1950–1952

K. von Meyenn (Ed.)
Wolfgang Pauli: Scientific Correspondence IV, Part II: 1953–1954